AF772433

Women in Engineering, Science and Technology:
Education and Career Challenges

Aileen Cater-Steel
University of Southern Queensland, Australia

Emily Cater
Bupa Health Assurance, UK

ENGINEERING SCIENCE REFERENCE

ENGINEERING SCIENCE REFERENCE
Hershey • New York

Director of Editorial Content: Kristin Klinger
Director of Book Publications: Julia Mosemann
Acquisitions Editor: Lindsay Johnston
Development Editor: Joel Gamon
Publishing Assistant: Keith Glazewski
Typesetter: Keith Glazewski
Production Editor: Jamie Snavely
Cover Design: Lisa Tosheff
Printed at: Yurchak Printing Inc.

Published in the United States of America by
Engineering Science Reference (an imprint of IGI Global)
701 E. Chocolate Avenue
Hershey PA 17033
Tel: 717-533-8845
Fax: 717-533-8661
E-mail: cust@igi-global.com
Web site: http://www.igi-global.com/reference

Copyright © 2010 by IGI Global. All rights reserved. No part of this publication may be reproduced, stored or distributed in any form or by any means, electronic or mechanical, including photocopying, without written permission from the publisher.

Product or company names used in this set are for identification purposes only. Inclusion of the names of the products or companies does not indicate a claim of ownership by IGI Global of the trademark or registered trademark.

Library of Congress Cataloging-in-Publication Data

Women in engineering, science, and technology : education and career challenges / Aileen Cater-Steel and Emily Cater, editors.
p. cm.
Includes bibliographical references and index.
Summary: "This book discusses increasing the participation of women in science, engineering and technology professions, educating the stakeholders - citizens, scholars, educators, managers and policy makers - how to be part of the solution"--Provided by publisher.
ISBN 978-1-61520-657-5 (hardcover) -- ISBN 978-1-61520-658-2 (ebook) 1. Women in the professions. 2. Women scientists. 3. Women--Employment. I. Cater-Steel, Aileen, 1954- II. Cater, Emily, 1982-
HD6054.W66 2010
502.3--dc22

2009043103

British Cataloguing in Publication Data
A Cataloguing in Publication record for this book is available from the British Library.

All work contributed to this book is new, previously-unpublished material. The views expressed in this book are those of the authors, but not necessarily of the publisher.

Editorial Advisory Board

Jenine Beekhuyzen, *Griffith University, Australia*
Lyn Brodie, *University of Southern Queensland, Australia*
Edel Chadwick, *Athlone Institute of Technology, Ireland*
Jo Devine, *University of Southern Queensland, Australia*
Sue Durbin, *University of West England, UK*
Kate Kaiser, *Marquette University, USA*
Yan Li, *University of Southern Queensland, Australia*
Jacquie McDonald, *University of Southern Queensland, Australia*
Charmaine Ryan, *University of Southern Queensland, Australia*
Eileen Trauth, *Penn State University, USA*
Mara Wasburn, *Purdue University, USA*

List of Reviewers

Jenine Beekhuyzen, *Qld Institute of Business & Technology, Australia*
Lyn Brodie, *University of Southern Queensland, Australia*
Aileen Cater-Steel, *University of Southern Queensland, Australia*
Edel Chadwick, *Athlone Institute of Technology, Ireland*
Jo Devine, *University of Southern Queensland, Australia*
Sue Durbin, *University of the West of England, UK*
Shelly Grist, *University of Southern Queensland, Australia*
Jess Guth, *Bradford University Law School, UK*
Kate Kaiser, *Marquette University, USA*
Yan Li, *University of Southern Queensland, Australia*
Birgit Loch, *University of Southern Queensland, Australia*
Ann Manuel, *Oxford University, UK*
David Maree, *University of Pretoria, South Africa*
Maslin Masrom, *University Teknologi, Malaysia*
Jacquie McDonald, *University of Southern Queensland, Australia*
Julie Prescott, *University of Liverpool, UK*
Petrea Redmond, *University of Southern Queensland, Australia*

Shirley Reushle, *University of Southern Queensland, Australia*
Charmaine Ryan, *University of Southern Queensland, Australia*
Ruta Sevo, *Consultant, USA*
Haifa Takruri-Rizk, *University of Salford, UK*
Natasha Veltri, *University of Tampa, USA*
Mara Wasburn, *Purdue University, USA*
Gwen White, *Gateway Community and Technical College, USA*
Elizabeth Yost, *University of Alabama at Birmingham, USA*

Table of Contents

Detailed Table of Contents

Alexandra Winter describes the policy and outcomes adopted by the Queensland Government to encourage women to participate in Science, Engineering and Technology (SET). The Queensland Government's Smart Women – Smart State Strategy (the Strategy) was an innovative response to address the under-representation of women in fields traditionally under-represented by women. From 2005 – 2009, the Office for Women (OFW) coordinated the delivery of the Strategy which encompassed the integration of policy analysis and development with program delivery. OFW is recognised as leading policy development and implementation to support the participation of women in SET in Australia, and have been approached by other jurisdictions seeking to develop similar policy initiatives. The Strategy used strategic leverage points with industry and Government to promote science, engineering and technology (SET) education and careers to women, profile women achieving in SET, and to influence employers in SET industries to consider strategies to attract and retain women.

Ruta Sevo and Daryl Chubin focus on Bias Literacy by reviewing the concepts in research on gender discrimination and the U.S. context. It offers a quick digest of the evidence for discrimination, especially with reference to women in science and engineering in the U.S., explains common terminology and lists relevant legislation and national policy initiatives. The chapter summarizes the difference between tradition and bias, conscious and unconscious discrimination, overt and covert discrimination, and personal versus institution bias. Drawing on research in psychology and social science, it summarizes core concepts including: gender schema, accumulative advantage, stereotype threat, implicit bias, glass ceiling, mommy track, occupational segregation, statistical profiling, climate study, and the value of diversity in learning. A short section lists some U.S. national and international approaches to measuring whether discrimination is occurring and how improvements are benchmarked. There is a list of major organizations working for diversity in the U.S., with links.

Natasha Veltri and her colleagues report on GETSMART, an Academic-Industry partnership encouraging female participation in Math, Science and Technology careers. The GETSMART (Getting Everyone To Study Math and Related Technologies) program is a model mechanism for influencing socio-cultural factors affecting female IT career choice (role models and mentors). They describe the development a formal social mechanism for communication among female IT role models (industry executives and recent college graduates), higher education IT academicians, and female middle school and high school students, and report the attitudes of GETSMART program participants towards a career in STEM. This feedback from girls indicates the importance of teachers, parents, role models and early STEM experiences in formation of interest in STEM, and serves as a feedback loop to the program executive and academic leaders.

Increasing the number of women in science is the goal of the next author, Gwen White. Although recently women have had more opportunities to enter these fields, previously stereotypes, misinformation and ignorance shut women out of these important career choices. This chapter explains that although there are a number of programs available that encourage females to enter the fields of engineering and technology; they just need to move up on the priority lists of corporations, guidance counselors and parents.

Elizabeth Yost and her colleagues from Birmingham Alabama examine the impact of mentoring on self-efficacy for female graduate students and post doctoral fellows in STEM fields. Using data from a national study of selected U.S. academic institutions, recommendations are made in order to enhance mentoring practices that will reduce the barriers women face within STEM fields. Quality mentoring programs represent a viable way to enhance institutional change that may result in increased numbers of women in STEM fields.

This chapter presents a case study by Jacquie McDonald and her colleagues about a current project to support students and academics in an Australian University. The Go WEST - Supporting Women in Engineering, Science and Technology project has built a mentoring and support network among female Science, Engineering and Technology students, staff and industry professionals. As well as the context and history of the project, the chapter describes the activities undertaken and the challenges faced in making the project sustainable. Factors critical to the success of the project are identified and include securing funds and commitment from senior management, having a multi-disciplinary team with strong leadership and effective support, and using information technology to enhance personal networks and to promote activities.

This chapter examines a new and exciting industry: computer games development. Julie Prescott and Jan Bogg focus on the current position and experiences of women working within the computer games industry, the Information and Communication Technology (ICT) sector and the wider context of the Science, Engineering and Technology industry (SET). They review global data collected as part of a larger quantitative study on women who are currently working in the male dominated computer games industry, in relation to the 'long hours' culture associated with the sector. In addition, the lack of females, especially females with families in the gaming industry, skills shortages, work life balance and flexible working is discussed in the games industry context. The discussion is related to the question of attracting and retaining women in the games development workforce of the future.

Jessica Guth and Fran Wright explore women's perceptions and experiences of promotion of academic women. The chapter focuses on the promotions process and criteria as one important issue emerging from that research. At the UK university studied by the authors, the Human Resources department has sought to make the promotion process more transparent and, officially at least, it no longer depends purely upon research achievements. However, these changes will not necessarily result in easier progression for women academics. The authors' study confirms that there is still a mismatch between what women think the criteria for promotion are, what the formal criteria are and how those criteria actually operate.

David and Marinda Maree discuss the structural and individual barriers responsible for the under representation of women in SET (Science, Engineering and Technology). Self-efficacy as a requirement for success in SET is discussed. The results are illustrated with a report on a study done with a sample of 29 women in successful SET careers. Strong self-efficacy beliefs which can be associated with resilience and cognitive hardiness came to the fore. These women believe that they can achieve their goals and they do to a large extent.

Anne Manual addresses the career challenge of the gendered academic research culture and asks if Internet technologies make a difference. The use of Internet technologies (ITs) by academics in their research might be enabling women to meet this particular challenge and whether women in a scientific academic setting have a different experience in this respect from colleagues in social sciences and humanities. Findings are presented from 25 in-depth interviews and content analysis of 750 academic web profiles. Results suggest that although there are areas where the gendered research culture is being circumvented by the use of ITs for women in all disciplines, there are areas where women are not taking full advantage of the potential of the web to increase their visibility and research profile. Moreover, the opportunities presented by IT are not necessarily tied to discipline and women in SET disciplines appear to be no more or less likely to take advantage of them.

Focusing on women engineers in the UK, Haifa Takruri-Rizk, Natalie Sappleton and Sunrita Dhar-Bhattacharjee explain that retention of female engineers is falling, and there has been minimal progress on breaking down vertical segregation in the engineering workplace. In this chapter they explore the situation of women engineers in the UK. They examine, using qualitative and quantitative data collection methods, whether women in engineering do indeed suffer from the application of descriptive and prescriptive stereotypes in the workplace, and how they are able to circumvent them. In particular, we instigate the factors, such as self-promotion and confidence, which may aid or hinder the progression of women's careers. Our findings indicate that confidence and self-promotion should be high on the agenda of women engineers to survive and progress in the engineering industry.

In the final chapter, Mirella Moro, Taisy Weber and Carla Freitas provide an overview of women in the Brazilian Computer Society research community. The authors believe that the first step in order to solve the problem of girls not choosing technology and science related areas is to understand the current situation. They present the first study to identify which areas of Computer Science have more and less feminine participation. In order to do so, the chapter considers the gender balance in program committees of the Brazilian conferences in those areas. The study evaluates the 2008 and previous instances of such conferences. Some Brazilian initiatives to bring more girls to Computer Science are also discussed.

Foreword

In June 2008 a report was published by Harvard University Press that turned heads. The Athena Factor: Reversing the Brain Drain in Science, Engineering and Technology (Hewlett et al. 2008) analyzed the career trajectories of women in the science, engineering and technology (SET) labor force in selected private sector companies around the world. The report documents the unequivocal seriousness of the gender imbalance in the SET fields. The findings of this report claim that women hit an apparent 'breaking point' in their mid to late 30s, which has resulted in 52 percent of highly qualified SET women quitting their jobs. Similar reports have emanated recently from other sources. This 'breaking point' is but one indication of the challenges confronting those who are seeking a greater gender balance in the scientific and technological work force.

The issue of female under representation in the SET disciplines has implications at both the individual and the societal levels. At the individual level it is an issue of fairness in the pursuit of a career; at the societal level it is an issue of economic opportunity. The fairness issue is about the inherent right of an individual in a free society – regardless of gender, race, nationality or other characteristics – to participate in a profession for which one has the interest and the capability. That is, women and men should have equal opportunity to participate in the SET disciplines. The economic opportunity issue is about innovation and competitiveness in the new economy. That is, a country, community or company cannot afford to limit occupational participation on the basis of gender stereotypes. To do so would be to lose the potential contribution of half of the population. Talent knows no gender.

In order to reduce this gender imbalance there is a dire need to produce knowledge about what we know and what works – and to share this knowledge. There is a need for research to understand and frame 'the problem' of gender under representation in science, engineering and technology. Equally important, is a need for publicizing exemplars of interventions that can address these problems. These interventions need to occur in the classroom, the workplace and society. Finally, there is a need for public policy and industry level support for interventions. It is for these reasons that we need the sort of dialogue that occurs in this book. The chapters flow in accordance with a pipeline metaphor that is often used to characterize the range of challenges involved in redressing the gender imbalance. The book starts with an examination of gender policy issues in two representative countries: Australia and the United States. It then moves to the beginning of the pipeline by considering issues and interventions associated with recruiting secondary school females into science and technology fields. It continues with an examination of the situation for university women in SET. The book concludes with chapters about women currently working in scientific and technological jobs. Throughout the book, the scholarly is balanced with the applied through the mechanism of 'career spotlights' that consider practical interventions that contribute to successful career progression of female students and graduates.

There is a varied audience for the messages contained in this book. It includes those from the academy, the workplace, the research community and society at large. University administrators who are concerned about the recruitment and retention of women preparing for careers in the science and technology professions, and faculty from these disciplines who must address the gender imbalance in the classroom want an understanding of interventions that might increase female participation in SET courses of study. Another audience for this book consists of practitioners and managers who seek practical ways to affect the retention of women in the workplace. Fellow researchers who want to remain current on gender research in the SET fields make up a third audience. Given the range of academic disciplines from which this research emanates and its global representation, it is difficult to keep up with all of the journals publishing gender and SET research. Hence, books such as this are a handy way to maintain an overview of current work. Finally, those who want equal opportunity for their daughters, sisters and spouses, and policy makers who desire both equal opportunity and economic opportunity in SET will benefit from learning more about the issues and efforts to address them.

If we are serious about increasing the participation of women in science, engineering and technology professions, then all of the stakeholders – citizens, scholars, educators, managers and policy makers – need to be part of the solution. But in order to be effective, we need to share knowledge about how to understand the challenges and best practices for addressing them. This is the contribution of this book.

REFERENCES

Hewlett, S.A., Luce, C.B., Servon, L.J., Sherbin, L., Shiller, P., Sosnovich, E., and & Sumberg, K. (2008). *The Athena Factor: Reversing the Brain Drain in Science, Engineering, and Technology.* Boston, MA: Harvard Business Review Research Report 10094.

Eileen M. Trauth
The Pennsylvania State University, USA

Eileen M. Trauth *is Professor of Information Sciences and Technology, Women's Studies and International Affairs at The Pennsylvania State University. Trauth's research is concerned with societal, cultural and organizational influences on information technology and the information technology professions with a special focus on the role of diversity and social inclusion. During 2008 she held the University of Klagenfurt (Austria) Fulbright Distinguished Chair in Gender Studies. She has lectured about and investigated gender under representation in the information technology professions in Austria, Australia, Finland, Greece, Ireland, New Zealand, Romania, South Africa, Spain the UK and the United States. Trauth was also the recipient of a Fulbright Scholar award to investigate socio-cultural influences on the emergence of Ireland's information economy. She has lectured about and analysed cultural, economic, infrastructure and public policy influences on the development of the information technology sectors in The Netherlands, Egypt, Romania and various locations in the U.S. Trauth's research has been supported by grants from the Fulbright Foundation, the National Science Foundation, the Australian Research Council and Science Foundation Ireland. She has published 9 books and over 100 scholarly papers on her work on gender and social inclusion, the information economy, qualitative research methods, critical theory, global informatics, information policy, information management, telecommunications policy and information systems skills.*

Preface

INTRODUCTION

Despite improvements in workplace and employment conditions, women remain seriously under-represented in some specific disciplines of engineering, science, and technology, and furthermore, are not well-represented at the most senior levels in all disciplines. There has been academic research into the under-representation of women in these non-traditional fields and changes to government and organisation policy but the situation does not appear to be improving.

In recent years, some women in Australia have received national recognition in the disciplines of engineering, science and technology. Role models include Julie Hammer who became the first female President of Engineers Australia (EA) in 2007. In the same year, EA announced it had achieved a significant milestone: membership of females has broken the 10 percent barrier. Australia's Chief Scientist is also a woman, Professor Penny Sackett. In the field of information technology, Professor Shirley Gregor was made an Officer of the Order of Australia in the Queen's Birthday Honour's list in June 2005 for services as an educator and researcher in the field of information systems and in the development of applications for electronic commerce in the agribusiness sector.

This book provides a collection of recent high-quality empirical studies related to the education and careers of women in engineering, science and technology disciplines. As such, this book provides a valuable contribution as it highlights innovative approaches to attracting girls and women to non-traditional disciplines and well as supporting students through their university experience. The transition to the workforce and challenges faced by graduates are also examined. The profiles of successful girls and women included as 'career spotlights' provides a practical, rather than a theoretical discussion of factors which contribute to successful career progression of female students and graduates. The women profiled are inspirational role-models to other women in these non-traditional fields.

In achieving the overall objective of presenting a collection of recent empirical studies related to attracting, educating and retaining women in engineering, science and technology disciplines, it is hoped this book is relevant to educators and well as business managers as each chapter investigates and reports on a particular issue. The contributions include cases of innovative approaches to attracting and supporting women to engineering, science and technology training courses. It also includes a discussion of the effectiveness of mentoring, role models, government policy, and initiatives by professional bodies.

Although the target audience of this book is professionals and researchers working in the field of education and career development of women in engineering, science and technology, the book also provides insights and support for academics concerned with attracting and retaining women in these disciplines where women are traditionally under-represented. The book also provide valuable information to executives and members of professional bodies representing engineers, scientists and technology workers who wish to encourage women during their transition from education to the workforce, and during their career progression.

As well as the editors' home locations of Toowoomba (Australia) and Bristol (UK), the book attracted submissions from authors based in South Africa, Malaysia, Brazil, five different states in the USA, and Salford and Manchester in the UK. This interest from the international community proves the issues are of global, not local relevance.

The call for chapter proposals was widely disseminated though the use of women's networks and mail lists subscribed to by academics in engineering, science and technology. However, the Moderator of the software engineering mail list 'SEWorld' refused to distribute the call for proposals with the excuse that 'the list is about software engineering research and the message has a broader scope'. As a result, there were no proposals related to the challenges experienced by women in software engineering!

STRUCTURE OF THE BOOK

This book is divided into four main parts: government initiatives, schools, college and university, and finally, careers. In each section, profiles of outstanding women are provided in the form of career spotlights.

Section 1 sets the scene with two chapters related to government initiatives and legislation. These two chapters relate to the general situation without solely focusing on school, university or career aspects.

In the opening chapter, Alexandra Winter describes the policy and outcomes adopted by the Queensland Government to encourage women to participate in Science, Engineering and Technology (SET). The Queensland Government's Smart Women – Smart State Strategy (the Strategy) was an innovative response to address the under-representation of women in fields traditionally under-represented by women. From 2005 – 2009, the Office for Women (OFW) coordinated the delivery of the Strategy which encompassed the integration of policy analysis and development with program delivery. OFW is recognised as leading policy development and implementation to support the participation of women in SET in Australia, and have been approached by other jurisdictions seeking to develop similar policy initiatives. The Strategy used strategic leverage points with industry and Government to promote science, engineering and technology (SET) education and careers to women, profile women achieving in SET, and to influence employers in SET industries to consider strategies to attract and retain women.

The next chapter by Ruta Sevo and Daryl Chubin focuses on Bias Literacy by reviewing the concepts in research on gender discrimination and the U.S. context. It offers a quick digest of the evidence for discrimination, especially with reference to women in science and engineering in the U.S., explains common terminology and lists relevant legislation and national policy initiatives. The chapter summarizes the difference between tradition and bias, conscious and unconscious discrimination, overt and covert discrimination, and personal versus institution bias. Drawing on research in psychology and social science, it summarizes core concepts including: gender schema, accumulative advantage, stereotype threat, implicit bias, glass ceiling, mommy track, occupational segregation, statistical profiling, climate study, and the value of diversity in learning. A short section lists some U.S. national and international approaches to measuring whether discrimination is occurring and how improvements are benchmarked. There is a list of major organizations working for diversity in the U.S., with links.

Section 2 is focused on encouraging girls at school to consider engineering, science and technology careers.

Natasha Veltri and her colleagues report on GETSMART, an Academic-Industry partnership encouraging female participation in Math, Science and Technology careers. The GETSMART (Getting Everyone To Study Math and Related Technologies) program is a model mechanism for influencing socio-cultural factors affecting female IT career choice (role models and mentors). They describe the development a formal social mechanism for communication among female IT role models (industry executives and recent college graduates), higher education IT academicians, and female middle school and high school students, and report the attitudes of GETSMART program participants towards a career in STEM. This feedback from girls indicates the importance of teachers, parents, role models and early STEM experiences in formation of interest in STEM, and serves as a feedback loop to the program executive and academic leaders.

Increasing the number of women in science is the goal of the next author, Gwen White. Although recently women have had more opportunities to enter these fields, previously stereotypes, misinformation and ignorance shut women out of these important career choices. This chapter explains that although there are a number of programs available that encourage females to enter the fields of engineering and technology; they just need to move up on the priority lists of corporations, guidance counselors and parents.

Section 3 is focused on College and University students.

Elizabeth Yost and her colleagues from Birmingham Alabama examine the impact of mentoring on self-efficacy for female graduate students and post doctoral fellows in STEM fields. Using data from a national study of selected U.S. academic institutions, recommendations are made in order to enhance mentoring practices that will reduce the barriers women face within STEM fields. Quality mentoring programs represent a viable way to enhance institutional change that may result in increased numbers of women in STEM fields.

The next chapter presents a case study by Jacquie McDonald and her colleagues about a current project to support students and academics in an Australian University. The Go WEST - Supporting Women in Engineering, Science and Technology project has built a mentoring and support network among female Science, Engineering and Technology students, staff and industry professionals. As well as the context and history of the project, the chapter describes the activities undertaken and the challenges faced in making the project sustainable. Factors critical to the success of the project are identified and include securing funds and commitment from senior management, having a multi-disciplinary team with strong leadership and effective support, and using information technology to enhance personal networks and to promote activities.

Section 4 presents eight chapters and is focused on women who work in engineering, science and technology disciplines, either as academics or in the wider work force.

The first chapter in Section 4 examines a new and exciting industry: computer games development. Julie Prescott and Jan Bogg focus on the current position and experiences of women working within the computer games industry, the Information and Communication Technology (ICT) sector and the wider context of the Science, Engineering and Technology industry (SET). They review global data collected as part of a larger quantitative study on women who are currently working in the male dominated computer games industry, in relation to the 'long hours' culture associated with the sector. In addition, the lack of females, especially females with families in the gaming industry, skills shortages, work life balance and flexible working is discussed in the games industry context. The discussion is related to the question of attracting and retaining women in the games development workforce of the future.

Jessica Guth and Fran Wright explore women's perceptions and experiences of promotion of academic women. The chapter focuses on the promotions process and criteria as one important issue emerging from that research. At the UK university studied by the authors, the Human Resources department has sought to make the promotion process more transparent and, officially at least, it no longer depends purely upon research achievements. However, these changes will not necessarily result in easier progression for women academics. The authors' study confirms that there is still a mismatch between what women think the criteria for promotion are, what the formal criteria are and how those criteria actually operate.

The next chapter is from South Africa. David and Marinda Maree discuss the structural and individual barriers responsible for the under representation of women in SET (Science, Engineering and

Technology). Self-efficacy as a requirement for success in SET is discussed. The results are illustrated with a report on a study done with a sample of 29 women in successful SET careers. Strong self-efficacy beliefs which can be associated with resilience and cognitive hardiness came to the fore. These women believe that they can achieve their goals and they do to a large extent.

Maslin Masrom and Zuraini Ismail provide a Malaysian perspective on the access of women to computers and the internet. Since Malaysia gained independence in 1957, Malaysian women have participated and contributed towards the social and economic development of the country. This was made possible by the policy standpoint of the Malaysian government which considers women are an important pool of resource that can be mobilized towards achieving the national development agenda. In Malaysia, where the use of computers is widespread, women have been found to have benefited from modern technologies. They have greater access to computers and Internet. This chapter emphasizes accessibility, delivering access to the computers and internet info-structure to women in Malaysia. A framework for analyzing Malaysian women accessing computers and the Internet is presented and discussed.

The next chapter, by Sue Durbin presents a case study of senior female scientists in the UK. It focuses upon the careers of thirteen senior female scientists in a UK public sector organisation engaged in scientific research and development. This chapter explores how female scientists have achieved senior status in a field where men dominate and where the sector is culturally and structurally gendered. Although these women's experiences of gender relations were variable and a limited career path has reduced their strategic leadership options, they have sustained a positive orientation, built around dedication to their work and a strong desire to reputation build, working within a generally supportive work environment.

The next chapter by Anne Manual addresses the career challenge of the gendered academic research culture and asks if Internet technologies make a difference. The use of Internet technologies (ITs) by academics in their research might be enabling women to meet this particular challenge and whether women in a scientific academic setting have a different experience in this respect from colleagues in social sciences and humanities. Findings are presented from 25 in-depth interviews and content analysis of 750 academic web profiles. Results suggest that although there are areas where the gendered research culture is being circumvented by the use of ITs for women in all disciplines, there are areas where women are not taking full advantage of the potential of the web to increase their visibility and research profile. Moreover, the opportunities presented by IT are not necessarily tied to discipline and women in SET disciplines appear to be no more or less likely to take advantage of them.

Focusing on women engineers in the UK, Haifa Takruri-Rizk, Natalie Sappleton and Sunrita Dhar-Bhattacharjee explain that retention of female engineers is falling, and there has been minimal progress on breaking down vertical segregation in the engineering workplace. In this chapter they explore the situation of women engineers in the UK. They examine, using qualitative and quantitative data collection methods, whether women in engineering do indeed suffer from the application of descriptive and prescriptive stereotypes in the workplace, and how they are able to circumvent them. In particular, we instigate the factors, such as self-promotion and confidence, which may aid or hinder the progression of women's careers. Our findings indicate that confidence and self-promotion should be high on the agenda of women engineers to survive and progress in the engineering industry.

In the final chapter, Mirella Moro, Taisy Weber and Carla Freitas provide an overview of women in the Brazilian Computer Society research community. The authors believe that the first step in order to solve the problem of girls not choosing technology and science related areas is to understand the current situation. They present the first study to identify which areas of Computer Science have more and less feminine participation. In order to do so, the chapter considers the gender balance in program committees of the Brazilian conferences in those areas. The study evaluates the 2008 and previous instances of such conferences. Some Brazilian initiatives to bring more girls to Computer Science are also discussed.

Interspersed between the chapters throughout the book are career spotlights of successful women. These profiles include women who are students, and women working in academia as well as the public and private sectors. The achievements of these women have been recognised with awards in the Queensland Government's Smart Women - Smart State program. These women are role models and it is hoped their personal stories will inspire other women to achieve success in the fields of engineering, science and technology.

BOOK DEVELOPMENT PROCESS

A double-blind review process was used for all chapters submitted to the editors. Authors of selected chapters were invited to act on the reviewers' comments and resubmit their chapters to the editors. Chapters were checked and final revisions applied. The Career Spotlights were sourced from recent winners of the Queensland Government Smart State Smart Women awards. The winners were given the opportunity to provide a brief biography and revise their award submission document to suit the style of this book.

We have enjoyed the process of compiling this book and in particular working with the contributors who provided such wide-ranging research on the topic of women in engineering, science and technology. It is up to you, the readers to decide if the perspectives offered here are relevant to your research or to the practical application of the concepts in your organisations. We would be delighted to hear your feedback on the usefulness of this collection.

DISCLAIMERS

No product or service mentioned in this book is endorsed, nor are any claims made about the capabilities of such product or service.

All trademarks are copyrighted to their respective owners.

Aileen Cater-Steel
University of Southern Queensland, Australia

Emily Cater
Bupa Health Assurance, UK

Acknowledgment

The authors who contributed to this book deserve our heartfelt thanks for their contribution, patience and co-operation throughout the long and complex process of compiling this book. All the chapter contributors are listed with biographical sketches in a section of the book. Thanks to the Queensland Government Office of Women who helped by contacting the Smart State Smart Women award winners and providing the photos. The inspiration for the book came from the GoWEST team at USQ and their support and encouragement is much appreciated. Sincere thanks also to Eileen Trauth for writing the Foreword and for her consistent and valuable contribution to researching the role of women in the fields of information systems and technology.

The reviewers also played an essential role and we know the authors were very appreciative of the valuable comments provided by the reviewers. We sincerely thank the reviewers for taking the time to read and comment on the original submissions. Their contribution was an essential ingredient necessary to improve the content and presentation of the chapters.

A special note of thanks to the staff at IGI Global who provided the necessary processes, templates, reminders and project management of the entire process from our first proposal to this final publication. In particular we wish to express our thanks to Dr. Mehdi Khosrow-Pour, Executive Director of IGI Global, and Elizabeth Ardner, Editorial Assistant.

Finally, Aileen dedicates this book to Rick Glassey (the loveable rogue) for his encouragement, charisma and joie de vivre; and Emily dedicates it to Amanda Connell for being the smartest woman in science, the universe and everything; and to Mrs. Dr. Levy for being proof it can be done - for Emily and for so many others.

Section 1
Government Initiatives

Chapter 1
The Smart Women – Smart State Strategy:
A Policy on Women's Participation in Science, Engineering and Technology in Queensland, Australia

Alexandra Winter[1]
Queensland State Government, Australia

ABSTRACT

Despite growing demand for skilled workers, women remain under-represented in science, engineering and technology industries. Limited opportunities for women not only impacts upon individual women's access to the workforce and economic security, but also means that the State does not have the appropriate skills for future growth. In Australia, the Queensland State Government developed and implemented a strategy to promote science, engineering and technology to women and girls, and to retain women in these fields of education and work. The strategy involved a range of methods, including policy and program delivery, to engage women and girls at different points of their lives. This chapter gives an overview of the Smart Women – Smart State Strategy and the initiatives delivered as part of the strategy, and finishes by highlighting the need for ongoing strategies to effect social and attitudinal change. This overview highlights the importance of policy development and delivery that is gender sensitive, as well as the broader social and economic benefits of women's equitable participation in science, engineering and technology.

INTRODUCTION

The Queensland Government's *Smart Women – Smart State Strategy* (the Strategy) was an innovative response to address the under-representation of women in science, engineering, and technology in Queensland, a state of Australia. From 2005 – 2009, the Office for Women (OFW) coordinated the delivery of the Strategy which encompassed the integration of policy analysis and development with program delivery. The Strategy utilised strategic leverage points with industry and Government to promote science, engineering and technology (SET) education and careers to women, profile women achieving in SET, and to influence employers in SET industries to consider strategies to attract and

DOI: 10.4018/978-1-61520-657-5.ch001

Copyright © 2010, IGI Global. Copying or distributing in print or electronic forms without written permission of IGI Global is prohibited.

retain women. OFW are recognised as leading policy development and implementation to support the participation of women in SET in Australia.

Research on women's participation in SET, and OFW's expertise in social policy and gender analysis, demonstrated the need to understand the issue of women and girls' under-representation in SET as embedded in gendered structures of education, training and workforce participation. Developing and implementing initiatives to address the participation of women and girls in SET was therefore not just focussed on promoting opportunities in areas in which women are under-represented, but also needed to be sensitive to the social dimensions of study and work in some male-dominated areas of SET. The Strategy needed to recognise the gendered dimensions of women and girls' participation in SET, which are economic and social in cause and effect. The Strategy therefore recognised some of the possible economic benefits to women and girls (for example, employment in well-paid areas of skills shortage) as well as some of the challenges for women studying and working in SET industries (for example, women employed on research grants can lack access to maternity leave provisions). The Strategy identified the need to promote social and structural change in education, in the workforce, in leadership, and in perceptions of appropriate work and study for women and girls.

While the extent of systemic change sought by the Strategy was ambitious, during the five years of its implementation (2004 – 2009) a number of successful outcomes that assisted, and will continue to support, women and girls' participation in SET were achieved. These included: profiling women's achievements in SET at OFW's *Smart Women – Smart State Awards*; supporting retention of first year female engineering students at *Women Engineering Queensland* workshops; developing information resources to assist women in SET, and employers; and raising awareness of the barriers to women's participation in SET education and employment within industry, academia and across government. The Strategy extended the Queensland Government's *Smart State Strategy* by applying gender-sensitive policy analysis to ensure that women had the opportunity to contribute to, and benefit from, Queensland's strategic policy direction to address long-term skills needs and promote economic development.

BACKGROUND

The *Smart State Strategy (2005 – 2015)*, launched by the then Premier, the Honourable Peter Beattie, promoted the development of a knowledge economy for Queensland, a State that has historically been focussed on primary industries and resources. The *Smart State Strategy* envisioned a State in which economic growth and prosperity are driven by knowledge, creativity and innovation, particularly in biotechnology, information technology, nano-technology and communication technology. The Strategy recognised that in order to meet the Smart State objectives there was a need to expand the SET workforce. Because of the under-representation of women in SET, increasing the participation of women and girls in SET education and employment was one important way to boost the State's SET workforce while also providing new opportunities for women to benefit from entering into SET study and employment.

Prior to the launch of the *Smart State Strategy*, Dr Lesley Clark, former Member of Parliament for Barron River, and Parliamentary Secretary to the Premier, developed the *SET for Success in the Smart State* report, which documented the barriers to, and under-representation of, women in SET as an issue to be addressed under the *Smart State Strategy*. A number of Dr Clark's recommendations were implemented. Specifically an awards program to recognise and promote the achievements of women in SET commenced in 2004 (the *Smart Women – Smart State Awards*), and a taskforce was established in 2005 (the *Smart Women – Smart State Taskforce*).

The *Smart Women – Smart State Taskforce* (the Taskforce) was established to provide expert advice to the Minister for Women on strategies to assist the Government to increase women's participation and leadership in Queensland's SET industries. Membership consisted of women representing SET industries, academia, and government, as well as an identified youth representative. Members brought not only specialist expertise from engineering, biochemistry, technology and physics, but leadership and personal experience in their chosen fields. The content knowledge, as well as the personal experiences, of these women was important in shaping the activities carried out under the Strategy. Representatives of the Queensland Government agencies provided a cross government perspective on the issues, and played a role in supporting the progression of a number of initiatives. The structure of the Taskforce enabled a forum that brought together these key stakeholders to share industry knowledge with expertise in Government policy development to collaboratively identify and implement opportunities to support women and girls' participation in SET.

Drawing on their combined knowledge and experience, the Taskforce developed a twelve point action plan to address the imbalances for women's participation in SET. In order to better understand the underlying issues, a concept paper, *Supporting Women's Participation in Emerging Industries: Science, Engineering and Technology* (Office for Women, 2006a), outlining statistics on the participation of women and girls in SET was researched and written by OFW. The concept paper provided an analysis of national and international data and research on women's participation in SET, and was also informed by consultation with Government departments, industry and educational institutions.

The concept paper provided an evidence base and gaps analysis that was utilised by the Taskforce in developing the *Smart Women – Smart State Science Engineering and Technology Action Plan(2006 – 2009)* (the Action Plan). The Action Plan proposed twelve initiatives to be undertaken by industry bodies and the Queensland Government, and received Cabinet endorsement in 2006. The Action Plan and concept paper were published on the OFW website (www.women.qld.gov.au).

In recognition of the numerous factors influencing the participation of women and girls in SET education and employment, the Strategy aimed to look at the most significant points of engagement and disengagement with SET. A "life cycle" approach was adopted and four "crunch points" at which women and girls were at risk of disengaging from SET education and work were identified. These included engagement with girls at school; subject selection at year 10; retention at university; and, leadership and retention in the workforce (particularly in relation to career breaks) (Office for Women, 2006a). Aligned with these crunch points, the Action Plan highlighted four priority areas for action (outlined below) to support women and girls' participation in SET.

- Developing awareness: Raising awareness of SET careers to girls at school was identified as an important factor in attracting women to SET industries.
- Recruitment and retention: Once a SET career has been chosen, the need for retention strategies for women in undergraduate courses (especially engineering students) and careers (especially scientists) was noted.
- A partnership approach: The building and maintenance of partnerships, both across Government and with industry and educational bodies, was seen as essential. The establishment of collaborative, strategic alliances between sectors facilitated the development and implementation of the Action Plan, and benefitted SET industries as well as girls and women in SET study and careers.

- Monitoring and progress: Gaps were identified in the availability of accurate quantitative and qualitative data to provide a clear picture of women's participation in SET industries and to aid in the monitoring of progress of women in SET.

In order to address the above priority areas the actions in the Action Plan were grouped under four key goals: profiling, partnerships, pathways and progress. Through raising the profile of science, engineering and technology, promoting pathways into SET education and careers, fostering cross-sectoral partnerships, and measuring success, the Action Plan aimed to help advance women's roles in SET.

OUTCOMES

In the short term, the Strategy delivered some very successful outputs (detailed below). However, because it was recognised that increasing women's participation in SET will require long term systemic change, the initiatives delivered were also designed to encourage long term outcomes that would outlast the period of the Action Plan. Initiatives addressing workplace culture in SET, for example, encouraged reform in the recognition that change will need to keep occurring beyond the life of the Strategy. Activities also encompassed different forms of delivery and scale, from public events to targeted information provision and dissemination. Activities delivered under the Strategy were also pitched at a range of different target audiences (including young women, university students, women in the SET workforce, and employers). The range and scale of the initiatives delivered under the Strategy highlight the breadth of the issues pertaining to women and girls in SET, as well as the spheres (government, education and industry) through which women's participation in SET can be supported. OFW's work in delivering the Strategy therefore included policy development and commentary, large scale events, (such as the Awards), and industry engagement.

Some of the key outputs of the Strategy are outlined below. These activities are grouped under the following themes: profiling women's achievements in SET; sharing good practice to engage women in SET; promoting pathways for women in SET; retaining women in SET; and partnerships. These themes and accompanying examples give an overview of the range of activity that was auspiced by the Strategy and show some of the ways in which the priority action areas outlined above, as well as patterns in women's educational and career trajectories, were addressed.

Profiling Women's Achievements in SET

The *Smart Women – Smart State Awards* (the Awards) were first delivered in 2004 and have been delivered over six consecutive years since. The Awards have been very successful in celebrating the achievements of women in SET industries and promoting careers in SET to Queensland women and girls. The Awards reward women's work in SET and thereby recognise both individual and team contributions made by women and girls in Queensland. By profiling successful women, the Awards also encourage other women and girls to pursue non-traditional career paths in emerging industries. Many of the nominees and winners have gained confidence and new career opportunities have arisen as a direct result of their participation.

Winning the award has opened doors locally and regionally in regards to being a natural invitee to many community forums and as such has extended my ability to form partnerships with different organisations and groups in regards to a diverse range of projects. My resume certainly looks impressive with the inclusion of the award and has also been helpful in adding credence to my proposals to businesses and government organisations. – Samantha Faint, 2006 Smart Women – Smart State Award Winner, Clermont

With the money I won, I am able to volunteer on a conservation project in Africa, so in January I am off to Africa for 7 months to work with Mongooses. I have been volunteering on animal related projects around Australia to get some experience, and writing papers. – Alecia Carter, 2007 Smart Women – Smart State Award Winner, Brisbane

Coming from a poor rural background with very limited opportunities, these awards should inspire other women to strive to be the best in any field they choose. I congratulate the Queensland Government on having initiated this scheme. – Dr Kirsten Heimann, 2008 Smart Women – Smart State Award Winner, Townsville.

The Awards have involved industry engagement through sponsorship and judging. The Awards have therefore enabled organisations to promote their support for women in SET and the benefits of employing women in SET industries.

The success of the Awards is also reflected in the significant media and community interest that has increased each year since their inception. Media coverage has again provided an important opportunity to profile women's achievements in SET, and raise community awareness of girls' and women's contributions in these fields. For example, in 2008 the Awards received approximately 50 media mentions, including 30 print articles profiling the awards and showcasing the award winners. These articles were published in a range of State-wide publications including *The Courier Mail* and *The Sunday Mail*, and regional publications such as *The Gold Coast Bulletin, The Townsville Bulletin, The Cairns Post*, and *The Mackay Daily Mercury*. National television coverage was received through ABC's *Landline* program, and a story aired on ABC Television's Queensland news.

Securing media coverage of the Awards is particularly significant given the important role media plays in either reinforcing, or challenging perceptions about gender roles. Representations of a diverse range of women working and achieving in SET demonstrate that women and girls can develop successful careers in SET, and challenge broader community preconceptions about the appropriateness of SET education and work for women. The range of coverage is also a positive outcome for the Awards as it is important the information regarding the Awards and SET reaches women in regional and rural Queensland.

Sharing Good Practice to Engage Young Women in SET

Programs to Promote SET to Young Women

In order to profile opportunities in SET careers to young women, OFW delivered panel events at schools and careers expos at a number of locations throughout Queensland. School-based programs often provide a good opportunity to give young women a taste of the previously unknown opportunities in SET fields. OFW delivered a pilot event, *Bright Futures – Shining Stars* (see case study below), that has been used as a model to deliver similar events in regional Queensland. Hearing first hand from women in SET, and being able to ask questions, is potentially a far more direct and engaging way of reaching young women than other forms of communication. Feedback from young women who attended the event showed that it changed their perceptions of what SET careers can offer and influenced some young women in their career choices.

Bright Futures, Shining Stars

Bright Futures – Shining Stars (BFSS) took place at Parliament House on Wednesday 20 September 2006 and preceded the 2006 *Smart Women – Smart State Awards*.

BFSS was aimed at year 10 to 12 female high school students interested in or enrolled in SET subjects to inspire and motivate them to pursue further study or a career in SET. The event provided an opportunity for students to interact with a panel of three female role models, one from each of the SET industries, chosen for their successful and inspiring careers.

The event received positive feedback from students and teachers, and formed the basis of a model replicated in a number of regional locations in Queensland:

I think the Bright Futures event had just the right flavour. Congratulations. It allowed the girls to ask the questions they really wanted answers to and the three presenters had such good honest stories. I would strongly encourage as many repeats of this kind of event as we can manage. I would be surprised if out of that one event we do not have some of the girls thinking seriously about SET careers.

The BFSS event is a good initiative. Comments from students to her indicated that it was good to hear the pathways the speakers had taken and that they hadn't gone straight down the traditional road. It was influential on student decisions.

It inspired me also to keep trying my best and never give up.

It is while at school, particularly during the middle years of high school, that decisions regarding subject selection are made, and these can have a significant impact on future study and employment options. In order to support the provisions of information to young women regarding opportunities in SET, OFW developed a resource profiling programs that targeted young women's participation in SET.

OFW undertook an environmental scan of international, Australian and Queensland programs promoting SET to girls in school. While the scan identified a range of national programs promoting SET to young people, a limited number of programs that specifically target women were apparent. A need to share and promote existing models of good practice to engage young women in SET was identified, and based on the environmental scan, OFW put together a resource – *A directory of science, engineering and technology programs for girls in Queensland schools* – which profiles five successful models and provides a directory of other available programs. The initiatives profiled in the resource represent a range of ongoing and one-off programs, developed by a variety of educational, industry and government groups, and include enough information that these programs can be replicated or adapted by other organisations. The resources profiled the following Queensland based initiatives, and covered different scales and locations:

- Go Go GIDGITS Program;
- Scientist in Residence Program;
- Secondary Schools and Queensland University of Technology Engineering Activity Kits (SQUEAK);
- Technology Takes You Anywhere Event; and
- Sunshine Coast Science, Engineering and Technology Expo 2008 (see summary case study below).

The development of the case studies highlighted an ongoing need for the development of gender specific programs that encourage engagement with SET, and future education and employment in these areas. OFW will continue to promote the resource that has been developed to assist in this process. In addition to promoting programs that encourage the participation of girls in SET through the distribution of the above resource, OFW will also continue to highlight and promote relevant events, programs, sponsorships and awards via industry networks and the OFW website.

The Sunshine Coast Science, Engineering and Technology Expo Get SET for the Future
The Sunshine Coast Science, Engineering and Technology Expo *Get SET for the Future* program consisted of two days – one day for Year 5 students and one day for Year 9 students. It was delivered by OFW in partnership with Education Queensland ('Science Centres of Innovation and Professional Practice' (SCIPP) and 'Girls in ICT'), the University of the Sunshine Coast, and Engineers Australia Queensland Division (Sunshine Coast committee).

In addition to the student's Expo, a Parent's Information session was planned to address stereotypes held by some parents that SET careers were not appropriate for girls and women to pursue.

The main goal was to educate and inspire regional students to further their studies in science and ICT by:

- providing dynamic and interactive information relating to Science, Engineering and Technology;
- showcasing opportunities available in the SET industries by showcasing men and women already working/studying in these fields;
- challenging stereotypes of SET careers as 'male-only' domains; and
- promoting SET career pathways to parents/caregivers.

Successes/Achievements
Participant surveys indicated that the satisfaction level of all students was very high, with many rating the Engineering Challenge as a favourite. Most importantly, the responses regarding intentions to continue science and ICT studies in Year 10 were heartening. Analysis demonstrates that at the beginning of the Expo 71% of Year 9 girls rated science positively, and 57% of Year 9 girls rated ICT positively. Following the Expo 96% of Year 9 girls said they would now consider choosing a science subject in Year 10, and 69% of Year 9 girls said they would now consider choosing an ICT subject in Year 10.

Quotes from Student Participants

Want to come back

Fantastic

Learnt a lot

Liked meeting new people

Best science expo I've been to

Want to become a forensic scientist!

Promoting Pathways for Women in SET

Women Engineering Queensland Workshops

The *Women Engineering Queensland Workshops* (WEQ) aimed to increase the retention of Queensland women in undergraduate engineering courses by building networks among women in engineering and showcasing opportunities in the engineering profession. By encouraging networks and providing information about future opportunities WEQ was a strategy to promote women's long term participation in engineering. The workshops were a collaborative venture and were delivered in partnership with key university and industry stakeholders. A working group was established that included representatives from industry and all Queensland universities offering engineering. The input of these stakeholders was invaluable in developing a program that included practical, academic, and informal information and resources to attendees. The generous sponsorship from industry meant that students from throughout Queensland were able to attend free of charge.

The WEQ program provided attendees with information about working in engineering, and included speakers who had recently graduated from engineering alongside senior women engineers. The forum enabled students to ask practical questions about work, as well as raise some long term issues pertaining to work life balance and family responsibilities. WEQ also included practical exercises to develop women's skills in dealing with workplace issues (particularly issues that may arise in a male dominated workforce) and their networking skills. An evening function with industry representatives enabled attendees to put their networking skills into practice and to meet senior engineering representatives.

A pilot WEQ was run in 2007, and was attended by over 50 first year students from all Queensland universities offering engineering. Feedback from participants indicated that the 2007 workshop was very successful with 100% of participants strongly agreed or agreed that the workshop was useful, and 99% strongly agreed or agreed that the workshop helped them feel more confident in their decision to study engineering.

Building on the 2007 pilot, a much larger WEQ was held in 2008. The 2008 workshop was held from 18 – 19 July in Brisbane, and was attended by 140 female students from all regional universities. The working group, made up of Taskforce members, OFW staff, and industry and university representatives, was successful in securing sponsorship from 13 separate industry and Government bodies. This level of industry support testifies to the importance of such an event to retain women in engineering.

Feedback from participants indicates that the workshop was, overall, a very positive experience for female students:

- 98% of participants strongly agreed or agreed that the workshop was useful;
- 96% of participants agreed or strongly agreed that meeting other women studying engineering at the workshop was valuable; and
- 83% agreed or strongly agreed that the workshop assisted in making them feel more confident in their decision to study engineering.

Feedback from students revealed that the event was particularly successful in showcasing inspiring female engineers and providing students with important information about career paths and the opportunities associated with various engineering disciplines. The opportunity to make contact with industry representatives at the evening networking function was also highly valued by participants, and was useful to establish friendships with other students and contacts with industry representatives.

I think this was a superbly organised event. It has been wonderfully enlightening and motivating for me as a young female engineering student. I think the networking skills I learnt will be very valuable in many aspects of my life and career. The chosen speakers were excellent. Thank you.

After attending I felt so much better about a career in engineering and I'm definitely inspired to follow a career path similar to those of speakers here. Everyone was really friendly and welcoming and it was an awesome experience to get to know other female engineers. LOVED IT.

[b]y listening to engineers in industry I have felt inspired and this has affirmed my choice of studying engineering.

- Quotes from Women Engineering Queensland workshop participants 2008

As part of the workshop evaluation, students were asked to rate various elements of the workshop. Figure 1 illustrates some of the areas covered by WEQ, and participants' evaluation of these.

Figure 1. Summary of student feedback on key components of the Women Engineering Queensland workshop 2008

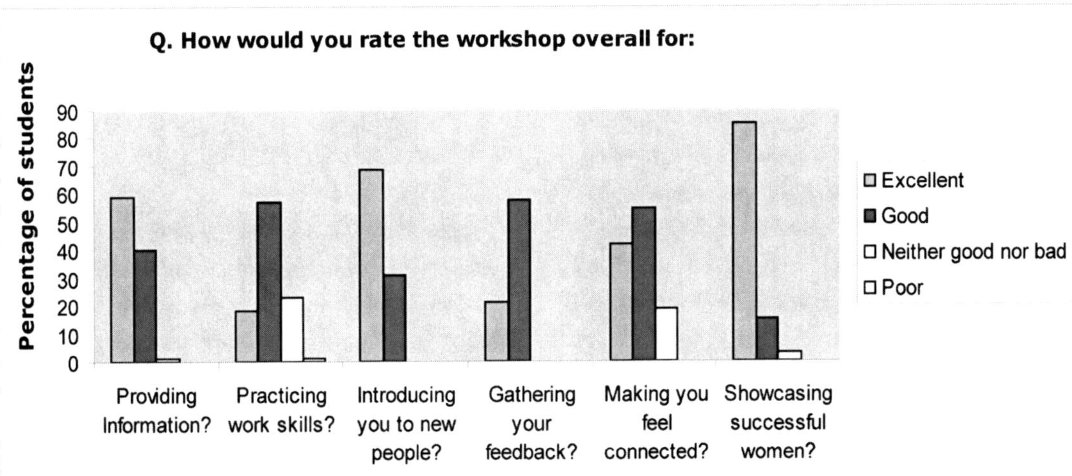

Smart Women – Smart State Board Shadowing Pilot

It is well established that women are under-represented in leadership positions and on boards across industry sectors (Equal Opportunity for Women in the Workforce Agency, 2008), and that this is heightened in SET industries in which there are fewer women at all levels. OFW developed a pilot "board shadowing" program to mentor emerging women leaders in SET industries that was delivered in partnership with the Taskforce. The board shadowing program aimed to:

- draw on the expertise of Taskforce members to enhance the capacity of emerging women leaders in SET industries;
- provide practical experience in board processes and operations;
- provide an opportunity to develop networks for participants with women leaders in SET industries; and
- demonstrate leadership to other government boards and committees by supporting the development of women's leadership skills.

The results of the evaluation of the board shadowing program indicated that participants gained valuable insight into how a board operates, and appreciated the opportunity to observe female leaders in the SET industries interacting on the Taskforce. Both participants and mentors were in agreement that the concept of a board shadowing program is a good way to support the development of women's leadership skills. OFW has published this case study online (Office for Women, 2007) in order to promote this as a model of good practice in supporting women's leadership aspirations to other Government agencies and industry.

Retaining Women in SET

Returning to the SET Workforce Following a Career Break

Taking a break from the workforce is an issue for women in all industry sectors, but can be exacerbated for women working in SET as a result of industry specific challenges. The Strategy recognised that many women temporarily exit the workforce at least once in their professional lives to raise children, care for ageing parents, and meet family responsibilities or other commitments. For women working in SET, time out of the paid workforce can have a significant impact on their career trajectories, their pay, and ultimately their long-term financial security. Despite the fact that career breaks can have a significant impact on women's ability to return to the SET workforce, a significant gap in information and support for women returning to the workforce following a career break was apparent. The loss of qualified women from the SET workforce can have negative consequences for organisations, which lose valuable skills and experience, and has a significant economic cost due to the need to recruit and retrain new staff.

Research was undertaken by OFW to investigate women's experiences of temporarily exiting, and returning to, the SET workforce. This research informed the development of strategies to assist women to successfully transition back into the SET workforce. General research on taking a career break was undertaken, and in-depth interviews conducted with eight women who had recently taken a break from, and returned to, the SET workforce. In response, a resource was developed to assist women working in SET with career planning both before and after career breaks, and to link them with networking, skills development opportunities and support services. The resource – *Taking a Career Break: A Resource for Women in Science, Engineering and Technology* – was developed and was made available on the OFW website (www.women.qld.gov.au). It covers the following key areas to assist women take a career break:

- how to assess a workplace for supportive policies and practices;
- an overview of legal rights and entitlements during a career break;
- planning for a break;
- negotiating flexible working arrangements;
- connecting with other workers who have taken a career break;
- securing a mentor;
- establishing future child care and elder care arrangements;
- networking and staying connected to the SET industry; and
- reviewing and re-negotiating a return to work plan.

In general, it was found that the issues facing women in SET are similar to issues facing women in other industries, and include access to parental leave, flexible working arrangements, workplace culture (including entrenched behaviours and attitudes), and opportunities for career progression. These can, however, be experienced more acutely in industries in which there are fewer women. For women working in SET, the rapidly changing, and fast-paced nature of these industries can further exacerbate the difficulties associated with taking a career break. For example, fast technological change in IT industries makes it difficult to maintain up to date knowledge and skills, while the reliance on publications and research grants in science means that a period out of the workforce can impact significantly on future opportunities. In engineering, project based work that entails long hours, rosters and on-site work can influence opportunities to return to the workforce. Taking a break and returning to the SET workforce remains a significant issue that needs to be addressed for many women in all industry sectors. However, specific and further work in SET is needed, both with industry, and within organisations, to continue the development and implementation of strategies that assist women to take career breaks and successfully transition back into SET industries.

Flexible Working Arrangements in Science, Engineering and Technology: Information for Employers

Retaining women in SET industries balances efforts to engage women and girls in SET education and training. Without ensuring that women are able to obtain and remain in SET work, educational efforts will do little to influence long term change. Retaining women in SET requires engagement with, and support from industry. Flexible working practices that assist women to balance their work, caring, and other responsibilities are part of organisational retention strategies, and alleviate costs associated with high staff turnover. Some of the challenges associated with working in SET industries include managing work and family or other responsibilities, and social issues, including gender stereotyping and masculine workplace cultures. The Association of Professional Engineers, Scientists and Managers Australia's (APESMA) 2007 *Women in the Professions Survey* found that women are leaving the engineering profession at a rate of 38% faster than their male colleagues, and found that for 21.1% of the female engineers, scientists, and computer professionals surveyed, flexible hours of work would be a significant factor in influencing their next career move (APESMA, 2007).

Flexible, family-friendly work practices are an important strategy for SET organisations that wish to attract, motivate and retain quality workers. Both male and female employees face increasing difficulties managing their work, family and other commitments, especially as is widely acknowledged that the intensification of work in Australia has been increasing. Work practices that are accommodating, such as flexible work hours, job sharing and telecommuting, enable women to continue working and building their career in SET, while meeting these responsibilities.

In order to identify the best way to promote the importance of flexible working practices to SET employers, and work towards addressing some of barriers to women's ongoing participation in SET, consultation was undertaken with a number of key industry associations, including Engineers Australia, the Federation of Australian Scientific and Technological Societies (FASTS), the Women in Science Enquiry Network (WiSENet), and APESMA. Consultation with these groups identified a number of distinct challenges in the implementation of flexible working arrangements in SET industries. Commonalities were evident, however each sector identified some specific issues for women seeking to work flexibly.

In science, the grant-based, competitive nature of many academic contracts was seen to be the most significant barrier to employees being able to work flexibly. It was suggested that many employers may need to make changes to the laboratory environment (for example, by providing a distinct space for parents) in order to implement fully effective family friendly policies. In the information technology sector, entrenched cultural practices, including an expectation that employees work a core 40 hour week plus additional hours, were identified as a key issue. Similarly, the 'male-dominated' culture of the engineering sector, and the nature of the work undertaken in this industry (for example, on-site work and fixed working hours) were identified as barriers to the implementation of flexible working practices such as telecommuting and job-share arrangements.

It is important to recognise the positive outcomes for business that flexible working practices can bring. OFW identified that there are a range of business benefits for organisations including higher levels of employee loyalty, and reduced absenteeism and employee turnover. Industry consultation reiterated the need for data and research to demonstrate the strong business case for flexible workplaces, and to give confidence to organisations considering the implementation of flexible working arrangements. The Boston College Center for Work and Family, for example, outlines the benefits to organisations of implementing flexible working practices that impact upon all aspects of an organisation (including finance, operations, and employees) and its engagement with customers and its social reputation (Van Deusen, Ladge, James and Harrington, 2008).

Drawing on information gathered through the consultation process and international and Australian research regarding the business benefits of flexible working arrangements, an information sheet was developed for SET employers that focused on building the business case for flexible employment practices, and providing organisations with practical information about how to develop and implement policies to support a flexible, family-friendly workplace. The resource, *Flexible Working Arrangements in Science, Engineering and Technology: Information for Employers* was developed and made available to SET organisations via the OFW website. It covers the following key areas:

- why flexible, family-friendly workplaces makes economic sense;
- facts about flexible work arrangements;
- flexible work in the current financial climate;
- information on what a flexible workplace is;
- the importance of organisational culture;
- hints on how to develop and implement flexible working policies;
- good practice in science, engineering and technology; and
- links to further information.

Underpinning the need for flexible working practices is not only a cost saving measure for industry, but the need for a diverse and appropriately qualified workforce. Flexible working practices and work/life balance were identified by the *SET Fair Report* as one of the organisational strategies to achieve a diverse workforce, alongside the introduction of transparency of pay and promotion procedures; gender balance in decision making processes; and research to achieve a better understanding of scientific and employment cultures (Greenfield, 2002).

Partnerships

The scope and success of the outcomes delivered under the Strategy represent a significant and quantifiable body of work. It is also important to recognise that one of the significant achievements of the Strategy includes the development of relationships and partnerships that have proved invaluable. Though the impact of these is harder to establish, the importance of industry engagement and support in increasing women's participation in SET is critical. The relationships established by OFW include those with members of the Taskforce, whose networks and industry knowledge were immensely valuable, and partnerships with organisations and industry bodies. Partnerships assisted with information gathering, promotion of outcomes, and mutual support for programs and policies. Members of the Taskforce have also applied knowledge gained through the Strategy to other areas in which they have influence. In this way the intent of the Strategy continues in new forums. Examples of partnerships include: the industry, university and government input into, and support for, the development of the Action Plan; delivery of the *Smart Women – Smart State Awards*; and the *Women Engineering Queensland* workshop, which brought together representatives of academia, industry and government to deliver a workshop program that was able to take into account the university experiences of women studying engineering, as well as their future employment challenges and needs.

OPPORTUNITIES AND CHALLENGES

The promotion of a knowledge economy through the *Smart State Strategy* presented an opportunity to develop gender sensitive policy to ensure that women benefited from, and were able to, contribute to SET industries. The development and delivery of the Strategy also coincided with a resources and economic boom, and resultant skills shortages. Queensland experienced record unemployment rates of less that 4%, and a number of Government and non-government initiatives were developed to address Queensland's skills needs. These included the *Queensland Skills Plan*, various Skills Formation Strategies, the Advanced Engineering Capability Network, Engineers Australia's *Year of Women in Engineering* in 2007, the Queensland Resources Council's *Women in Resources Action Plan*, skilled migration programs, and significant investment in infrastructure development to house research institutes. This context was favourable in that several industries, including engineering, mining and construction, specifically recognised the need to attract and retain women in these industries in order to meet skills needs. IT, however, remains an industry in which the participation of women is decreasing. The Strategy was able to build on and further the momentum in government and non-government organisations in order to work towards increasing the participation of women in SET.

At the time of writing, the full impact on women of the global economic downturn that began in 2008 has yet to be fully felt in Queensland. While job losses in mining and construction in particular have been well documented, and the impact of these on their predominantly male workforce is considerable, the effect on women is still to be fully realised and documented. There is a risk that initiatives targeting women will be linked to good economic times, and that measures to increase diversity will be shelved in the too hard basket while the downturn lasts and that progress made on attracting and retaining women in SET, implementing flexible working policies and practices, and promoting women's leadership will be lost. It is important to identify ways to maintain momentum in an economic downturn, not only in terms of the employment of women, but also the retention of women already employed in SET industries. Continuing to promote women's participation in SET will require a mixture of policy initiatives and industry commitment.

The attraction and retention of women must be regarded as more than a solution to skills shortages during economically prosperous times, but as having a wider social and economic benefit. The Strategy has worked towards addressing the bigger and more unwieldy issue of gendered occupational segregation. Australian Bureau of Statistics (ABS) reveals, for example, that in Queensland in 2006 women in the mining sector comprised: 20.45% of professionals; 1.5% of intermediate production and transport workers, 0% of tradespersons, labourers and related workers;[2] and 100% of advanced clerical and service workers. In Queensland in 2006 women in the construction industry comprised 1.6% of tradespersons and related workers, 98.88% of advanced clerical and service workers, and 4.95% of labourers and related workers. Such occupational segregation has also been associated with vertical segregation within these industries, and relatively few women therefore hold leadership positions in SET. To draw on the example from mining again, only 9.09% of women held managerial and administrative positions in 2006.

Occupational segregation is also associated with systemic pay inequity, in which male dominated industries receive higher wages than female dominated industries, despite adjusting for education, training and experience. Therefore strategies to address occupational segregation not only ensure that women have true choice and opportunity in any industry in which they study and work, but also have access to work that is well paid. According to ABS data, in 2008, in Queensland:

- average weekly ordinary time earnings (for award, agreed or standard hours only) for full time adults were $976.10 for women and $1,157.30 for men (a gap of 15.7%);
- average weekly total earnings (ordinary

times plus overtime earnings) for full time adults were $993.60 for women and $1231.70 for men (a gap of 19.3%); and

- average weekly total earnings for all wage and salary earners (including full time, part time and casual) were $685.80 for women and $1033.80 for men (a gap of 33.7%).

ABS data also shows that women working in the mining industry earn significantly more, on average, than women employed in many other sectors. In 2006 women in the mining sector earnt an average weekly wage of $1335.40, compared to women working in the retail trade sector, who earnt an average $726.10 per week. Despite this relatively high income, women in the mining sector earnt just 77.35% of the average weekly wage of their male counterparts, who received an average $1726.50 per week. In contrast, women working in the retail trade sector during the same period earnt 91.08% of the average weekly wage of their male colleagues, who received an average $797.20 per week (ABS, 2006).

Above and beyond the current downturn, it must be recognised that women's participation in SET, and the related issues of occupational segregation and gender pay inequity, are also related to deeply entrenched norms regarding women's roles, both in the workforce and in their social and cultural identity. The impact of gender role stereotyping, and perceptions of appropriate behaviour and work for women, has been well documented and is not an issue restricted to SET industries. It has impacts on the perceived choices that women have regarding their study and work options, and therefore on women's participation in a whole range industries.

Increasingly the argument is being made that workforce diversity has a number of benefits for organisations: it supports better decision making within an organisation, appeals to consumers, and supports the attraction and retention of the best possible staff. Enabling women to enter, and remain in, SET study and work furthers workforce diversity and benefits organisations and industries (see for example Van Deusen, Ladge, James, and Harrington, 2008 and Joy, Carter, Wagner, and Narayanan, 2007).

GOING FORWARD

The *SET Fair Report* (Greenfield, 2002) stated that the key reasons for including women in SET are to: promote competitiveness of SET industries by engaging the best people to drive innovation; ensure that the return on investment and training is realised by retaining people in SET industries; reaping the benefits of a diverse workforce to enhance the quality of SET; and, missing market opportunities and skills by not including women in the design and development of policy and products (Greenfield, 2002). These arguments remain current, and are still compelling in the face of an economic downturn, and perhaps are even more so given that diversity may give organisations a business or research edge.

The work undertaken to promote women's participation in SET, not only in Queensland, but in many countries, needs to be regarded as long term strategy that requires systemic and cultural change. As noted above, attitudinal and behavioural change regarding perceptions of women's and men's social roles impact upon education and workforce participation, and addressing this in the broadest possible terms is a task that needs to continue comprehensively into the future. In Queensland, OFW continues to develop and implement policy that promotes the attraction and retention of women in non-traditional industries through its *Women in Hard Hats* initiative, which builds on and extends the *Smart Women – Smart State Strategy*.

Finally, dedicated initiatives such as the *Smart Women – Smart State Strategy* demonstrate the importance of policy development and implementation that is sensitive to gender. The work undertaken by OFW has highlighted the importance of examining women's participation in SET in a framework that addresses education and employment alongside social policy and economic policy. Although the scope of this is challenging, a systemic view is necessary in order to comprehensively address the under-representation of women in SET and support their participation at every stage of their lives. In a state-wide policy context, attention to the impacts on and outcomes for women should lead to better outcomes not only for women, but for organisations, for the State, and for government. Developing and implementing the *Smart Women – Smart State Strategy* under the auspices of the *Smart State Strategy* meant that comprehensive approach was taken to ensure that a group historically under-represented in SET industries was acknowledged, and steps taken to redress this under-representation, while Queensland began working to build a knowledge economy to benefit the State as a whole.

REFERENCES

Advanced Engineering Capability Network. (n.d.). Retrieved August 26, 2009 from http://www.engineeringcapability.net/

Association of Professional Engineers. Scientists and Managers Australia. (2007). *Women in the Professions Survey Report*. Retrieved 11 February 2009, from http://www.apesma.asn.au/women/survey/women_in_the_professions_2007.pdf.

Australian Bureau of Statistics. (2006). *Calculations based on data in: Average weekly Ordinary time earnings (AWOTE) - Full time adults, 6302*, Australian Bureau of Statistics, 1984-2006. Retrieved 30 January 2009 from http://www.abs.gov.au/AUSSTATS/abs@.nsf/DetailsPage/6302.0Nov%20 2006?OpenDocument

Australian Bureau of Statistics. (2008). *Average Weekly Earnings Australia, August 2008*, 'Table 13C Average Weekly Earnings, Queensland (Dollars) - Original', data cube: Excel spreadsheet, Cat. no. 6302.0. Retrieved 30 January 2009 from http://www.abs.gov.au/AUSSTATS/abs@.nsf/DetailsPage/6302.0Aug%202008?OpenDocument

Australian Bureau of Statistics. (2009). *Employed Persons by Sex, Industry, Occupation, State, 6291.0.55.003 E09_aug86*, 1986-1996 and 1996-2007. Retrieved 2 March 2009, from http://www.abs.gov.au/AUSSTATS/abs@.nsf/DetailsPage/6291.0.55.003Feb%202009?OpenDocument

Australian Computer Society – Women. (n.d.). Retrieved 26 August 2009 from http://www.acs.org.au/acswomen/

Equal Opportunity for Women in the Workplace Agency. (2008). *Australian Census of Women in Leadership*, Australian Government. Retrieved 2 March 2009 from http://www.eowa.gov.au/Australian_Women_In_Leadership_Census/2008_Australian_Women_In_Leadership_Census/Media_Kit/EOWA_Census_2008_Publication.pdf

Greenfield, S., J. Peters, N. Lane, T. Lees and G.l Samuels. (2002). *SET Fair: A Report on Women in Science, Engineering, and Technology*. Retrieved February 12, 2009 from http://www.ukrc4setwomen.org/downloads/Government/setfair.pdf

Hooker, C. (2004). Irresistible Forces: Australian Women in Science. Carlton, Victoria: Melbourne University Press.

Joy, L., & Carter, N. M. Wagner, H. M. & Narayanan, S. (2007). *The Bottom Line: Corporate Performance and Women's Representation on Boards*. Retrieved March 5 2009 from http://www.catalyst.org/publication/200/the-bottom-line-corporate-performance-and-womens-representation-on-boards

Lord, L., Preston, A., & Crosbie, R. (2006). Unearthing New Resources: Attracting and Retaining Women in the Australian Minerals Industry. Canberra, Australia: Australian Government Office for Women and Minerals Council of Australia.

Office for Women. (2006a). *Smart Women – Smart State Concept Paper: Supporting Women in the Smart State*. Brisbane: Queensland Government. Retrieved from http://www.women.qld.gov.au/work-and-life/smart-state-strategy/documents/set-concept-paper.pdf

Office for Women. (2006b). *Smart Women – Smart State Strategy Science, Engineering and Technology Action Plan (2006 – 2009)*. Brisbane: Queensland Government. Retrieved from http://www.women.qld.gov.au/work-and-life/smart-state-strategy/documents/set-action-plan.pdf

Office for Women. (2007). *Women's Leadership in Science, Engineering and Technology: The Smart Women – Smart State Taskforce's Board Shadowing Program*. Brisbane: Queensland Government. Retrieved from http://www.women.qld.gov.au/leadership-and-community/women-on-boards/documents/board-shadowing-case-study.pdf

Panteli, N. (2006). Returning to IT. *Labour & Industry*, *16*(3).

UK Resource Centre for Women in Science. Engineering and Technology. (2004). *Work Life Balance – A SET Perspective*. Retrieved on 12 February 2009 from http://www.ukrc4setwomen.org/html/resources/ukrc-publications/

UK Resource Centre for Women in Science. *Engineering and Technology*. (n.d.). Retrieved on 26 August 2009 from http://www.ukrc4setwomen.org/

Van Deusen, F., Ladge, J., James, J., & Harrington, B. (2008). *Building the Business Case for Work-Life Programs*. Retrieved on 12 February 2009 from http://www.bc.edu/centers/cwf/research/meta-elements/pdf/BCCWF_Business_Case_EBS.pdf

Women in Science Enquiry Network (WISENet). (n.d.). Retrieved on 26 August from http://www.wisenet-australia.org/

Women in Technology (WiT). (n.d.). Retrieved on 26 August from http://www.wit.org.au/

ENDNOTES

1. I would like to acknowledge Cara Rothery's assistance in developing this chapter.
2. The number of female tradespersons, labourers and related workers are so small that they cannot be regarded as statistically valid.

Profile 1
Engineering Career Spotlight:
Senior Engineer for Major Projects at Goonyella Riverside Mine

Melanie Gordon, Moranbah

Melanie Gordon (nee Webber) graduated from St Peters Lutheran College in Brisbane in 1996. She went on to complete a Bachelor of Mechanical Engineering at the University of Queensland's St Lucia campus and a Masters of Project Management at the University of Southern Queensland. Now Melanie is a Project Manager for BHP Billiton and has become an ambassador for women in the engineering and resources field. Her efforts have been recognised with two prestigious awards during the past six months. Melanie took out the top gong in the resources sector winning the Resource Award for Women in March 2008 and more recently winning the Smart Women Smart State Award for Engineering in September 2008. Melanie is currently the Maintenance and Engineering Manager for an underground coal mine in the Bowen Basin area. She also recently captained 176 people at Goonyella Riverside Mine as Project Manager for a Major Dragline Maintenance Project achieving fantastic results in terms of safety and budget. Melanie was the first female Project Manager ever appointed to such a project. Melanie loves her work because of the challenges, variety and freedom to try new innovations. Outside of work Melanie is married to Maclean who she met at university and who is the maintenance superintendent at Goonyella Riverside Mine.

Copyright © 2010, IGI Global. Copying or distributing in print or electronic forms without written permission of IGI Global is prohibited.

Melanie's Role

Melanie's current title is Senior Engineer – Major Projects. Her role is to manage a group of engineers to successfully deliver major projects at the Goonyella Riverside Mine. Melanie's most recent role was as Project Manager for the Dragline 27 Shutdown. She is the first female in BMA to ever fill the role of Project Manager for a dragline shutdown. Shutdown sites are traditionally a man's domain so Melanie being the Project Manager has come as a shock to the crews however she has received positive feedback from them regarding her work ethic and dedication. In this role, Melanie was responsible for captaining 176 people to execute the shutdown scope of work safely, with high quality outcomes and within the defined budget and schedule constraints.

The shutdown project was completed in March 2008 and has been declared a success. There were many challenges during the project, in particular the floods which occurred in January and February and resulted in the project site being closed for 15 days. Besides the obvious impacts on the schedule from the floods, the additional safety risks introduced were of the highest consideration. The safety performance during the project was the biggest success with there being no medical treatment or recordable injuries. The project was completed with a 5% underspend on the approved budget. The dragline was returned to work with high reliability and the key customers, the dragline operators, were happy with the result, in particular the new kitchen which had many features only a female could dream up.

Melanie continues to implement innovative projects and to look for a better way to do things. She is currently managing a long-life bearing trial which could potentially see dragline shutdown frequencies extended by 2-3 years resulting in a significant cost saving for the business. She is also trialling a whole body vibration dose meter in mobile equipment. The dose meter provides live feedback to the operator allowing them to modify their operating techniques to reduce their exposure to vibration and the associated long term health effects.

Other important aspects of Melanie's role are to advance and mentor the mechanical and electrical engineers in her team, develop and manage the 15 year maintenance plan and develop and manage the 5 year budget.

Typical Day at Work

Melanie starts work at 6am every weekday. There is no typical day at work as every day is different and her plans can change from hour to hour – this being the main attraction and love of the job for her. Tasks that she is required to complete include managing projects, assisting with problems identified in the workplace, managing contractors and consultants, providing presentations to site personnel, investigating safety problems, reviewing and approving engineering designs, scheduling, facilitating risk assessments, managing budgets, writing specifications and writing submissions for project approvals .... just to name a few! She could never complain about being bored and the variety in her day makes life exciting. Melanie spends a significant amount of time in the field and working one on one with people.

Community Benefit

Melanie has always wanted to be in a position to help improve the self-esteem of other women and to help them achieve success. She has been actively involved in the community for a long time however her role as a female Project Manager and winner of the 2008 Resource Award for Women has catapulted her into the lime light and a very good position to be able to help others. Melanie was the initiator to setup a regional networking group for women in the mining industry, she has spoken at schools, conferences and career expos, she mentors several younger and older women, she has appeared on television, radio, magazines and newspapers. Her message to women is to reach for the stars and that they can achieve anything that they want. She also actively promotes careers in the mining industry and uses her own experiences as inspiration to others as an example of what can be achieved.

Melanie was asked by the Queensland Government's Office for Women to speak at two of the "Girls in Hard Hats" forums, at the Queensland Resources Expo in Rockhampton and careers expo in Mackay. The sessions began with a presentation by two or three females working in male dominated fields. The presentations detailed how they began their career, their qualifications and real experiences in the workforce. The second half of the session was an interactive forum where females in the audience were able to meet and ask the presenters questions. The forums were well attended by school students, teachers and parents.

The interactive forums were valuable to the attendees who were able to ask questions which could potentially have been preventing them from considering a career working in the mining sector. Melanie had several young ladies talk to her after the session and say that they were now interested in working at the mines and how her presentation had influenced them. The sessions were widely publicised on the television (channel 7 news) and several newspapers, gaining more opportunities to communicate the message to females who could potentially work in the mining sector. Melanie's profile has been used on the Office for Women internet page to continue to promote a career in the mining industry for women.

In March, Melanie represented BMA at the Skirts in Engineering Pink Tie Ball. Skirts in Engineering are a society at the University of Queensland for female engineering students. At the dinner Melanie spoke about how to apply for a job in the mining industry, why they should consider working in the mining industry and personal stories of her experiences in her career to date. These female engineers have the opportunity to work in any sector due to the world wide shortage of engineers so it is important to "market" the idea of working in the Queensland mining industry to these ladies before they leave university or start applying for jobs.

Melanie is currently driving the initiative from Moranbah to set up a networking forum with "Women In Mining & Resources Queensland" (WIMARQ) for women working at mines in the area. Melanie has contacted companies in the Moranbah area to discuss the potential of them being involved and sponsoring the event. She has also spoken to BMA management who have agreed to assist with sponsorship. The first meeting was held in February 2008 and the theme focused on "Work and Lifestyle balance". The purpose of the forum is to provide an opportunity for women working at mines to network together, make some new friends and discuss opportunities to contribute to the community and encourage more females to join the industry. Currently, women working in the mines do not support each other and sometimes the opposite, they compete with each other. It can be a lonely journey for a professional woman in mining and it would make life much more fun to have some friends in a similar position to relate to. Over 80 women attended the first forum in February and it was declared a huge success. More events are planned for July and October.

Melanie was asked if her profile could be used by the QRC for a presentation, titled "Jobs You Won't Hear About From Your Mum", which was to be shown to school children in Queensland. The presentation shows the human side of a female working in the mining industry starting from primary school, university, her first job, getting married and up to where she is today in her career. The presentation gets the message out to females at a young age that jobs in the mining industry are a real option for them and that the females who currently work in the industry are normal everyday people.

Melanie's contribution and success in the mining industry has provided a role model to other women in the minerals industry, within BMA and to younger people who may one day work in the industry. Melanie's key message has been "*If you have any interest in working in the resources industry then give it a go. Ask heaps of questions, listen to the advice that you're given and try everything at least once. Nobody expects you to be an expert straight away and we all have to start somewhere. Certainly don't ever think that females don't belong in the resources industry because I know many amazing women who have achieved amazing things and have stories similar to mine.*"

Chapter 2
Bias Literacy:
A Review of Concepts in Research on Gender Discrimination and the U.S. Context

Ruta Sevo
Consultant, USA

Daryl E. Chubin
AAAS Capacity Center, USA

ABSTRACT

The chapter offers a quick digest of the evidence for discrimination, especially with reference to women in science and engineering in the U.S. It explains common terminology and lists relevant legislation and national policy initiatives. The chapter summarizes the difference between tradition and bias, conscious and unconscious discrimination, overt and covert discrimination, and personal versus institution bias. Drawing on research in psychology and social science, it summarizes core concepts including: gender schema, accumulative advantage, stereotype threat, implicit bias, glass ceiling, mommy track, occupational segregation, statistical profiling, climate study, and the value of diversity in learning. A short section lists some U.S. national and international approaches to measuring whether discrimination is occurring and how improvements are benchmarked. There is a list of major organizations working for diversity in the U.S., with links. Many of the concepts are more fully described in the recent U.S. national report Beyond bias and barriers (2007), which inspired this literacy effort.

INTRODUCTION

The idea of literacy is applied to many types of knowledge and skills. It usually refers to basic competency in a skill set (e.g., reading or writing) or a profession (e.g., computer technology). Literacy is a set of expectations for knowledge among people who share a culture -- what any insider knows.

Bias Literacy is a timely construct because many individuals and organizations are engaged in understanding the problem of discrimination in society and in the workplace, but have not anchored their thinking in reflection and research on discrimination. When – in what contexts -- do we claim that

DOI: 10.4018/978-1-61520-657-5.ch002

Copyright © 2010, IGI Global. Copying or distributing in print or electronic forms without written permission of IGI Global is prohibited.

someone is discriminating and is biased? How do we arrive at that assessment?

We also believe that literacy is the first step toward action on bias. This is not an intellectual exercise. Once we understand the dynamics and impacts of discrimination, we should understand what to do, and what others are doing successfully, in order to make bias transparent where it has been hidden or unacknowledged, and to control illegal bias. We are all victims of discrimination when our society or profession or group is built on a false sense of equity.

Here, we use the enterprise of science and engineering as the social context from which to draw examples and research. We are concerned particularly with the extent to which women and minorities are treated with or without bias and discrimination in the American system of producing and employing scientists and engineers. We hope the dialog is improved by isolating the basics.

BACKGROUND

The Vocabulary

There are a few key words and phrases that occur often in the discourse about discrimination.

Bias. The definition of bias is inherently negative. It is to favor a view or group over others and to be unfair or partial to a view or a group. Regarding ethnicity, bias may be rooted in racism, which is to consider members of one race or group to be intrinsically superior to others. The *motive* for bias may be unconscious and unintended, however. Bias can be rooted in tradition or prejudice, such that the socialization of children and new members encourages members of a group to be biased against members of another group. ("Asians are superior to whites." "Women do not have a brain for math.") Bias can also be unobservable. We discuss these dimensions in the next section.

Discrimination. "Discrimination" is a synonym for "discernment." It is not inherently negative to analyze differences among things or people, and to make distinctions. It is negative when it is acted upon: treating an individual based on a class or category – such as a stereotype or statistical profile – to the detriment of the individual. ("A woman who becomes a mother will lose interest in her scientific work.")

Prejudice. When a person holds or expresses an adverse, preconceived judgment or opinion about someone else, they are prejudiced. The meaning of the word emphasizes a lack of evidence, and a lack of interest in seeking evidence to sustain this attitude.

Bigotry and Misogyny. These two words are placed together because they both mean irrational suspicion or hatred of a particular group, race, or religion. The common aspect is hatred. Extreme prejudice and extreme discrimination are behind bigotry and misogyny. They imply not only a lack of information or a lack of sophistication, but deep emotional roots and conviction. ("No woman is going to get tenure here.")

Intolerance. Intolerance is a version of bias and discrimination. It stands for the unwillingness to recognize and respect differences in others. There is not necessarily a feeling of superiority on the part of the intolerant person, but possibly a feeling of impatience, discomfort, or dislike (i.e., lighter versions of bigotry and misogyny).

To Be Fair. What does it mean to be without bias and without prejudice, and to be fair? We suggest that *social tolerance is the capacity to recognize different beliefs, practices, and life experience in others, and to respect others in spite of their difference from you. It means holding stereotypes at bay. It means letting the merits of an individual trump preconceived, unsubstantiated negative assumptions about the individual's group. It recognizes that an individual may not represent the group – especially not a gross statistical profile or reductionist view of a group.*

Social Justice. The cause of social justice is about making society fair in its distribution of rewards and burdens. The "cause" is not necessarily rooted in some universal notion of what is fair, and how society should be made fair. Although everyone embarked on the mission of social justice feels a kinship and an affinity to a struggle, it pays to ask fellow-travelers which injustice is the target, and whether you agree on the methods for achieving a remedy. Sometimes the "movement" for equality on the basis of race is itself sexist, and the "movement" for equality on the basis of sex is racist. For example, the U.S. nation's Founders implicitly excluded "women" and "slaves" when they established the principle that "all men are created equal."

Research. Research is a systematic, controlled, and empirical investigation and process of inquiry. Its goal is to discover, interpret, or revise facts, events, behaviors, or theories, and to develop or contribute to credible, knowledge that can be generalized. A scientific method will often employ a hypothesis and the gathering of evidence to test or reject the hypothesis. Social science research may investigate the phenomenon of discrimination in society. Research on discrimination and characteristics of groups and their differences is *not the same as advocacy for social justice*.

Advocacy and Activism. Advocacy is the active support and promotion of an idea or a cause for the sake of changing public opinion, acceptance, and behavior, for example, the cause of social justice. Its methods of influence are wide and varied, ranging from public education, social marketing, and lobbying. Advocacy is inherently driven by values – notions of right and wrong.

Advocacy is not motivated by the same goals and constraints as research (that is, empirical investigation validated by expert judgment). Advocates may use research to make their case for an idea or a cause. ("Because X is true, Y is the right course of action.") Conversely, researchers may advocate for social change. ("X is true, and it is socially wrong or bad.") Advocates summarizing research may be biased (favoring a viewpoint) and selective (e.g., omitting contrary results) in making a "research-based" argument. Researchers may fail to be "scientific" in crossing the boundary toward advocacy for a particular social change. Many published articles on discrimination, especially in the media, confuse research and advocacy. A critical reader should ask themselves if a particular author is making a research case ("this is what we know about X") or whether the author is offering a prescription ("this is the way things should be"). Sometimes both appear in the same piece of writing.

Policy. Policy is a commitment to achieve a valued objective. It may suggest details as to delegated actors, stakeholders, processes, and timetables. Public policy is synonymous with government legislation and investment. Outside government, policies are more community-specific. For example, a university may hold the policy that " faculty diversity is important." Leadership, resources, and detailed action plans are necessary to act on the policy. Since the objective is a subjective value, the process of deciding, supporting, and achieving the policy can be highly politicized.

Diversity. The word represents the value on having the composition of a group, organization, or workforce reflect the distribution or composition of groups in the greater society and population, by race and ethnicity, gender, culture, disability, and other social differences. There are disputes about what constitutes sufficient diversity in a team, group, organization, etc. Diversity is a "problem" when groups are not represented to the extent that they "should be," given the customer base, the population base, and the society in which they exist. Some regard diversity as a condition to be improved (that is, a starting point), while others consider it a solution to a problem retarding achievement of an organization's goals (an end point).

Law. A law is public policy that has been adopted as mandatory practice through a legislative process. It is a rule or body of rules of conduct given authority by a government. It is enforced by designated oversight or accountability agencies. For example, notions of "fairness" can be legislated, disputed and interpreted in courts, and enforced. Individuals or organizations that "break the law" can be punished.

The Legal Context of the U.S.

When a group of boys in school take over a scarce computer to engage their interests (educational or not), pushing girls aside, they are expressing a preference for each other's company and for the pleasure of sharing a passion with like-minded friends. When a white, middle-aged engineer teams with his white male peers on a project he is expressing a preference for collegiality, friendship, and familiarity in tackling the complexities and risks of his project in a socially comfortable environment.

Preferences in both contexts have consequences beyond the immediate situation. In the school setting, the lack of access to a computer may mean that girls do not learn as much as the boys. They are put at a disadvantage. They may be discouraged from using certain tools. They may associate the tools with a select, narrowly defined group (computer=boy thing). They may not enroll in technology classes because they find them unpleasant and unwelcoming. They may fail to develop skills that are foundational to advanced study and preparation for certain careers.

In the second case, the elite team is choosing not to seek out or enlist younger researchers or researchers who are not in the personal network of the group, though they may be more appropriate to the project or offer needed expertise. They are placing a higher value on social comfort and familiarity over wider brain power and intellectual inputs, over the development of new talent, and over the extension of their network to a broader community.

Generally, in the U.S., we may subscribe to the principles that schools should "leave no child behind," that all kinds of students should be exposed to every feasible publicly funded educational experience, that the workforce needs the best minds and more people, especially in advanced science and engineering. The behaviors (on the part of the boys in school and their teachers; on the part of senior researchers and their administrators and funders) are not consistent with expressed values and national policy. The behaviors are unthinking, traditional, or uncontested. They are not technically illegal, unless they are part of a pattern of exclusion, and their impact adds up to measurable discrimination. They are merely unfortunate, lamentable, counter-productive – and maddening to the excluded.

United States's Core Values. The United States Constitution expresses the fundamental values of our country. Use of the phrase "illegal bias or discrimination" is only meaningful with reference to our legal system. Other countries may not have codified "fairness" to the same extent as the United States, nor in the same ways. Here is the U.S. statement of "truths we hold to be self-evident:"

...all men are created equal. (Declaration of Independence, 1776)

We should not assume that other countries share our definitions of rights and privileges, nor enforce them in the same ways as in the U.S.

Here are legal constraints to exclusion and discrimination in the U.S., as contained in:

- **Equal Pay Act of 1963.** Abolishes differential pay based on sex.
- **Civil Rights Act of 1964.** Title IV outlaws racial segregation in schools. Title VI prohibits employment discrimination on the basis of race, color, religion, sex or national origin and establishes the **Equal Employment Opportunity Commission** as enforcer.
- **Title IX of the Education Amendments of 1972**. "No person in the United States shall, on the basis of sex, be excluded from participation in, be denied the benefits of,

or be subjected to discrimination under any educational program or activity receiving Federal financial assistance."

- **Americans With Disabilities Act of 1990.** Prohibits employers from discriminating against qualified individuals with disabilities in job application procedures, hiring, firing, advancement, compensation, job training, and other terms, conditions and privileges of employment.
- **Civil Rights Act of 1991.** Strengthens Federal civil rights laws. Title II set up the **Glass Ceiling Commission** to study discrimination (1991-1996).

Compliance with legal mandates can be soft. For 30 years, Title IX was applied mostly to the issues of inequitable provision of sports programs and sexual harassment in K-12 schools, though this law explicitly identifies neither "sports," sexual harassment, nor primary and secondary schools as its focus. There is room for interpretation as to whether a program that has few girls enrolled is designed to keep them away. With sports, the improved availability of high school programs and the provision of programs designed to appeal to girls led to a ten-fold increase in participation over 30 years. At the college level, the increase was 5-fold (National Coalition for Women and Girls in Education, 2002; Blumenthal, 2005). Could the same be achieved in engineering? Other countries have shown that there is nothing inherently "male" about engineering, for example. Why, then, the gender imbalance in participation?

The U.S. Department of Education Office for Civil Rights bears the responsibility for monitoring compliance with Title IX for the nation's schools, and the Office of Science and Technology Policy oversees the compliance among institutions funded by the federal research and development (R&D) agencies, including Health and Human Services (National Institutes of Health), National Aeronautics and Space Administration, and National Science Foundation. In the case of compliance reviews in higher education, very little activity has been reported publicly.

There have been national policy initiatives to promote diversity and equity in America:

- **The Perkins Act of 1978** (reauthorization) required each state to hire a sex-equity coordinator to make the vocational training system more equitable. (However, it provided little funding and its reauthorization in 1998 eliminated endorsement of the position.)
- **The Equal Opportunities for Women and Minorities in Science and Technology Act of 1981.** States that it is U.S. policy and in the national interest to encourage all groups to participate in science and engineering. Mandated that the National Science Foundation report statistics on underrepresented groups and initiate a suite of programs to influence diversity in the science and engineering workforce. To this day, NSF is the only R&D agency with this mandate.
- **Commission on the Advancement of Women and Minorities in Science, Engineering, and Technology Development 1998**. A two-year congressional commission created to research and recommend ways to improve the recruitment, retention, and representation of women, under-represented minorities (namely, African Americans, Hispanic Americans, and American Indians), and persons with disabilities in science, engineering, and technology (SET) education and employment. A follow-up to what became known as the Morella Commission was a public-private partnership, BEST (Building Engineering and Science Talent) that published several reports in 2002-2004.
- **U.S. Government Accountability Office Report on Gender Issues 2004**. Found

Title IX enforcement inadequate, especially pertaining to higher education in science and engineering, including at institutions with research funded by the U.S. Department of Energy and NSF.

Bias is illegal when members of society are denied access to resources and opportunities on the basis of their membership in a group subject to discrimination. They are disadvantaged and cannot participate in the benefits of the society or contribute equally. America values equal opportunity in education and employment. America needs the participation and contribution of its best talent.

CONCEPTS

Tradition vs. Bias

The first college in the United States, Harvard, was founded in 1636. Only elite young men attended college at that time. Women and people of color (of any class) were excluded. Oberlin College was the first to admit women, in 1833, and the first to admit students of all races. This means that *traditionally*, for nearly 200 years in America's history, women and people of color did not have access to higher education. There was no legal mechanism in place to contest the tradition. The first women to get degrees found it hard to gain employment, as *traditionally,* they were not welcome in the workplace. The early female and "colored" pioneers in higher education might have protested, but in those days, they did not sue in court.

Highly-rated institutions continued to exclude all women as late as 1970 (e.g., the University of Virginia), to exclude them from certain fields of scientific study, and to deter them from graduate study in certain fields of science, such as physics and astronomy. Exclusion from certain fields of science in the 1960's partially explains the low numbers of women available to be nominated to the National Academy of Sciences until recently.

Once women entered the workforce in great numbers, there was social acceptance for the break with the traditional idea that "a woman's place is in the home." However, some occupations were considered more appropriate than others. The "helping" roles of nurse, secretary, primary school teacher, and assistant were generously open to women, due to a number of traditional assumptions: women were not primary income providers for families and did not need high pay; their salaries were disposable income; they were not building careers; they were not interested in power and leadership. Both men and women commonly held these views. There was great social pressure to conform to tradition. There was little public debate about assumptions, and little support for women who had other ideas about work and career. It is true that many women did not feel that they suffered bias, because they found satisfaction in their traditional role and in their conformity with the majority.

Regarding ethnicity, class, and sexual orientation:

Until the 1960s, the great majority of American college and universities were overwhelmingly white. Minority students were rarely seen; their cultural traditions were almost entirely excluded from both the formal curriculum and the informal campus culture. Many institutions sustained quotas on specific religious groups. Other campuses either excluded or subordinated women while virtually every institution denigrated homosexuality. For students from less privileged economic and/or cultural backgrounds, class was--and remains still—a difficult and often painful issue (Smith et al., 1997, p. 8).

The struggles to change tradition can be long and hard, especially when life styles of the advantaged classes are dependent on the free or cheap labor of a group. For example, slavery was abolished in America only in 1865, and women were given the right to vote only in 1920. Other countries were on a different timetable – some earlier and some later. Australia, for example, gave non-aboriginal women the right to vote in 1902, but excluded aboriginals (men and women) until 1962. Discrimination dies hard and fades with much scrubbing. Legal milestones do not automatically bring about social acceptance and conformity with laws. We might ask historians how long it usually takes for new legal rights that challenge tradition to be assimilated – 100 years or more?

Conscious vs. Unconscious Discrimination

When the MIT Report on the Status of Women was issued in 1999, Dean Robert J. Birgeneau added among his introductory comments:

I believe that in no case was this discrimination conscious or deliberate. Indeed, it was usually totally unconscious and unknowing. Nevertheless, the effects were and are real. (Massachusetts Institute of Technology, 1999)

This is also called subtle prejudice. An in-group holds unconscious beliefs and associations about an out-group. In the minds of the in-group, their differences are exaggerated (linguistic, cultural, religious, sexual). Members of the out-group are portrayed as outsiders, worthy of avoidance and exclusion or disadvantage. The in-group may justify the negative attitude and treatment by blaming the out-group for choosing to be deficient. If the in-group is a majority, in society, and in command of most of the resources (jobs, positions of authority, leadership, ownership), it might unconsciously favor its own members, and portray its own members more favorably in many contexts. For example, media portrayals of adventure heroes or doctors or heads of countries will show the in-group "in its element," with the out-group invisible or in subordinate or negative roles.

It is very hard to detect and protest subtle, unconscious discrimination. It may be manifest in petty slights, nuanced visual clues, tone of voice, lack of eye contact, and so on. A TV show that omits the presence of women and minorities may seem normal. The victim's claim that "You are not treating me as an equal" might be hard to substantiate (without a five-year MIT study by senior scientists!).

Overt vs. Covert Discrimination

It is possible for *intended* discrimination to be *covert*—unobservable and possibly uncontestable. For example, an academic committee may choose among candidates for a fellowship and discuss their belief that a woman should not take the place of a man, and they do not document their discussion, nor report it outside the meeting. If the committee members are unaware of their bias, for example assuming that a woman who just married or just had a child probably lost interest in graduate school, their discrimination is *unintended* and *covert*.

Personal vs. Institutional Bias

Sometimes it is not the personal behaviors of individuals that disadvantage a group, but *rules and policies* of an institution. This is also called "structural discrimination." For example, a highly qualified couple may find that one is offered a position at a university in a remote small town. There are no other potential employers for the other spouse within a reasonable radius. Because the university's nepotism rules preclude hiring the "trailing spouse," that spouse must remain unemployed or under-employed, and usually out of their professed field. Another example are rules for qualification for tenure that are unforgiving of temporary leave-of-absence to bear a child, thus posing a disadvantage to a woman who chooses to have children in the first 7-10 years of her career, which are optimal years for childbearing.

Another example of a structural bias is to require a test as a qualification for a job (or other opportunity) that is known to disadvantage a certain group. An English language test – beyond the requirements of a job -- can be used to screen out immigrants. An aptitude test was used to prevent African Americans from obtaining jobs. Until the Civil Rights Act of 1964, flight attendants had to meet height, weight, age, and vision (no glasses) requirements, and be single -- all attributes irrelevant to their primary work function, which was passenger safety. Fire departments historically used excessively challenging physical tests to eliminate women from eligibility. "Women seeking admission to the New York State College of Agriculture in the 1970s needed SAT scores 30 to 40 points higher than those of men." There were quotas for women in college admissions (National Council for Women and Girls in Education, 2002).

Structural bias is impersonal – it is embedded in rules and procedures. It makes it easier for the insider to say "Sorry, I don't make the rules" to the outsider. It may also be unintended—more tradition than policy. And it is sometimes hard to measure the effects, though we are learning how (see "Climate Study" below).

RESEARCH SNAPSHOTS

Gender Schema

All people are exposed to assumptions about gender from birth on. A girl and a boy will have different experiences in any society from the first minute their sex is known through the rest of their lives. They will be differentiated, and learn to differentiate, on aspects of appearance, activities, toys, roles in the family, and occupations. Like other cultural values and assumptions, assumptions about gender are internalized and are likely to become unconscious.

Gender schema theory (Martin & Haverson, 1981; Valian, 1998) holds that gender plays a major role in how children organize information. Gender provides a structure that constitutes an implicit theory. There are probably differences in how strongly gender typing acts between individuals. Younger children tend to have simpler constructs that they hold more rigidly. It is postulated that their initial simple and rigid constructs provide consistency and stability as they make sense of the world.

The gender labels that children learn can evolve into stereotypes. Stereotypes are not inherently bad. They simplify our world and offer rules of thumb about what to expect in social interactions. Gender schema theory is one of several psychological approaches to explaining how children develop identity, particularly gender identity.

Stereotypes across cultures for men and women are not uniform. A majority of cultures expect men to be dominant, competitive, and achieving, and women to be co-operative and supportive. However, expectations (especially implicit, unconscious expectations) and actual behavior may not be consistent. Our interpretations of the normal range of acceptable behavior will vary by culture and by individual, as will our tendency to assign a negative weight to deviations. One example of "gendering" is to consider the behaviors and clothing that are considered gay in America today. In some countries, men hold hands and use expressive gestures, and they are not labeled "gay."

According to Virginia Valian (1998), schemas are hypotheses and unconscious beliefs, shared by both men and women, that can skew our perceptions of their respective abilities. We tend to over-rate men, and under-rate women. In one study, people making a judgment about the height of a man and a woman of equal height, for example, perceived the man as taller. In a study of the peer review system of the Swedish Medical

Research Council for post-doctoral fellowships, female applicants had to have more credentials than men to get the same competence rating from reviewers (Wenneras & Wold, 1997). A study of the influence of gender in the review of curricula vitae for faculty found that both men and women favored male job applicants. One study took actual curricula vitae, created male and female versions, and sent them for review as candidates for faculty positions. It found that both men and women preferred male job applicants (Steinpreis, Anders, & Ritzke, 1999). Another study compared letters of recommendation for medical faculty candidates and found that the letters differed "systematically" in preference toward men, in terms of length, "doubt-raising" language, and references to status (Trix & Psenka, 2003).

Other studies show that we expect men to be capable of independent action, and that men do things for a reason (in contrast with women). We expect women to be more nurturing, expressive, and caring about the group or community than men. In situations where we evaluate the professional competence of men and women, and where there is much room for interpretation, men will have a significant advantage due to unconscious assumptions. Our schema for males is a better fit for professional success, and especially for high-intensity scientific and engineering careers. The theory of schemas explains the disparities between men and women in salary and advancement across professions.

Accumulative Advantage

Due to gender schema and other deep-rooted schemas about race, class, sexual orientation, there are many expressions of unconscious and undetected bias against excluded groups. Accumulative advantage is the idea that many small advantages cumulate over time and add up to a larger advantage. Disadvantages work the same way, negatively. The idea of cumulated disadvantage explains why even small disadvantages are important. They can over time grow into large disparities in salary, promotion, and status. "The rich get richer..." "Like interest on capital, advantages accrue, and ... like interest on debt, disadvantages also accumulate" (Valian, 1998, p. 3). A computer simulation postulated an organization staffed with equal numbers of men and women at the lowest level, having 8 levels of promotion potential. At each level of promotion in the simulation, men were given a mere 1% advantage over women. At the beginning, the starting group was comprised of 50% men and 50% women. After eight minutely advantaged moves, the top level of the organization was 65% male (Martell, Lane & Emrich, 1996). Even small imbalances over time add up to major differences. When a woman or minority is denied access to a course, given a slightly lower grade, denied academic assistance, excluded from a study group, not encouraged, denied a fellowship, a job, or a promotion because of discrimination, they suffer an increasing gap in the competition for opportunity. The rich not only get richer, but the poor poorer. To recover competitive ground, the disadvantaged would have to gain more than 1% in qualifications or opportunity to reach the same level as members of the advantaged group, in the simulation by Martell, Lane & Emrich.

Stereotype Threat

Both men and women think and act according to gender schema automatically and unconsciously, for the most part. Children develop gendered stereotypes from early elementary school and by late elementary school show strong stereotypes. At these ages, girls and boys show equal primary abilities for mathematics. Sex differences in quantitative skills emerge after elementary school (Steele, Reisz, Williams & Kawakami, 2007).

The theory of stereotype threat holds that when an individual is made aware of his or her identity in a group that is negatively stereotyped for an activity, their performance will be impaired (thus confirming the negative stereotype). For example,

African Americans taking an IQ test who are told that their intelligence is being measured will do worse than those told it is not measuring their IQ (Steele & Aronson, 1995). Another example is female students taking a math test and being made aware that "women are not as smart as men in math." The effect is stronger in those who care about the domain (Steele et al., 2007). The effect is triggered by raising awareness of negatives just prior to testing. When women are told that they usually do worse on a math test than men before the test, they tend to do worse than the men. When white men are told they usually do worse on a math test than Asian men, they tend to do worse than the Asians. The effect has been shown in African Americans, Latinos, low socio-economic status students, and the elderly (regarding memory), especially in testing environments. The effect can be triggered without *explicit* reference to the stereotype. For example, exposing individuals to strong cues about their group identity (gender, race, age, etc.) prior to the test can trigger the effect (e.g., filling out a form that asks subjects to identify their group).

Girls have often heard in school, from peers, teachers, and parents: "What are you doing in the physics class?" or "Do you think you can handle the material?" The effect can be mediated a number of ways. Strongly optimistic (and non-judgmental) relationships between students and teachers help. Making the negative stereotype conscious, and refuting it, helps. Awareness of positive role models – individuals from the group who counter the stereotype – helps. Individually, self-affirmation and raising a strong sense of self-adequacy can help foster psychological resistance. If the subject is made aware that they are possibly anxious about anticipated negative expectations, and not really anxious about their ability, they tend to do better. "Teaching about stereotype threat inoculates against its effects" (Committee on Maximizing the Potential of Women in Academic Science and Engineering, 2007, pp. 46-47).

A related idea is the Pygmalion Effect. In studying the effect of teachers' expectations on students, Rosenthal and Jacobson (1968) found that high expectations can result in better performance. This is also known as a self-fulfilling prophecy, applied to many types of relationships such as doctor-patient or manager-employee. The dynamic is different than stereotype threat – it is interactive, between an authority figure and a student or subordinate, and, it usually refers to *positive* expectations. Yet, similarly, the student internalizes the expectation and changes to meet it.

In education literature, the phrase "soft tyranny of low expectations" is often used to blame schools for the poor performance of disadvantaged students. It is another variation on discrimination – the projection of not necessarily a stereotype but rather a negative, low expectation, which research shows will affect response.

Implicit Bias

Can we tell if we are prejudiced? Project Implicit at Harvard University recognized that people are either unwilling to report their attitudes and beliefs, or, unable to report them because those attitudes and beliefs are unconscious (Project Implicit, 2007). They designed an online test designed to uncover unconscious thinking and feeling. The test prompts for associations between words. It measures not so much the patterns of associations that the subject makes, but rather the time (in milliseconds) it takes to respond to a sequence of rapid prompts. Longer times suggest a difficulty making an association automatically. Among the Implicit Association test suite of experiments are tests of a subject's preference for young versus old people, preference for white versus black people, sexuality, gender and career, and gender and science. For example, it often reveals an association between "liberal arts" and "females" and between "science" and "males."

People taking the tests have been surprised at their implicit associations that indicate bias and discrimination. Some are shocked and dismayed at their scores because they have spent their careers and lives committed to fighting discrimination; their identity as a "fair" person is threatened. Subjects often want their results kept private because they consider them professionally damaging. The site warns: "If you are unprepared to encounter interpretations that you might find objectionable, please do not proceed further." The Implicit Association tests are showing that that even the best *conscious* intentions and *conscious* reflection can cannot easily "undo" our *unconscious* schemas. The test can also measure change in attitudes and beliefs after a subject has learned to make different associations.

Glass Ceiling

Why can't women and minorities "get to the top" or "move up the organization"? Because of the glass ceiling, which is a metaphor for an invisible barrier. It does not matter what the qualifications and ambitions of an individual who is "up against the ceiling," if they are a member of certain groups, they will be blocked from advancement. It is possible to identify *where the ceiling is*, in an organization or in a profession. In an equal and fair system of promotion and advancement, every individual has an equal chance of getting ahead. If members of certain groups never or rarely advance, then their group membership is the variable that explains their lack of success. Statistics about the characteristics of people at each level of the organization can show that a disproportionate number of the "disadvantaged" people drop off. *Why* they drop off can have many reasons and mechanisms.

Title II of the U.S. Civil Rights Act of 1991 created a 21-member, bipartisan Federal Glass Ceiling Commission. The Commission's mandate was to study the barriers to the advancement of minorities and women within corporate hierarchies. Their definition was: "artificial barriers based on attitudinal or organizational bias that prevent qualified individuals from advancing upward in their organization into management-level positions" (U.S. Glass Ceiling Commission, 1995).

The case of Asian Americans in science and engineering illustrates the glass ceiling effect (Chen & Farr, 2007). Asian Americans are a population minority in the United States – about 5%. They are over-represented among students and professionals in the fields of science and engineering, that is, Asians are more likely than other groups to enroll and graduate. More than 15% of all doctorates in science and engineering are held by Asians (National Science Foundation, 2003). As a group they have higher educational attainment than whites. Chen and Farr offer four criteria for a glass ceiling: the inequality represents a gender or racial difference that is not explained by other job-relevant characteristics of an employee (such as education, training, field, location), the inequality is greater at higher levels, it is an inequality of opportunity and not merely an inequality in numbers of people at high levels, and finally, the inequality increases over the trajectory of a career. The authors conducted statistical analyses of data crossing seven years, between 1993 and 1999. They found a glass ceiling effect for Asian Americans (both men and women) at all stages of their careers in science and engineering, and confirmed the effect for all women (regardless of race) in science and engineering.

A significant monograph on Asian Americans across all occupations, not just science and engineering, was produced for the Glass Ceiling Commission (Woo, 1994). Other studies have found the effect for African Americans. Catalyst, Inc. updated the Commission report in 2000, with an emphasis on women in management (Catalyst, 2000a).

Mommy Track

The "Mommy Track" is the socio-economic phenomenon that working women who get pregnant and have children experience lower expectations at work, lower opportunity, and lower pay than those who do not have children.

The "Mommy Track" is a complex dynamic. We cannot definitively say that it is the result of systematic discrimination against women who choose to raise children. If anything, it is structural discrimination in our society generally. The tremendous rise in the participation of women in the workforce over the 20th Century has collided with tradition that treasures women's total dedication to the home. We cannot blame any single employer for traditions and stereotypes that make women our primary care-takers in the family. Nor can we expect any single employer to offer the ideal answer to staying competitive (i.e., having a workforce that works long and hard and at reasonable cost by competitive standards) and at the same time allowing for employees to "have a second job" at home in order to fulfill tradition's ideal family scenario. Instead, we tend to deny that the "second job at home" exists. Mothers returning from maternity leave (a "stolen time" of three days to three months) are supposed to "return to normal" and go back to "giving their soul" to work(Crittenden, 2001, p. 35).

The Job of family. The work of raising children and maintaining a family is significant. There is a large literature on the time spent on household work, showing the effects of education, marriage, ages of spouses, presence of children, and so on, and trends in the last century. Roughly, married partners together spend about 53-55 hours per week on household work. Typically two-thirds of that burden falls on the woman. Working women tend to spend 25-30 hours a week maintaining the family. In a study of business school graduates, "the presence of children adds from three to ten hours per week to the workload of male graduates and from ten to 20 hours per week to the workload of female graduates" (Bujaki & McKeen, 1998), p. 114). Since children are not as easily ignored as laundry and cleaning, the workload constitutes *at least the equivalent of a part-time job.*

The Family and home is women's work. "American culture generally stereotypes care-giving as feminine work" (Committee on Maximizing the Potential of Women in Academic Science and Engineering, 2007, p. 174). Tasks and responsibilities at home have not shifted significantly to men, in America. The pressure to maintain the family and the home falls unequally on women.

Changing expectations. Our expectations for how to raise children and how much time to spend raising them have increased. Although there are more supports for child care than there used to be, such as child care centers, pre-school, kindergarten, after-school programs, and community centers, there is a greater expectation that children need constant enrichment and attention in order to thrive. Popular media feature the importance of reading to children every night, exposing them to varied experiences, food, people, and providing much close physical contact. Research shows that children who are not stimulated at home are disadvantaged when they enter school. The research is behind programs like Head Start to provide enrichment in early years to disadvantaged populations. The cost-benefit to society of rich early family environments have been calculated (Heckman, 2006). In disadvantaged children, early stimulation of cognitive, linguistic, social and emotional competencies reduce the need for later social programs and interventions such as reduced pupil-teacher ratios, public job training, convict rehabilitation programs, tuition subsidies, or expenditure on police. We don't think we should compromise parenting and child care, we don't want to, and we recognize the later costs of early neglect. Old authoritarian approaches to child raising – "stay out of sight and out of mind" and "play amongst yourselves" – are out of vogue.

Family-friendly workplace? The definition of a family-friendly workplace is one where an employer acknowledges the demands of family responsibilities and adopts policies that allow for flexibility, particularly in the time and location of work, absence from work, and allowances for family care. In spite of the Family and Medical Leave Act of 1993, the United States "lags far behind all wealthy countries" with regard to family-oriented policies, and lands in a group with Lesotho, Liberia, Swaziland and Papua New Guinea in not guaranteeing some form of paid maternity leave (The Associated Press, 2007). Among established policies are paid maternity and paternity leave, paid sick days, support for breast-feeding, child care facilities, and limits on the maximum work week. The U.S. is a leader in providing equal opportunity to work, but has not helped the family. We can only speculate why American society, relatively, has "left the family behind." An enormous literature in the late 20th Century explores the problem, identifies solutions, and offers model implementation plans and policies. Good employer behavior is frequently highlighted and recognized via national awards and media exposure. An example is the Sloan Work and Family Research Network (2007). A common intervention for improving on retention of female faculty in science and engineering is to provide relief from the traditional tenure process which coincides with prime child-bearing years (National Science Foundation, 2007a).

Women and career interest. Women's expectations that they will work and have careers have risen phenomenally and rapidly. In 2006, adult women with B.A.s held 52% of those degrees, 53% of Masters degrees, 36.5% of professional degrees, and 31.5% of Ph.Ds (U.S. Department of Education, 2006, Table 9). Women with degrees are more likely to work than women without them (Coontz, 2007). Over the last century, the participation of women in the workforce has gone from 18.4% to 46.4% (Rutgers University Center for Women and Work, 2002).

Higher expectation for advancement in career. Legal challenges to discrimination have opened more positions, more kinds of positions, and higher positions, than ever before, as well as more equal pay. Young women graduate from college now perceiving no barrier to their ambitions based on gender. Research reports monitor the changes in opportunity for women in business, closely, and are encouraging (Catalyst, 2007).

Opting out. The tension between working (typically full time) and raising a family has led more women to opt out of careers in spite of investing in long and intensive career preparation - confirming old stereotypes that "women just want to get married and have a baby." Early in the 20th Century,

The first group of American women to receive a higher education – those graduating around 1910 – virtually had to renounce motherhood and family life if they wanted a career. More than half of the female college graduates of that generation never had children. Theirs was a clear-cut either-or choice: career or family. (Crittenden, 2001 quoting Goldin, 1995, p. 33)

The next generation, graduating around 1933, had a "*job then family*" pattern. The graduates in the 1950s (the post WW II decade) "had substantially higher marriage and fertility rates than their predecessors" and went to work after their children were older. Their pattern was *family then job*, for almost all the doors to the professions were still closed to them with the exception of teaching" (Crittenden, 2001, p. 33). "Fewer than 20 percent of college-educated baby-boomer women [graduating between 1966 and 1979] managed to achieve both motherhood and a career by their late thirties or forties. ... Women without children have been *twice* as successful in achieving a career as the women with children. Fully *half* of the women who had attained a career by midlife were childless" (Crittenden, 2001, p. 32). A study of female M.B.A.'s in high management positions found that fewer than half have children, compared with 84 percent of male peers (Crittenden, 2001 quoting Catalyst, 2000b, p. 35).

Many statistics show that women are disproportionately concentrated in less pressured career positions – lower levels of management, community and four-year colleges rather than research universities, non-profit associations, and staff jobs. Did they opt out?

Or pushed out? Employers are in the position of pressure to hire without bias regarding family status (single or married, children or no children) and to invest in professional development fully, regardless. At the same time, they may find prized employees walk away from the investment because of family pressure. (It is a joke for politicians – "I quit in order to spend more time with my family.") Some of this experience may reinforce old biases against committing to women and making allowances for their "home distractions." A study of high-achieving women – graduates of Harvard's professional schools between 1971 and 1981 – found that 25 percent of the M.B.A.s, who were "some of the most expensively trained and highly motivated people in the country, had left the workplace entirely by the early 1990s. Many said they had been forced out of the best jobs once they became mothers" (Crittenden, 2001, p. 36 quoting Walker & Swiss, 1993, p. 34). According to the study, the women were taken by surprise at falling in love with the new baby, and, at the inflexibility and even hostility they encountered in the workplace.

Studies of the female brain drain found that the cost of replacing professionals could exceed the cost of a year's salary. Some employers respond by providing flexible schedules and other allowances, and not punishing women who elect to enjoy them (Crittenden, 2001, p. 36). Inevitably, some male managers and supervisors will consider these benefits unfair advantages, and unconsciously shy away from hiring or promoting women. When a woman has a child, she signals a choice to be available to family over career. Very few people would question her new priority of responding to family needs (sick child?) over work demands (critical presentation?). Peer perception in the workplace is likely to expect the woman, rather than a man, to stay home with a sick child. A 1992 study of government executives found that "women with children have received fewer promotions than women without children and than men, even after controlling for length of service, education, and other factors"(Crittenden, 2001, p. 41 quoting Merit Systems Protection Board, 1992).

Parents and productivity. There are a few studies that show that people who have children and no one at home full time *devote a few hours less to work per week* than those who do not (Committee on Maximizing the Potential of Women in Academic Science and Engineering, 2007; Crittenden, 2001). Employers and peers can exaggerate the negatives associated with "family distractions." Some states have laws prohibiting discrimination against parents. In academics, there are studies that show a gender gap in the number of papers published, but they tend not to isolate the effect of parenthood (Committee on Maximizing the Potential of Women in Academic Science and Engineering, 2007, pp. 117-123). Even so, the practice of counting papers as a measure of productivity is controversial. Measuring work productivity across all types of work is problematic, so the question of whether parents are less productive cannot be answered objectively. That is unfortunate, because the assumption is common.

The Dilemma persists. It is hard to separate out how much the Mommy Track is due to women's withdrawal from intensive career pressures, and how much it is due to being "pushed out" or "parked" by employers. There are signs that the conflict is a taboo subject in professional life. Crittenden (2001) calls it a "conspiracy of silence." Academics are even reluctant to fill out questionnaires about their needs for accommodation for family life, lest it have repercussions on their career advancement. Meanwhile the subtle effects of workplace marginalization and the female brain drain cumulate, reducing the income and status of women, and wasting everyone's investment in the development of human capital. There is a joke that Ginger Rogers did everything Fred Astaire did, only backwards and wearing high heels. There is research evidence that a baby on a woman's psychological hip can be a career drain and a career buster.

Occupational Segregation

Occupational segregation is a sociological label for the phenomenon whereby certain groups are concentrated in certain jobs. In the United States, the Census data currently identifies 449 broad occupations (Bureau of Labor Statistics, 2007). The distribution of men and women across such a wide spectrum is, however, surprisingly uneven. In 1997, 75% of women were concentrated in mainly ten (10 of 449!) occupations: secretary, cashier, administrator, nurse, sales supervisor, nursing aide, bookkeeper, elementary school teacher, waitress, and retail sales clerk (Fain, 2007a). There are many, many more jobs where there are a few women, but mostly not: specifically, 66 occupations in which they are less than 25% of the workers in the occupation category. (Fain, 2007b) These are often called non-traditional occupations.

Affirmative action has opened many job opportunities that were formerly closed due to blatant exclusion or special rules that screened applicants by gender. For example, now there are more males among nurses, secretaries, and airline stewards. There are more females among police officers, carpenters, construction workers, and security officers.

Vocational education in the United States before Title IX restricted women's access to non-traditional careers such as automotive, aviation, food, and maritime trades, all normally higher-paying than the careers open to women, such as health services, teaching, graphic arts, office technology, and cosmetology. Efforts to open vocational programs are still ineffective (National Council for Women and Girls in Education, 2002). The National Alliance for Partnerships in Equity (2006) tracks annually the statistics on non-traditional careers, showing those in which males or females are more than 75% of the current occupation group. Collapsing classifications of jobs from multiple sources, they list 204 jobs that are *non-traditional for women* (women comprise less than 25% of workers), and 62 jobs that are *non-traditional for men* (men comprise less than 25%). Many reports have observed that, generally, women are still concentrated in the lowest paying jobs, and in relatively few "pink" jobs.

There is horizontal segregation or uneven distribution across types of work specializations. There is also vertical segregation, which is the concentration of one gender in higher, more advanced positions in a given occupation.

As the gender gaps in university attendance narrow, and as more women enter the workforce than ever before, and more egalitarian views enter the public consciousness, and legal barriers to discrimination are assimilated – is there less sexual segregation in the workforce, especially in developed countries? Not really. "Occupational ghettoes" persist, because of deeply rooted cultural assumptions across many cultures that men and women are fundamentally different (Charles & Grusky, 2004). Women are considered suited to provide service and nurturing, and men are considered suited to do physical labor, technical tasks, and abstract calculation or analysis. Thus women cluster in service-oriented occupations that don't involve manual labor, and men are clustered in physically demanding occupations often involved with manual labor. When both participate in occupations that do not require manual labor, there is still vertical segregation due to views that men are more competent and status-worthy than women. The segregation is perpetuated, especially in affluent economies in which a wide variety of careers and occupations are available, as males and females adapt their interests to conform to traditional models, and employers project their fundamental assumptions about differences.

Occupational segregation by gender is a self-fulfilling prophecy on a societal level. Gradually, as men work as nurses, administrative assistants, and in care-taking professions, and as women work as soldiers, security guards, economists, CEOs, and civil engineers, the public may adjust its view of differences from real experience and allow for a less segregated profile. During World War II, women were called to take on jobs normally excluding them, and showed they were quite capable of performing them well much to the surprise of many. This demonstrates that strong biases about differences can be suspended.

Statistical Profiling

Statistical profiling is a legitimate and powerful technique made possible by the availability of huge banks of electronic data (such as the national census), ubiquitous computing, and software that put it feasibly in the hands of anyone who knew statistics. It made it possible in marketing to analyze and target consumer behavior – "if a person with a median income likes to barbeque they will probably buy lawn care products." Statistics made it possible to analyze voter behavior, and the characterization of highly refined subsets of population such as soccer moms and their likelihood of holding certain positions on issues. Unfortunately, it also can promote simplistic, and reductionist thinking: "women don't want demanding jobs." The logical fallacy is to mistake an individual who shares characteristics with a profiled group as having all the characteristics, and attributing a determinism to the profile. For example, "Soccer moms will support investment in education."

Statistical profiling can thus create and reinforce false stereotypes and assumptions by making them appear to be a truth about individuals.

Climate Study

Sociologists have characterized the culture of organizations. Universities promote themselves on the basis of unique traditions and constellation of values. Professions such as physics, geology, and mathematics develop a pride of professional identity. Accordingly, academic departments which are a microcosm of a disciplinary field can reverberate and reinforce characteristic identity as they bring new students into the profession. Unfortunately, certain fields (electrical engineering), like certain occupations (airline pilot) are gendered -- especially strongly associated with male dominance, and thus women are not perceived by themselves or others as a "good fit."

There is no wide consensus about the definition of organizational or professional culture (Powell, Bagilhole, & Dainty, 2007). Cultural life encompasses psychological predispositions, beliefs, values, and learned ways of behaving. Organizations can be characterized as hierarchical, patriarchal, sexist, resistant to change, or competitive as well as collaborative, egalitarian, open, and dynamic. "Historically, the image of engineering has been tough, heavy and dirty, and to do with machinery" (Powell, Bagilhole & Dainty, 2007, p. 50). When women go against cultural dictates and undertake "male work," everyone is uncomfortable. They can experience a hostile environment of bullying, stalling, sabotage, harassment, and spite. The behaviors can be small instances of conscious or unconscious, intended or unintended, overt or covert bias and resentment. A culture can also be simply indifferent, uninteresting, or uninviting.

There is a research literature devoted to climate study -- the challenge of determining whether a culture is experienced as friendly by women and whether it has embedded structural, or institutional barriers. The subject of a climate study is the atmosphere and environment for students or professionals, both intellectual and social, and interactions between individuals, the community, and the system of the institution. The survey tool typically addresses perceptions and experiences of individuals in various roles (e.g., students, faculty, teaching and lab assistants, peers), perceptions of typical experiences (classroom, laboratory, advising, tutoring, conferences), and personal characteristics (confidence, perception of the field, preparation for the field). Data can be disaggregated by gender, ethnicity, year in school or year in profession, and citizenship status. Most climate studies focus on an educational institution or a department.

Two major studies of students found that aspects of a negative climate explain the loss of qualified students in their first years of college study. An ethnographic study about students leaving science and engineering was conducted by Elaine Seymour and Nancy Hewitt (1997). They conducted a three-year, multi-institutional qualitative survey to identify the reasons why undergraduate students, with an emphasis on females, leave S&E majors for non-science fields. With no significant differences in academic preparation or performance between leavers and non-leavers, it was found that leavers: lose interest in S&E; are drawn to a non- S&E major; recognize poor teaching; and feel overwhelmed by the fast pace and workload of their coursework. Women were more likely than men to be discouraged by the competition in S&E departments. Seymour and Hewitt also found that, for many women, experiencing engineering education as a distinct minority automatically puts them at a psychological disadvantage, eroding their confidence.

In another large-scale effort, the *Women's Experiences in College Engineering Project* (WECE) conducted a cross-institutional, longitudinal examination of undergraduate women's experiences and persistence in engineering majors at 53 universities (Goodman & Cunningham, 2002). Consistent with Seymour and Hewitt (1997), over 30 percent of the exiting students cited *school climate factors* as a reason for leaving. WECE also noted losing interest and attraction to another field as key factors for leavers.

Brainard and Carlin (1997) conducted a six-year longitudinal survey of female students in science and engineering at the University of Washington. Factors that influenced the decisions of women to persist in, or switch out of, degree programs in engineering and science were: a significant loss of self confidence in their first year of study; a loss of interest, discouragement, lack of confidence, poor advising, and an attraction to another field. Persisters continued because they had a positive view of their coursework, understood career opportunities, and had positive experiences with faculty and with a Women-In-Engineering program.

Hartman and Hartman (2006) conducted a longitudinal survey of engineering students to determine characteristics of stayers and leavers at Rowan University. Involvement in student activities was a major factor in student retention. Leavers were less likely to be involved in discipline-specific professional societies, academic enrichment programs, and interactions with faculty.

Three case studies of physics departments yielded a set of characteristics that produce a high percentage of female graduates in physics (Whitten, 2003; Whitten, 2004; Whitten, 2007).

- Faculty support structure
  - More women on the faculty
  - Family-friendly policies, addressing child care, family leave, and the helping the partner of a hiree find a job (called the "two-body" problem)
  - A tolerant atmosphere (no sexist remarks, invited speakers include females)
  - Good team work
  - Support for junior faculty, especially in the first year
- Recruiting and outreach
  - An inviting departmental web site
  - Community presence at science fairs, summer programs, school visits
  - Department and admissions office work together
  - Offer bridge programs
- Role of alumni and alumnae
  - Department maintains relationships with former students
  - Use alums to illustrate career options
  - Highlight alum accomplishments

    - Bring alums back to give talks as role models and career awareness
- Departmental culture
    - Students have an active role in maintaining supports for others
    - Strong sense of community
- Introductory courses and early course experience
    - Courses are interactive
    - There is a value and use of team work
    - Courses include learning "the culture of the profession"
    - Offer special math skills development, such as 3-D spatial skills
    - Courses convey applications of the field to the environment and to social issues
    - Students participate in research with faculty
    - Safe labs
- Student care
    - Mentoring for four years
    - Student lounge
    - Tutorial service
    - Lab assistants
    - Seminars on career aspects
    - Club and social activities

A climate study can be one stage in an institutional process to change its culture in order to recruit and retain more diverse students in science and engineering. Data and discussion about a local context is more persuasive to faculty than national statistics on the issues. An example of such an intervention is the University of Wisconsin-Madison Climate Workshops for Department Chairs (Committee on Maximizing the Potential of Women in Academic Science and Engineering, 2007).

Value of Diversity in Learning

Many people dismiss the pressure to diversify as rampant political correctness. Personally, they assume that their profession or group's system of recruitment and advancement is fair, and that members who are brought into the group are evaluated objectively. They consider the pressure to include individuals from under-represented groups as an effort to favor less qualified people and to compromise standards. The research evidence for discrimination is not usually considered, in their argument, nor is the evidence for *the value of diversity in learning and work.*

The question gained critical importance in 2003 in the Supreme Court's deliberation of a suit against the University of Michigan for using race-sensitive criteria in admitting students to law school and to its college (University of Michigan, 2003). The University's intention was to ensure that it had a racially diverse student body in order to provide the best possible learning environment for students. It was exercising affirmative action by artificially reversing a historical trend of low enrollments by African American students. The Court found *for* the University's procedures in the law school (allowing continued use of existing criteria) but *against* them in the college (finding the criteria were biased) (Alger, 2003).

Many organizations supported the University's position by submitting briefs summarizing substantial research on the educational benefits of diversity. One brief represented 34 associations in higher education (American Council on Education et al., 2003). The summary of a key brief reads:

Research studies show that student body diversity can promote learning outcomes, democratic values and civic engagement, and preparation for a diverse society and workforce—goals that fall squarely within the basic mission of most universities. Several studies demonstrate that student body diversity broadens the range of intellectual opinions on university campuses and improves classroom learning environments, that diverse learning environments promote thinking skills, and that cross-racial interaction has positive effects on retention, college satisfaction, self-confidence, interpersonal skills, and leadership. Diverse learn-

ing environments challenge students to consider alternative viewpoints and to develop tolerance for differences, and can promote participation in civic activities. Studies further show that student diversity better prepares students for an increasingly diverse workforce and society. (American Educational Research Association et al., 2003)

Within the briefs were hundreds of citations for research studies and policies behind the value. Some verified empirically positive learning outcomes for students exposed to diversity, such as thinking ability, sophistication of opinions, higher self-confidence, and improved interpersonal and leadership skills. A special case for the value of diversity in science and engineering was among them (Massachusetts Institute of Technology et al., 2003). Prior events, especially California's Proposition 209 and the legal decision in Hopwood versus University of Texas, prompted an earlier published literature review (Smith, 1997) and the development of a web site to consolidate and update knowledge on diversity (Association of American Colleges and Universities & University of Maryland, 2007). Findings are generally that "environmental and experiential components of a diverse campus have positive effects on retention, overall college satisfaction, college grade point average, and intellectual and social self-confidence" (Smith, 1997, p. 28).

MEASURING SOCIETAL PROGRESS

Most countries gather data about their population through a census. The type of data gathered and its ability to show social and economic trends depends on the theories behind the design of the survey. Disaggregation by sex is common for census data, but not always common for data about occupations or education and other domains of life that are assumed to be neutral and independent of differences due to sex.

International comparisons are problematic due to uneven investments in data collection, lack of infrastructure to gather data, and no universal conceptualization of the issue, for example, women's empowerment. Nevertheless, the international development community has worked to adopt and refine measures related to discrimination and inequity. The United Nations uses a Human Development Index to measure a country's achievements in longevity (life expectancy at birth), knowledge (adult literacy rate and education enrollments), and a decent standard of living (GDP per capital). A **Gender Development Index** is based on the Human Development Index and measures disparities in achievement between men and women. Following an international conference on sustainable development in 1994 and the 1995 Beijing conference on women, the use of a **Gender Empowerment Measure** was introduced. The Gender Empowerment Measure also estimates gender inequality in economic and political spheres of activity. Economic participation and decision making is measured by the percentage of female administrators and managers, and professional and technical workers. Political participation and decision making are measured by the percentage of seats in parliament held by women. Power over economic resources is measured by Women's GDP per capita (PPP US$) (United Nations Economic Commission for Europe, 2007). Extensive studies have scrutinized the measure and its implementation (Malhotra, Schuler, & Boender, 2002) and refined it (Williams, 2005; Beteta, 2006).

In the United States, the U.S. Department of Labor, Bureau of Labor Statistics provides *Occupational Employment Statistics* (2007) which provides extensive, detailed information about employment by occupation, and wages, with analyses down to state and local levels. The web site used to offer breakdowns by sex, but those reports were removed under the George W. Bush administration and the information must now be specially retrieved or requested. Other organizations have made the effort to retrieve the data from BLS and present it for the analysis of occupational segregation by sex (National Alliance for Partnerships in Equity, 2007)

The Equal Opportunities for Women and Minorities in Science and Technology Act of 1981 mandated that the National Science Foundation report statistics on under-represented groups in science and engineering. The agency publishes *Science and engineering indicators* biannually, and a more detailed volume *Women, minorities and persons with disabilities* (National Science Foundation, 2007) The reports give statistics on enrollments in higher education by field at the undergraduate and graduate levels, disaggregated by race/ethnicity, disability status, and sex, as well as employment in science and engineering. From time to time, the National Science Board issues special reports with greater depth of analysis of statistical trends, for example, *Broadening participation in science and engineering faculty* (National Science Board, 2004).

The most detailed statistics on education at the K-12 level are available from the National Center for Education Statistics. Indicators on learning with breakdowns by race/ethnicity and gender are given in the publication of national statistics, for example, the National Assessment of Education Progress (NAEP). Also, special publications such as *Trends in educational equity of girls & women: 2004* assemble a series of indicators that examine the extent to which males and females have access to the same educational opportunities, avail themselves equally of these opportunities, perform at similar levels throughout schooling, succeed at similar rates, and reap the same benefits from their educational experiences (U.S. Department of Education, 2004). It shows, for example, the rates of enrollment in Advanced Placement courses, and compares median scores on national mathematics and science tests.

The Commission on Professionals in Science and Engineering (2007) monitors national statistics on human resources in science, engineering and technology. CPST regularly publishes *Professional women and minorities*, which is a detailed reference book of data on human resources presented in over 300 tables with breakouts by sex and minority status. *Professional women and minorities* is widely regarded as an authoritative source of data and trends in science and technology. Data on enrollments, degrees, and the general, academic, and federal workforce by field and subfield are included. It has been published biennially for nearly three decades.

A special project looked at the top fifty research departments in universities in science and engineering. It compared statistics on graduation rates, by sex and race/ethnicity and with profiles of the faculty (Nelson & Rogers, 2004; Beutel & Nelson, 2005). Comparing these two datasets showed that leading departments were not hiring women and minorities at nearly the rate as they were graduating. It also showed that less than 15% of full professors in top research departments are women or non-Whites. In the fields of biological sciences, psychology and social sciences, women were up to 15%, but they were much lower in physical sciences and engineering. In addition, Blacks and Hispanics made up 4.1% of faculty, with female Blacks and Hispanics at only 1%.

The American Association of University Professors (2006) looked at four indicators for "faculty gender equity" (across all fields in higher education, not just science and engineering):

1. Employment status (are women employed as faculty full-time tenure-track or other?
2. Tenure status (non-tenure track, tenure-track, tenured; by type of institution)
3. Full professor rank (percentage of women by type of institution)
4. Average salary (salaries by rank and across all academic ranks)

The National Science Foundation's ADVANCE program (NSF, 2007a) funded a number of institutions with the aim of institutional transformation in advancing female faculty. A number of projects have developed very detailed measurements to quantify progress. One evaluation (De Cohen & Clewell, 2006) for example, employed the following measures:

- Faculty in science and engineering by gender
- Faculty in tenured/tenure-track positions by gender and rank
- Faculty in non-tenure-track positions by gender and rank
- Faculty in administrative positions by gender
- Faculty in endowed/named chairs/professorships by gender
- Faculty on promotion and tenure committees by gender
- Tenure promotion outcomes (baseline and during grant) by gender
- Years in rank by gender
- Time at institution and differential attrition by gender
- Salaries of scientists and engineers (faculty) by gender (with controls)
- Space allocation by gender (with controls)
- Start-up packages of newly hired S&E faculty by gender (with controls)

ORGANIZATIONS IN ACTION

There are many motivations for working against bias and discrimination. Individuals and organizations may find themselves lumped together when their agendas are truly different, and actually narrowly intended:

- *Fairness and social justice* – harkening back to America's purest principles of equality and legal mandates
- *Wider educational opportunity* for the purpose of greater literacy in a complex society
- *Empowerment of women and minorities* to redress historical disadvantage
- *Rational human resource development* to stay responsive to demographic trends, a global economy, and diverse customers
- *National economic and technological competitiveness* – recruiting more good talent to build intellectual and economic capacities to maintain America's lead in productivity and growth

A growing constellation of organizations is engaged in promoting diversity in the U.S.:

- **Federal agencies** have diversity policies and programs. The National Science Foundation requires every grant application to address "broader societal impacts" (NSF, 2007b) which means pushing science to people it does not reach as well as it should. NSF has a division for human resource development (NSF, 2007c) and special initiatives. The National Institutes for Health in the U.S. Department of Health and Human Services (2007) asks grant applicants to include a "Recruitment and Retention Plan to Enhance Diversity."
- **The National Academy of Science (Committee for Women) and the National Academy for Engineering**
- **Professional associations** specializing in the issue:
  - American Association for Women in Science (AWIS)
  - American Association of University Women (AAUW)
  - Commission on Professionals in Science and Technology (CPST)
  - National Action Council for Minorities in Engineering (NACME)
  - National Alliance for Partnerships in Equity (NAPE)

- National Association of Diversity Officers in Higher Education (NADOHE)
- National Society of Black Engineers (NSBE)
- Society of Hispanic Professional Engineers (SHPE)
- Society of Women Engineers (SWE)
- Women in Engineering Professional Advocates Network (WEPAN)

- **Professional associations with a specializing unit**:
  - American Association for the Advancement of Science (AAAS), Center for Advancing Science and Engineering Capacity
  - American Association of Physics Teachers (AAPT), Committee on Women
  - American Council on Education (ACE), Office of Women in Higher Education
  - American Educational Research Association (AERA), Research on Women and Education Special Interest Group
  - American Institute of Physics (AIP)
  - American Physical Society (APS), Committee on the Status of Women in Physics
  - American Society for Engineering Education (ASEE), Women in Engineering Division
  - Association for Computing Machinery (ACM), Committee on Women in Computing
  - Association of American Colleges and Universities (AACU), Program on the Status and Education of Women
- **Centers and institutes**
  - Anita Borg Institute
  - National Center for Women in Information Technology (NCWIT)
  - Building Engineering and Science Talent (BEST)
- **Programs and centers embedded in universities**, for example:
  - Georgia Tech, ADVANCE Program
  - University of Washington, Office of Minority Affairs and Diversity
  - Arizona State University, Women in Applied Science and Engineering and Minorities in Engineering
- **Corporations** investing in diversity, for example:
  - L'Oreal
  - IBM
  - HP
  - Dupont
  - Raytheon
  - Texas Instruments
  - Intel

Accountability

There are many ways organizations are held accountable for good and bad performance regarding discrimination: publicity, awards, individual and class action lawsuits, and oversight by government organizations that are required, in their programs, contracts, and funding, to comply with the law.

Recently there was movement to use the provisions of Title IX to put pressure on universities to address inequities in science and engineering higher education. Senators Ron Wyden (Democrat-OR) and Barbara Boxer (Democrat-CA) sponsored legislation that mandated the General Accounting Office to audit Federal science agencies' oversight of Title IX in 2004. (Title IX applies to any organization receiving Federal funding; grantees are obligated to assure equal educational opportunity.) The Government Accountability Office (GAO) Report (2004) found that Title IX could provide the basis for reviewing research institutions and assessing whether they are discriminating in the treatment of students and faculty. Compliance reviewers could look at universities where, for example, a relatively low number of women go on to graduate study or where relatively few qualified women are hired even though they are available in the national candidate pool. National statistics reveal problem areas, and local statistics can show where universities are not making an effort to address and eliminate discriminatory policies and practices. There was little follow-up on the GAO Report. The potential for action does not go away as long as Title IX is the law.

CONCLUSION

The history of democracy and education in the United States illustrates a progression toward truer realization of equality. We have moved from a country run by a small handful of upper-class white men to one that gradually shared education and power with other groups. The abolishment of slavery, the right to vote to women, opening of education, and the myriad of laws and practices that these movements represent are not clean jumps over a transom.

Early philosophical assumptions about the inferiority of women and of other races are embedded in our religions and folklore, and they still persist. Now modern brain research is finding differences in cognitive preferences and in brain processing patterns between men and women. The differences that they identify do not indicate that men and women, or particular races, are determined from birth in capacity and interest for learning, work, and particular careers. The spectrum of interest and ability by sex and by race is not so different that it determines who should be in certain careers and who should be in power. From cross-cultural comparisons, we see that innate genetic differences do not explain global variations in who does what kind of work and who holds power. For example, we have evidence that Japanese girls can outperform Caucasian American boys in mathematics. That does not mean that we should direct Japanese girls into our advanced mathematics professions, and steer Caucasian American boys away, nor do we need to, in order to achieve excellence.

It is not easy to change our minds. It may be a function of our brains to resist locking on schema and stereotypes once they are learned:

Neuroscientists have recently shown that biases in thinking are built into the very way the brain processes information – all brains, regardless of their owner's political affiliation. In a study of people who were being monitored by magnetic resonance imaging while they were trying to process dissonant or consonant information about George Bush or John Kerry, researchers found that the reasoning areas of the brain virtually shut down when participants were confronted with dissonant information, and the emotion circuits of the brain lit up happily when consonance was restored. Researchers also showed that reading information that goes against your point of view can make you all the more convinced you are right. So there is a neurological basis for the observation that once our minds are made up, it is hard to change them. (Tavris & Aronson, 2007)

For profound personal change, we may have to depend on real-life, personal experiences that confront our bias. White soldiers in World War II who were rescued or saved by African American or Japanese conscripts saw them as valued peers for the first time. An older white male whose life was saved by a Black female physician changed his mind about his assumptions.

An advice columnist captured the problem of attitude, talking about a father who showed no interest in a daughter:

If he continues to put girls and boys in neat little unequal boxes ... then it's time to ask why. Pick any reason, that he's ignorant, 'old school,' controlling, angry, in a rut ... you still have the same common denominator. Fear. The clearer we draw our distinctions, the less gray we need to navigate, the fewer unknowns we need to face, the less we need to learn about ourselves or others, and the less we have to try. Trying, on the other hand, is scary. (Hax, 2007)

REFERENCES

Alger, J. (2003). *Summary of Supreme Court decisions in admissions cases, U.S. Supreme Court – June 23, 2003*. Downloaded October 2, 2007 from http://www.vpcomm.umich.edu/admissions/overview/cases-summary.html

American Association of University Professors. (2006). *AAUP faculty gender equity indicators 2006*. Downloaded October 2, 2007 from http://www.aaup.org/AAUP/pubsres/research/geneq2006

American Council on Education et al. (2003). *Brief of amici curiae American Council on Education et al., Nos. 01-1333, 01-1415, 0101418. Gratz et al. v Bollinger et al.* Downloaded October 2, 2007 from http://www.vpcomm.umich.edu/admissions/legal/gra_amicus/gra_ace.html

American Educational Research Association. Association of American Colleges and Universities, & American Association for Higher Education. (2003). *Brief of amici curiae in support of the respondents, No. 02-516, Gratz et al. v Bollinger et al.* Downloaded October 2, 2007 from http://www.vpcomm.umich.edu/admissions/legal/gra_amicus-ussc/um/AERA-gra.pdf

Association of American Colleges and Universities and University of Maryland. (2007). *Diversity web*. http://www.diversityweb.org/

Beteta, H. C. (2007). What is missing in measures of women's empowerment? *Journal of Human Development*, *7*(2), 221–241. doi:10.1080/14649880600768553

Beutel, A. M., & Nelson, D. J. (2005). The gender and race-ethnicity of faculty in top science and engineering research departments. *Journal of Women and Minorities in Science and Engineering*, *11*(4). doi:10.1615/JWomenMinorScienEng.v11.i4.50

Blumenthal, K. (2005, July 6). Title IX's next hurdle. *The Wall Street Journal*, p. B1.

Brainard, S., & Carlin, L. (1997). A longitudinal study of undergraduate women in engineering and science. In *Proceedings of 1997 ASEE/IEEE Frontiers in Education Conference.*

Building Engineering and Science Talent. (2002). *The quiet crisis: Falling short in producing American scientific and technical talent.* Downloaded January 9, 2008 from http://www.bestworkforce.org/PDFdocs/Quiet_Crisis.pdf

Building Engineering and Science Talent. (2004). *A bridge for all: Higher education design principles to broaden participation in science, technology, engineering and mathematics*. Downloaded January 9, 2008 from http://www.bestworkforce.org/PDFdocs/BEST_BridgeforAll_HighEdFINAL.pdf

Building Engineering and Science Talent. (2004). *The talent imperative: Diversifying America's science and engineering workforce*. Downloaded January 9, 2008 from http://www.bestworkforce.org/PDFdocs/BESTTalentImperative_FullReport.pdf

Bujaki, M. L., & McKeen, C. A. (1998). Hours spent on household tasks by business school graduates. *Women in Management Review*, *13*(3), 105–113. doi:10.1108/09649429810215848

Catalyst, Inc. (2000a). *Cracking the glass ceiling: Catalyst's research on women in corporate management #16133*. Downloaded September 10, 2007 from http://www.catalyst.org/xcart/product.php?productid=16133

Catalyst, Inc. (2000b). Women and the MBA: Gateway to opportunity. New York.

Catalyst, Inc. (2007). Retrieved from http://www.catalyst.org/

Charles, M., & Grusky, B. (2004). Occupational ghettos: The worldwide segregation of women and men. Stanford, CA: Stanford University Press.

Chen, T. T., & Farr, J. L. (2007). An empirical test of the glass ceiling effect for Asian Americans in science and engineering. In R.J. Burke & M.C. Mattis (Eds.), Women and minorities in science, technology, engineering and mathematics: Upping the numbers (pp. 128-156). Cheltenham, UK: Edward Elgar Publishing Lmtd.

Commission on Professionals in Science and Technology. (2007). *Workforce data.* Retrieved from http://www.cpst.org/

Commission on the Advancement of Women and Minorities in in Science. Engineering, and Technology Development. (2000). *Land of plenty: Diversity as America's competitive edge in science, engineering and technology.* Downloaded January 9, 2008 from http://www.nsf.gov/publications/pub_summ.jsp?ods_key=cawmset0409

Committee on Maximizing the Potential of Women in Academic Science and Engineering. (2007). Beyond bias and barriers: Fulfilling the potential of women in academic science and engineering. Committee on Maximizing the Potential of Women in Academic Science and Engineering / Committee on Science, Engineering, and Public Policy.

Coontz, S. (2007). Motherhood stalls when women can't work. *Hartford Courant*, May 13, 2007. Copyright Council on Contemporary Families. Downloaded September 10, 2007 from http://www.contemporaryfamilies.org/subtemplate.php?t=pressReleases&ext=motherhoodstalls

Crittenden, A. (2001). The price of motherhood. New York: Henry Holt & Co.

De Cohen, C. C., & Clewell, B. C. (2006). *External evaluation of the ADVANCE program Georgia Institute of Technology: Year four report.* Downloaded October 2, 2007 from http://www.advance.gatech.edu/files/GTYearFourReport.pdf

Fain, J. (2007a). *Occupational segregation home page, 20 leading occupations of employed women, 1997 annual averages.* Downloaded September 10, 2007 from http://spears.okstate.edu/home/jrfain/legacy/occseg/20_Leading%20Occupations.html

Fain, J. (2007b). *Occupational segregation home page, Non-traditional occupations for employed women in 1997.* Downloaded September 10, 2007 from http://spears.okstate.edu/home/jrfain/legacy/occseg/NontraditionalOcc.html

Goldin, C. (1995). *Career and family: Women look to the past* (Working Paper No. 5188). Cambridge MA: National Bureau of Economic Research.

Goodman, I. F., & Cunningham, M. L. (2002). *Final report of the Women's Experiences in College Engineering (WECE) Project.* Downloaded September 10, 2007 from http://www.grginc.com/WECE_FINAL_REPORT.pdf

Hartman, H. & Hartman, M. (2006). Leaving engineering: Lessons from Rowan University's College of Engineering. *Journal of Engineering Education,* January.

Harvard University. (2007). *Project Implicit.* Retrieved from https://implicit.harvard.edu

Hax, C. (2007, July 29). Tell me about it. *Washington Post, Sunday Source.*

Heckman, J. J. (2006, June 30). Skill formation and the economics of investing in disadvantaged children. *Science, 312*, 1900–1901. doi:10.1126/science.1128898

Malhotra, A., Schuler, S. R., & Boender, C. (2002). *Measuring women's empowerment as a variable in international development.* Prepared for the World Bank Workshop on Poverty and Gender. Downloaded on October 4 from http://www.aed.org/LeadershipandDemocracy/upload/MeasuringWomen.pdf

Martell, R. F., Lane, D. M., & Emrich, C. (1996). Male-female differences: A computer simulation. *The American Psychologist, 51*, 157–158. doi:10.1037/0003-066X.51.2.157

Martin, C. L., & Halverson, C. F. (1981). A schematic processing model of sex typing and stereotyping in children. *Child Development, 52*, 1119–1134. doi:10.2307/1129498

Massachusetts Institute of Technology. (1999). *A study of the women faculty in science at MIT.* Downloaded January 7, 2008 from http://web.mit.edu/fnl/women/Birgfnl.html

Massachusetts Institute of Technology. et al. (2003). *Brief of amici curiae in support of the respondents No. 02-241, 02-516.* Downloaded January 9, 2008 from http://www.vpcomm.umich.edu/admissions/legal/gru_amicus-ussc/um/MITfinal-both.doc

Merit Systems Protection Board. (1992). *A question of equity: Women and the glass ceiling: A report to the President and the Congress,* October 1992

National Alliance for Partnerships in Equity. (2006). *Non-traditional occupations SOC/CIP/Career Clusters/DOE Pathways crosswalks.* Downloaded September 10, 2007 from http://www.napequity.org/nape_publications.htm

National Coalition for Women and Girls in Education. (2002). *Title IX at 30: Report card on gender equity.* Retrieved September 10, 2007 from http://www.ncwge.org/pubs-reports.html

National Science Board. (2004). *Broadening participation in science and engineering faculty* (NSF 04-41). Downloaded October 2, 2007 from http://www.nsf.gov/pubs/2004/nsb0441/nsb0441.pdf

National Science Foundation. (2007a). *ADVANCE: Increasing the participation and advancement of women in academic science and engineering careers.* Downloaded from http://www.nsf.gov/funding/pgm_summ.jsp?pims_id=5383&from=fund

National Science Foundation. (2007b). *Merit review Broader Impacts criterion: Representative activities.* Downloaded January 7, 2008 from http://www.nsf.gov/pubs/gpg/broaderimpacts.pdf

National Science Foundation. (2007c). *Division of Human Resource Development.* Retrieved from http://www.nsf.gov/div/index.jsp?div=HRD

National Science Foundation. (2007d). *New formulas for America's workforce: Girls in science and engineering* (NSF 03-207). Downloaded January 9, 2008 from http://www.nsf.gov/ehr/hrd/Newformulas/newformulas.jsp

National Science Foundation. (2007e). *New formulas for America's workforce 2: Girls in science and engineering* (NSF 06-60). Downloaded January 9, 2008 from http://www.nsf.gov/ehr/hrd/Newformulas/newformulas.jsp

National Science Foundation. (2007f). *New tools for America's workforce: Girls in science and engineering* (NSF 06-59). Downloaded January 9, 2008 from http://www.nsf.gov/ehr/hrd/Newformulas/newformulas.jsp

National Science Foundation, Division of Science Resources Statistics. (2003). *Characteristics of doctoral scientists and engineers in the United States: 2003* (NSF 06-320). Arlington, VA: National Science Foundation. Project Officer, John Tsapogas. Table 3 downloaded on September 10, 2007 from http://www.nsf.gov/statistics/nsf06320/

National Science Foundation, Division of Science Resources Statistics. (2007). *Science and engineering statistics*. Retrieved from http://www.nsf.gov/statistics/

Nelson, D. J., & Rogers, D. C. (2005). *A national analysis of diversity in science and engineering faculties at research universities*. Downloaded October 2, 2007 from http://cheminfo.ou.edu/~djn/diversity/briefings/Diversity%20Report%20Final.pdf and later updates in http://cheminfo.chem.ou.edu/faculty/djn/diversity/Faculty_Tables_FY07/07Report.pdf

Powell, A., Bagilhole, B., & Dainty, A. (2007). The good, the bad and the ugly: Women engineering students' experiences of UK higher education. In R.J. Burke & M.C. Mattis, (Eds.), Women and Minorities in Science, Technology, Engineering and Mathematics: Upping the numbers (pp. 47-70). Cheltenham, UK: Edward Elgar Publishing, Ltd.

Rosenthal, R., & Jacobson, L. (1968). Pygmalion in the classroom. New York: Holt, Rinehart & Winston.

Rutgers University Center for Women and Work. (2002). *Fact sheet #1 Facts about working women*. Downloaded September 10, 2007 from http://www.cww.rutgers.edu/dataPages/FactSheet1.pdf

Seymour, E., & Hewitt, N. (1997). Talking about leaving: Why undergraduates leave the sciences. Boulder, CO: Westview Press.

Sloan Work and Family Research Network. (2007). *Work-family encyclopedia*. Retrieved from http://wfnetwork.bc.edu

Smith, D. G., et al. (1997). Diversity works: The emerging picture of how students benefit. Washington, DC: Association of American Colleges and Universities.

Steele, C. M., & Aronson, J. (1995). Stereotype threat and the intellectual test performance of African Americans. *Journal of Personality and Social Psychology*, *69*, 797–811. doi:10.1037/0022-3514.69.5.797

Steele, J. R., Reisz, L., Williams, A., & Kawakami, K. (2007). Women in mathematics: examining the hidden barriers that gender stereotypes can impose. In R.J. Burke & M.C.Mattis (Eds.), Women and minorities in science, technology, engineering and mathematics: Upping the numbers (pp. 159-182). Cheltenham, UK: Edward Elgar Publishing Ltd.

Steinpreis, R. E., Anders, K. A., & Ritzke, D. (1999). The impact of gender on the review of curricula vitae of job applicants and tenure candidates: A national empirical study. *Sex Roles*, *41*(7-8), 509–528. doi:10.1023/A:1018839203698

Tavris, C. & Aronson, E. (2007). Mistakes were made (but not by me). *AARP Bulletin*, June 2007.

The Associated Press. (2007, May 22). *U.S. workplace not family-oriented*. Downloaded January 8, 2008 from http://www.msnbc.msn.com/id/16907584

Trix, F., & Psenka, C. (2003). Exploring the color of glass: Letters of recommendation for female and male medical faculty. *Discourse & Society*, *14*(2), 191–220. doi:10.1177/0957926503014002277

United Nations Economic Commission for Europe. (2007). *Glossary*. Downloaded on October 4, 2007 at http://www.unece.org/stats/gender/glossary/g.html

University of Michigan. (2003). *Admissions lawsuits*. Downloaded October 2, 2007 from http://www.vpcomm.umich.edu/admissions/overview/

U.S. Bureau of Labor Statistics. (2007). *Standard Occupational Classification (SOC) System*. Downloaded September 10, 2007 from http://www.bls.gov/soc/home.htm

U.S. Bureau of Labor Statistics. (2007). *Occupational employment statistics*. Downloaded October 2, 2007 from http://www.bls.gov/oes/home.htm

U.S. Department of Education, Institute for Education Studies, National Center for Education Statistics. (2006). *Table 9. Number persons age 18 and over, by highest level of education attained, age, sex, and race/ethnicity: 2006.* Downloaded September 10, 2007 from http://nces.ed.gov/programs/digest/d06/tables/dt06_009.asp?referrer=list

U.S. Department of Education, Institute of Education Sciences, National Center for Education Statistics. (2004). *Trends in educational equity of girls & women.* Downloaded October 2, 2007 from http://nces.ed.gov/pubsearch/pubsinfo.asp?pubid=2005016

U.S. Department of Health and Human Services, National Institutes for Health. (2007). *Recruitment and retention plan to enhance diversity.* Retrieved from http://grants1.nih.gov/training/faq_diversity.htm

U.S. Glass Ceiling Commission. (1995). *A solid investment: Making full use of the nation's human capital, Final report of the Commission.* Washington, DC: U.S. Government Printing Office. Downloaded September 10, 2007 from http://digitalcommons.ilr.cornell.edu/key_workplace/120/

U.S. Government Accountability Office. (2004). *Gender issues: Women's participation in the sciences has increased, but agencies need to do more to ensure compliance with Title IX.* Downloaded January 9, 2008 from http://www.gao.gov/new.items/d04639.pdf

Valian, V. (1998). Why so slow? The advancement of women. Cambridge, MA: MIT Press.

Walker, J. P., & Swiss, D. J. (1993). Women and the work/family dilemma. New York: John Wiley & Sons.

Wenneras, C., & Wold, A. (1997). Nepotism and sexism in peer-review. *Nature, 387*, 341–343. doi:10.1038/387341a0

Whitten, B. (2003). What works? Increasing the participation of women in undergraduate physics. *Journal of Women and Minorities in Science and Engineering, 9*(3&4).

Whitten, B. (2004). 'Like a family': What works to create friendly and respectful student-faculty interactions. *Journal of Women and Minorities in Science and Engineering, 10*(3). doi:10.1615/JWomenMinorScienEng.v10.i3.30

Whitten, B. (2007). What works for women in undergraduate physics and what we can learn from women's colleges. *Journal of Women and Minorities in Science and Engineering, 13*(1). doi:10.1615/JWomenMinorScienEng.v13.i1.30

Williams, J. (2005). *Measuring gender and women's empowerment using confirmatory factor analysis.* (Research Program on Population Processes POP2005-01 and Population Aging Center PAC2005-01). Downloaded October 2, 2007 from http://www.colorado.edu/ibs/pubs/pop/pop2005-0001.pdf

Woo, D. (1994). *The glass ceiling and Asian Americans: A research monograph.* Washington, DC: U.S. Department of Labor, Glass Ceiling Commission. Downloaded September 10, 2007 from http://digitalcommons.ilr.cornell.edu/key_workplace/129/

ADDITIONAL READING

Bowen, W. G., & Bok, D. (2000). The shape of the river: Long-term consequences of considering race in college and university admissions. 2d ed. Princeton, NJ: Princeton University Press.

Chubin, D. E. (2006, February 13). Minding the student client. *Inside Higher Education.* Downloaded from http://insidehighered.com/views/2006/02/13/chubin

Chubin, D. E., & Malcom, S. M. (2008, October 6). Making a case for diversity in STEM fields. *Inside Higher Education*. Downloaded from http://www.insidehighered.com/views/2008/10/06/chubin.

Chubin, D. E., & Sevo, R. (2008). Work-Related discrimination in STEM workplaces. In P. Rayman et al. (Eds.), Working WISE: Intergenerational voices advancing research and policy for women in science, technology, engineering and mathematics (pp. 59-64). Lowell, MA: University of Massachusetts at Lowell, Center for Women & Work.

Corbett, C., Hill, C., & St. Rose, A. (2008). *Where the girls are: The facts about gender equity in education.* Washington, DC: American Association of University Women Education Foundation. http://www.aauw.org/research/whereGirlsAre.cfm

DePass, A. L., & Chubin, D. E. (Eds.). (2009). Understanding interventions that encourage minorities to pursue research careers. Bethesda, MD: American Society for Cell Biology.

Glazer, S. (2005, May 20). Gender and learning: Are there innate differences between the sexes? *CQ Researcher, 15*(19), 445–468.

Gornick, V. (2009). Women in science: Then and now. New York: The Feminist Press.

Gurin, P. (2002). Diversity and higher education: Theory and impact on educational outcomes. *Harvard Educational Review, 72*, 330–336.

Malcom, S. M. (2006). Diversity in physics. *Physics Today*, (June): 44–47. doi:10.1063/1.2218554

Malcom, S. M., Chubin, D. E., & Jesse, J. K. (2004). *Standing our ground: A guidebook for STEM educators in the post-Michigan era.* Washington, DC: American Association for the Advancement of Science and National Action Council for Minorities in Engineering. Downloaded from www.aaas.org/standingourground.

National Center for Women and Information Technology. (2008). *Programs-in-a-box*. Downloaded from. http://www.ncwit.org/resources.res.box.html

National Center for Women and Information Technology. (2008). *Promising practices catalog.* Downloaded from http://www.ncwit.org/resources.res.box.html

National Science Foundation, Research on Gender in Science and Engineering. (2003). *New formulas for America's workforce: Girls in science and engineering* (NSF 03-207). Arlington, VA. Downloaded from http://www.nsf.gov/ehr/hrd/Newformulas/newformulas.jsp

Page, S. E. (2008). The difference: How the power of diversity creates better groups, firms, schools and societies. Princeton, NJ: Princeton University Press.

Rosser, S. V. (2004). The science glass ceiling: Academic women scientists and the struggle to succeed. New York, NY: Routledge.

Society of Women Engineers. Assessing Women and Men in Engineering, & National Academy of Engineering Center for Advancement of Scholarship on Engineering Education (2009). *Applying research to practice resources*. Downloaded from http://www.engr.psu.edu/awe/secured/director/assessment/overviews.aspx

Summers, M. F., & Hrabowski, F. A. III. (2006, March 31). Preparing minority scientists and engineers. *Science, 311*, 1870–1871. doi:10.1126/science.1125257

Valian, V. (2001). *Tutorials for change: Gender schemas and science careers*. http://www.hunter.cuny.edu/gendertutorial/tutorials.htm

Profile 2
Science Career Spotlight:
Microalgal Research Fuels Biotechnology-Based New Industries

© 2008 Queensland Government Office of Women. Used with permission.

Associate Professor Kirsten Heimann, School of Marine and Tropical Biology, James Cook University, Townsville, Queensland, Australia

Kirsten Heimann is the director of the North Queensland Algal Identification/Culturing Facility within the School of Marine and Tropcial Biology at James Cook University, Townsville, Australia and leads a microalgal research team focusing on ecotoxicology and the development of microalgae for the commercial production of fertilizers, animal feeds, neutraceuticals, and biofuels. Kirsten has been fascinated by biology and strived to understand how organisms function from a very young age. Becoming a scientist was a logical progression to satisfy this curiosity. Her research led her to understand that microalgae can serve mankind in their quest for the development of new economically and environmentally sustainable industries. She has published extensively in diverse research areas in high ranking journals including Nature, Molecular Biology of the Cell, and Journal of Biological Chemistry. Kirsten's research has won several awards, the most recent

Copyright © 2010, IGI Global. Copying or distributing in print or electronic forms without written permission of IGI Global is prohibited.

the Queensland Smart Women – Smart State Award for Research Excellence in 2008.

Summary of Research Interests

My research focuses on microalgae and can be broadly divided into five categories: 1) applied phycology (biotechnological applications), 2) ecology (incl. development of genetic probes for the identification of toxic strains, physiology and nutrient utilisation), 3) cell biology (incl. signal transduction pathways triggered by environmental stimuli), 4) ecotoxicology, and 5) paleoclimate. I am especially interested in algal strain isolation to enable the use of the most productive native species for biotechnological applications (biodiesel production from microalgae, new microalgal strains for tropical aquaculture, and the potential isolation of novel bioactive compounds) and the use of microalgae as indicators of water quality.

Funding obtained through the Australian Research Council Linkage Infrastructure and Equipment Funds allowed me to develop the North Queensland Algal Identification/Culturing Facility (NQAIF), which presented a major breakthrough in my research career. Since its establishment (early 2004), my research facility has focused on establishing tropical microalgal cultures from the Great Barrier Reef (currently around 200 cultures), thereby creating the first systematic microalgal record for the region and directly enabling the research projects of 10 graduate students. The establishment of NQAIF in 2004 also enabled the set up of a large international collaboration, which resulted, for example, in a 2008 Nature publication (Moore et al. Nature 451:959-963, impact factor 28.75) describing a new phylum of a photosynthetic alveolate; a sister lineage to apicomplexans, colpodellids, and dinoflagellates. The establishment of GBR-specific microalgal cultures has significantly increased the breadth of current as well as collaborative research undertaken in my laboratory.

Current Research Projects

1. Microalgal potential for biotechnology (research category: applied phycology):
   a. Lipid content analysis of new microalgal strains to determine their usefulness as biofuels or as aquaculture feeds.
   b. Search for bioactive compounds from microalgae.
2. Algal blooms, invasiveness and toxicity (research categories: ecology, physiology):
   a. What are the factors / combination of specific factors controlling bloom establishment and durations?
   b. What are the environmental / biological determinants that promote invasive behaviour?
   c. Which biotic and abiotic factors influence levels of toxicity?
   d. Which organisms/ genetic strains produce these toxins?
3. Ecotoxicology (research categories: ecology, cell biology):
   a. Development of whole effluent toxicity tests using bioluminescence, esterase activity, photosynthetic quantum yields, and growth as toxicity endpoints.
   b. Sediment herbicide and break down product toxicity to benthic local microalgae and impact on community assemblages.
4. Ecology and cell biology of bioluminescent dinoflagellates:
   a. Ecological use of bioluminescence (burglar alarm *vs.* quorum sensing).
   b. Regulation of bioluminescence and signal transduction of mechanical stimuli.
   c. The function of the cytoskeleton in scintillon (bioluminescent organelle) and chloroplast circadian-controlled translocations.

Current Innovative Research and Future Directions

1. Microalgal potential for biotechnology – strategies to develop
   a. fossil fuel independence,
   b. non-arable land use for biodiesel and feedstock productions,
   c. remediation of nutrient-polluted waste water, and
   d. carbon neutral production practices

As a co-founder of the Australian Research Network "Biotechnological and Environmental Applications of Microalgae (BEAM) and director of NQAIF, I have a special interest in utilising microalgae for the greater good. Microalgae, are microscopic, plant-like, aquatic organisms, which, with the help of sunlight, draw down carbon dioxide from the atmosphere to produce oils, carbohydrates and proteins (for animal feeds). Therefore microalgae offer the potential to produce biodiesel in a carbon-neutral way. Microalgae also utilize phosphates and nitrates for growth and are therefore ideally suited to remediate nutrient-polluted waste water in brackish and marine applications, saving valuable freshwater resources. In addition, as aquatic microscopic organisms, microalgae can be cultured in ponds or bioreactors on waste land saving valuable arable land in Australia for the production of human food. Because microalgae grow many times faster than oil-producing crops (e.g. palm oil), they are likely the only viable alternative for sufficient biodiesel production to overcome our dependence on fossil fuels. However, more research is needed to address some problems that are currently hindering the economically sustainable production of biodiesel from microalgae.

I am addressing some of these problems, in a recently funded ***'microalgal biodiesel project'*** ($166,000 by the Queensland Government), a research and development project between MBD (a Melbourne based company) and James Cook University. A ***'lipid profiling research project'*** (funded by JCU) also started in 2008. My research analyses the lipid profiles of monoclonal cultures of microalgae isolated and established from the Great Barrier Reef. These analyses will establish the potential of selected microalgal strains as new feed organisms in tropical aquaculture, agriculture and their potential for biodiesel production. Starting in July 2008, I have established the suitability of selected microalgal strains for large scale culturing (e.g. photobioreactors or raceway applications) which is necessary for industrial applications. In addition, culture conditions are being explored, which will enhance lipid profiles and storage. Biodiesel production from microalgae is likely not economically viable on its own. Therefore my research will also explore the potential of selected and suitable microalgal strains as food additives (e.g. in aquaculture and agriculture after lipid extraction) and their production of other value-adding byproducts (e.g. phycobilins, essential amino acids, etc.).

2. Algal blooms, invasiveness and toxicity

In recent years elevated temperature, commonly addressed as "global warming", elevated nutrient loads, as well as other anthropogenic disturbances of estuarine systems are thought to have led to significant increases in bloom events, bloom durations, and invasion of disturbed habitats by algae. My contribution to a recent large international view point publication highlighted the potential for toxic bloom development if the proposed urea fertilization of the Sulu Sea was to be granted. As a consequence of this publication, governments globally decided not to fund these large-scale experiments and the International Oceanographic Commission has called for a procedure for international risk assessment of such future proposals.

My current and future research investigates the biology of bloom forming organisms by addressing major questions in the field with a focus on the effect of substrate availability and the interaction of chemical and physical parameters. For example, the *CRC-reef funded project* ($60,000) on a recently discovered bloom forming colonial chrysophyte, *Chrysocystis fragilis* in the Great Barrier Reef showed that bloom formation and duration is driven by elevated temperatures and substrate availability rather than elevated nutrients. Thus *C. fragilis* is an opportunistic invader colonizing disturbed coral substrates (i.e. damaged by cyclones or coral bleaching) when temperatures are high.

As part of a research team, I recently obtained funding ($160,000) from the Australian Government (DEW) through the Marine and Tropical Sciences Research Facility (MTSRF, manager RRRC) for evaluating the presence, distribution, and threats of toxic, invasive dinoflagellates to the Great Barrier Reef. During 2006 and 2007, the project conducted two literature reviews for the Australian Government. In addition, we hosted a workshop "Marine Invasive Species Research in Queensland" in December 2007, which was attended by industry and government representatives (stakeholders) (MTSRF/RRRC, Port of Townsville, the Queensland Museum, Sunfish, and Department of Primary Industries – Biosecurity Queensland). The workshop aimed to inform stakeholders on our current knowledge, research being conducted and to ensure that the research proposed will meet the needs of stakeholders. The experimental component of the project started mid 2007. As part of the distribution and abundance analysis, monoclonal cultures of toxin producing microalgae are being established and the influence of abiotic factors (pH, salinity, nutrient availability, light intensity and – quality), effects of phosphate and nitrate eutrophication, and impact of bacterial/ phytoplankton communities on levels of toxicity and densities will be analysed. The project also supports one new PhD student (stipend and research income), who will be involved with the establishment and validation of genetic probes, which can identify and quantify toxin producing dinoflagellates in field samples from the Great Barrier Reef. This latter part of the research project is particularly important in tropical and subtropical locations, due to reported increases of ciguatera poisoning events (tropical reef fish poisonings).

3. Ecotoxicology

An increase in coastal pollution has called for the development of early warning systems. Most toxicity assays still use lethality as an endpoint, despite the fact that molecular / enzyme-based endpoints offer far greater sensitivity. Bioluminescence in marine dinoflagellates is such an endpoint, where the effect of a pollutant on the light emitting machinery is easily quantified. My earlier ecotoxicological research has established the basic procedures to evaluate the usefulness of *Pyrocystis lunula's* bioluminescence in whole effluent testing (WET) application in collaboration with Lumitox Gulf, an American company and Dr. Paul Klerks, University of Louisiana at Lafayette.

Using bioluminescence as an endpoint, I have extended the applicability of the testing procedure by developing a "Flash Response Assay" to various metals, which reports metal / plasma membrane interactions in milliseconds. My data indicate that the metals tested lead to easily distinguishable 10-minute metal-characteristic bioluminescent flash signatures. This research also showed that copper and cadmium likely exhaust the bioluminescent system by interfering with the sodium / proton antiporter resulting in a decrease in cytosolic pH, whilst nickel and lead exert a different mechanism of interfering with the bioluminescent system.

My future research aims to focus on refining the system to compare the metal sensitivity, speciation and detoxification mechanisms of phototrophic and heterotrophic bioluminescent dinoflagellates. Currently, pilot studies are also being conducted to evaluate the use of the purified bioluminescence components in whole effluent toxicity tests. A successful outcome of these pilot studies will enable the development of an "off the shelf WET test", research that will certainly be funded by Research and Development Program grants.

These ecotoxicological studies have recently been extended to a more ecosystem-oriented approach with a comprehensive array of local benthic microalgae being used in toxicity screening of metals, herbicides and their break down products utilizing a variety of endpoints (i.e. esterase activity, photosynthetic quantum yields, and the traditional endpoint growth). This research showed that photosynthetic quantum yield is reduced by 50% in five benthic microalgae isolated from the Great Barrier Reef at concentrations significantly below those detected in the water column and in sediments. Future research will examine the effect of elevated herbicide levels in estuarine marine sediments on community structure of the microphytobenthos.

4. Ecology and cell biology of bioluminescent dinoflagellates

Marine environments exhibit vast bioluminescent displays with approximately 90% of marine organisms – ranging from bacteria to fish – thought to have luminescent capabilities. The ecological implications of being bioluminescent are poorly understood. In marine dinoflaglle-lates, bioluminescence is thought to serve as a burglar alarm, but the few studies provided very limited experimental evidence for this. In recent pilot studies, an honours student (Alana Owen) investigated the possibility of "quorum sensing" (the sensing and enumeration of self) as an ecological function of bioluminescence. Her results show that bioluminescent dinoflagellates may use bioluminescence in quorum sensing, which had previously only been shown for marine bacteria. Current research is focusing on identifying the possible inducer.

The circadian-controlled regulation of bioluminescence in dinoflagellates is diverse. While the bioluminescent system in some heterotrophic dinoflagellates appears not to be regulated by circadian rhythms, in phototrophic dinoflagellates two different modes of circadian regulation appear to exist: 1. diurnal degradation and biosynthesis of the scintillons (bioluminescent organelles) and 2. reciprocal organelle translocation processes. I am currently investigating the involvement of cytoskeletal structures such as filamentous actin and microtubules and their associated motor proteins in the diurnal redistribution of scintillons and chloroplasts from inactive to active locations (Heimann et al. 2009, Botanica Marina 52:170-177) and intend to extend this research to characterize the circadian clock components and the light perception machinery.

Current research also investigates the signaling pathways involved in relaying the mechanical stimulus from the plasma membrane to the scintillons. My work shows that calcium is involved in the signal transduction of the mechanical stimulus. Current research aims to establish the source of calcium, either internal stores (vacuole, mitochondria, chloroplast, endoplasmic reticulum) or the external medium via voltage-gated calcium channels.

Section 2

Initiatives to Reach Girls in Schools

Chapter 3
GETSMART:
An Academic-Industry Partnership to Encourage Female Participation in Science, Technology, Engineering and Math Careers

Natasha F. Veltri
The University of Tampa, USA

Harold W. Webb
The University of Tampa, USA

Raymond Papp
The University of Tampa, USA

ABSTRACT

This chapter reports on the development a formal social mechanism for interaction among female IT role models, such as industry executives and recent college graduates, higher education IT academicians, and female middle school and high school students. The GETSMART (Getting Everyone To Study Math and Related Technologies) program is designed to address the national issue of low female participation in the computer science and information systems fields. The goal of this initiative is to create an innovative educational and mentoring program that encourages women to pursue studies and careers in science, technology, engineering and math (STEM). The comments of the GETSMART participants serve as a feedback loop to the program executive and academic leaders and indicate the importance of teachers, parents, role models and early STEM experiences in formation of interest in STEM.

INTRODUCTION

Between 1950 and 2005, women's workforce participation in the United States has grown tremendously from 33.9 to 59.3 percent (Bureau of Labor Statistics 2007a). The role women play in the workforce has improved as well with women in the US holding 51 percent of all professional positions (NCWIT, 2007). However, women remain underrepresented in science, technology, engineering and math careers such as architecture and engineering (14.4 percent) and computer and mathematical jobs (25.6 percent) (Bureau of Labor Statistics 2007b).

At the same time, there is a worldwide shortage of engineering and Information Technology (IT)

DOI: 10.4018/978-1-61520-657-5.ch003

Copyright © 2010, IGI Global. Copying or distributing in print or electronic forms without written permission of IGI Global is prohibited.

workers and strong demand for STEM specialists. In the United States employment of computer scientists, engineers and mathematicians is expected to increase by 37, 11 and 10 percent respectively between 2006 and 2016 (Bureau of Labor Statistics). Increasing the female workforce is one possible solution to overcoming the worldwide shortage of engineering and IT personnel. Such a solution, however, is unlikely unless college enrollments of women in these fields can be increased. While nearly 60 percent of all undergraduate degree recipients in US colleges and universities in 2006 were female, only 21 percent of undergraduate degrees in computing and information sciences were awarded to women (NCWIT 2007). A miniscule 1 percent of all females taking the SAT reasoning test, a major college entrance exam in the USA, in 2006 indicated that they were interested in pursuing IT related degrees in college (NCWIT, 2007).

So what can be done to encourage women to pursue science and technology degrees and careers? In this chapter we describe an academic-industry partnership that aims to educate and excite middle and high school female students about math, science and technology. The purpose of the chapter is two-fold. First, we report on the development a formal social mechanism for communication among female IT role models (industry executives and recent college graduates), higher education IT academicians, and female middle school and high school students. The GETSMART (Getting Everyone To Study Math and Related Technologies) program is a model for influencing socio-cultural factors affecting female IT career choice (role models and mentors), the need for which has been suggested by Ahuja (2002), Adya & Kaiser (2005) and Trauth et al. (2008). Second, we report the attitudes of GETSMART program participants towards a career in STEM. The girls' comments indicate the importance of teachers, parents, role models and early STEM experiences in formation of interest in STEM, and serve as a feedback loop to the program executive and academic leaders.

The next section of the chapter discusses women's attitudes and perceptions about STEM careers. It is followed by a section that identifies gender differences that are salient in choosing a technical career with a special emphasis on role models and mentors. Finally, a section on academic-industry partnerships that provide a great opportunity for community outreach is included. Because role models are especially important for female career orientation, the GETSMART program is presented as an example of an academic-industry partnership that brings academic and industry mentors together to promote careers in math, science and technology among girls. Our experiences with GETSMART and recommendations for other programs are discussed in the final section.

WOMEN IN SCIENCE AND TECHNOLOGY

Researchers and practitioners alike have long wondered why women are significantly underrepresented in STEM fields. Various reasons have been proposed that can be grouped into three main categories: differences in aptitude, discrimination, and personal choice (Rosenbloom, Ash, Dupont, & Coder, 2008). In the field of computing this has been a topic of concern for decades with a series of articles published in a special issue of the Communications of the ACM in 1995 (Snow, 1995; Klawe & Levenson, 1995) and a follow-up review article on progress to date in 2009 (Klawe, Whitney, & Simard, 2009).

Some may argue that women simply are not wired the same way as men and they tend to perform worse on math, science and technology tasks because they have lower abilities when it comes to those areas. Women indeed report lower levels of computer self-efficacy (Johnson, Stone, & Phillips, 2008). Computer self-efficacy is the individuals' belief of their ability to perform across multiple computer related tasks (Marakas, Yi, & Johnson, 1998). In their interaction with computer technology women are more social than men and are more likely to view computers as responsible for their performance and believe that the computer determines the outcomes of their interactions with technology (Johnson, Veltri, & Hornik, 2008). Computer self-efficacy impacts attitudes towards IT which in turn influences IT career intentions. Similar patterns can be suggested for math, engineering and science and given women's low levels of self-efficacy they would be less likely to pursue STEM careers.

Women are different from men in their personalities as they tend to be more nurturing and agreeable and are more concerned with feelings rather than ideas (Costa, Terracciano, & McCrae, 2001). While a discussion of Jungian personality types is beyond the scope of this chapter, certain personality types are more suited to the pursuit of STEM careers. Women, as a whole, tend to have personalities that do not lend themselves to the pursuit of STEM careers with only 4.5 percent of women estimated to be categorized in the traditionally STEM focused NT personality types, those who are introspective, logical and on a quest for knowledge. This is in contrast to the male population, where 15.5 percent are estimated to have NT personality types (MyPersonality.info, 2009).

Social expectations shape women's perceptions of math, science and technology careers. Widely held gender stereotypes may be one of the reasons for women to underperform on STEM subjects. Girls are traditionally steered into liberal arts and away from mathematics and sciences (Ahuja, 2002). From an early age, girls are exposed to societal stereotypes and often miss the opportunity to learn and advance in sciences. Stereotype threat could lead women with high aptitude in STEM to underperform because of societal pressures (Rosenbloom, Ash, Dupont, & Coder, 2008).

Additionally, women's decision to use technology is more strongly influenced by subjective norm and perceived behavioral control (V. Venkatesh, Morris, & Ackerman, 2000), than by their attitude towards using new technology. Subjective norm is the individual's beliefs that people who are important to him or her think he or she should or should not engage in a particular behavior (Fishbein & Ajzen, 1975). Thus, a girl's perception of whether or not people of importance to her would support her interest in STEM can influence her subsequent decisions to take STEM classes and pursue a career in STEM. Perceived behavioral control is an individual's perception of ease or difficulty of performing a specific behavior (Ajzen, 1991). Girls' math and science abilities, prior experience in math and science classes, as well as societal influences can all play a role in forming a perception about ease or difficulty of science, math and technology.

Others may argue that women face discrimination and differential barriers to entry into STEM fields. For example, there are still disparities in salaries between males and females in both industry and academia for equivalent qualifications (Klawe, Whitney, & Simard, 2009). This is not a new disparity in the computing fields (Hemenway, 1995). Early studies indicate that students are treated differently in the classroom regarding science, math and technology, based on gender, as early as the kindergarten through high school (K-12) years (Klawe & Levenson, 1995). Additionally similar treatment of female students has been reported by the same authors in their analysis of both undergraduate and graduate computing programs. A review of studies from the 1980s and early 1990s indicates that females were discouraged from participating in classroom learning activities by both male and female teachers on a wide-spread basis. At the time of these studies females were reported to be less prepared for success in computing fields in college than male counterparts due to social stereotyping, lack of classroom encouragement by teachers, and even the non-availability of home computers to females (Klawe & Levenson, 1995).

On the other hand, however, a recent study reported that there is no biological evidence to clearly suggest that men and women are different in terms of math and science potential and their mathematical ability (Ceci, Williams, & Barnett, 2009). The authors of this study argue that women are making a rational choice when it comes to education and careers and simply are not as interested in STEM careers as men. Therefore, the gap between men's and women's participation in science, math and technology is not due to discrimination or lack of aptitude, but instead is the result of self-selection into fields that are perceived to better fit women's personalities. Indeed, women are much more focused on interpersonal communication (Tannen, 1990; Wood & Rhodes, 1992) and therefore look for career environments with high levels of perceived interpersonal communication. As a result, science and technology careers may be less attractive to women because they perceive them as focusing on the technical and not the interpersonal connections.

Women who have high math abilities are disproportionately more likely than men to also have high verbal competence. A combination of math and verbal abilities allows those women a greater choice of potential careers and very often they pursue careers in other fields that are of more interest to them (Ceci, Williams, & Barnett, 2009). Clearly, men and women value different aspects of their careers and women's occupational preferences differ from men's (Rosenbloom, Ash, Dupont, & Coder, 2008). In order to attract girls into STEM classes and women into STEM careers it is critical to better understand the specific factors that are important for women as they make their career choice.

CAREER CHOICE FACTORS FOR WOMEN

A variety of factors play an important role in female career choices including social expectations, work-family conflict, occupational culture, role models, computer training and computer attitudes (Ahuja, 2002). A recent study reported that women enter the IT profession because they seek job security and flexible work hours and perceive entry into the IT field as easy. On the other hand, men enter the IT profession because of their love of technology and computers and higher levels of comfort with their IT specific skills (McKinney, Wilson, Brooks, O'Leary-Kelly, & Hardgrave, 2008). Similarly, Papastergiou (2008) reported that females tend to pursue an IT degree more because of increased employment prospects, and not personal interest in IT. These findings indicate that women make a career choice not necessarily based on their aptitudes or interests but based on more practical factors such as overall fit with their life and family styles.

Women are more concerned about work-family balance because they traditionally focus more on family and their early career years coincide with child birth and child rearing (Ahuja, 2002). Many science and technology careers are demanding in terms of time and commitment and may be more stressful for a female who is trying to balance work and family. The underrepresentation of women in these fields could be due to the time demands of these careers. At the same time science and technology is a fast evolving field, and it may be difficult to stay current and maintain skills if an absence from the career is necessary due to family demands.

Computer training and computer attitudes is another factor that plays a role in IT career choice (Ahuja, 2002). Because women are more social in general, it is important to highlight the breadth of math and technology applications with emphasis on the human and social-oriented aspects of computing (Papastergiou, 2008) as well as systems development. Forming positive attitudes towards STEM, exposing girls to various applications of science and training them to use technology can help promote girls enthusiasm and interest in these areas.

A study of the socio-cultural environments affecting female recruitment and retention in IT careers argues for examination of cultural variation in gender and IT issues. Perceptions of women's role in society and socio-cultural moderators (gender stereotypes about aptitude, economic opportunity, social class, and gendered career norms) have been found to affect attitudes about working in IT (Trauth, Quesenberry, & Huang, 2008). Social-cultural moderators have been found to vary by both geographic region within the United States (Trauth, Quesnberry, & Yeo, 2008) as well as by country (Trauth, Quesenberry, & Huang, 2008). The practice-related findings of these studies suggest that IT workforce employers develop interventions to increase the underrepresentation of women by focusing on recruitment and retention. Other suggestions are that pre-college educators play a crucial role in increasing participation by females in the IT workforce. The collaboration of employers, IT educators, and primary and secondary institutions for the purpose of outreach programs for students and their parents is recommended (Trauth, Quesenberry, & Huang, 2008).

Adya and Kaiser (2005) present a theoretical model of factors influencing girls career choice that posits that a combination of social factors (family, peers, media, and gender stereotypes) interact with structural factors (teachers, access to technology, educational structure, and technology resources) to affect career choices (along with individual difference variables). They also posit that as technology diffuses throughout society and the presence of women in the workforce increases, the impact of these variables may be affected.

The Importance of Role Models and Mentors

A recent study of female Canadian IT professionals highlights the importance of female role models in making career choices. 41.7 percent of survey respondents identified shortage of female mentors as one of the challenges (Orser, Riding, Dathan, & Stanley, 2007). The role of mentors and role models is especially important for women who tend to be more social (Wood & Rhodes, 1992), network oriented (Tannen, 1990) and value personal aspects and relationships (Hofstede, 1980). Interacting with role models can give girls an opportunity to learn about science and technology careers and lift the persistent stereotypes of technology-heavy, difficult and male-dominant professions. Presence of role models in IT field has been reported to increase girl's computer self-efficacy (Papastergiou, 2008). Because self-efficacy is important in attitude formations and subsequent career choice, improving self-efficacy can be one way to encourage girls' choice of STEM fields of study and work.

Unfortunately, as females are underrepresented in math, science and technology fields, there are few female role models available to serve as mentors. This makes attracting young females to the field difficult, perpetuating a vicious cycle of low female participation. Reversing this trend requires a collaborative effort of industry professionals and academics using long-term strategies that focus on the next generation of female science and technology professionals (Papastergiou, 2008). These future professionals are currently students in middle and high schools. Involving these students in activities highlighting math, science and technology can improve their understanding and appreciation of those fields of study and hopefully lead to more involvement in STEM areas in their future studies and careers.

ACADEMIC-INDUSTRY PARTNERSHIPS

Collaborative approaches include various stakeholders and support problem solving through a comprehensive and coordinated effort (Siegel, 2008). Since both academia and industry are interested in gender balanced participation in

STEM, academic-industry partnerships can be a great way to bring both academic and applied perspectives together, combining educational resources with industry financial support and practical experience.

There are many examples of effective academic-industry collaborations that deliver outreach programs for the community, incorporate service and experiential learning into the college curriculum (Lazar & Lidtke, 2002; M. Venkatesh & Small, 2002), and bridge the digital divide through service-research partnerships (Lazar & Norcio, 2002). Such programs can also be useful for encouraging female interest in IT, increasing girls' IT skills through projects, and exposing them to successful role models and mentors in math, science and technology, especially when early-intervention efforts are critical (Klawe, Whitney, & Simard, 2009).

Over the past two decades a number of initiatives by government and professional associations have been in place to promote female participation in STEM. Examples of these include:

- National Science Foundation STEM initiatives,
- Legislation such as the America COMPETES act of 2007,
- Association for Computing Machinery Committee on Women in Computing,
- National Center for Women and Information Technology (NCWIT), and
- Society of Women Engineers.

NCWIT, for example, has established several alliances to address the underrepresentation of women in IT. These include an academic alliance consisting of over 70 colleges and universities, a workforce alliance with over 30 reported corporate members, and a K-12 (kindergarten through 12th grade) alliance that seeks to increase the number of girls who study computing at the K-12 level by improving the image and teaching of computing with over 20 members from industry, not for profit and academics, and an entrepreneurial alliance of 20 organizations (NCWIT, 2009).

While a large number of groups are addressing the problem of low participation by females in STEM fields, it appears that these are often compartmentalized in their approach to the problem. One area of interest of these groups, addressed primarily by industry, is mentoring female recruits and employees in the workplace, the current workforce. Key recommendations for industry in this area are to focus on recruitment, retention, and promotion of females (Klawe, Whitney, & Simard, 2009). A second area of interest is women currently enrolled in universities, the near term workforce. Academics mostly focus on this group by providing curricular, pedagogical, research support and social networking. Recommendations for increasing participation by female undergraduate and graduate students have been developed as well as career development recommendations for female computer science professors (Klawe, Whitney, & Simard, 2009). The third area of interest is the K-12 girls, the future workforce. This area is of interest to industry, professional groups, and to some extent, academics.

While there is some cross-over of various stakeholders into multiple areas of interest, it appears that an increased emphasis on academic-industry alliances could help address this long-term issue of women underrepresentation that continues to exist. Ahuja (2002) theorized that a lack of role models will negatively influence women's choice of IT as career. A coordinated approach by industry, IT academics in higher education, and secondary educators to address the socio-cultural influences on female students' attitudes toward IT careers is recommended (Trauth, Quesenberry, & Huang, 2008). One way to engage female students in STEM is to educate them about various applications of STEM and expose them to role models in the community. Simultaneously, female IT professionals should be made aware of the need for female IT role models and mentors. The following section discusses the GETSMART academic-industry partnership that aims to educate and mentor female students during their middle and high school years.

GETSMART

The GETSMART (Getting Everyone To Study Math and Related Technologies) program was founded in early 2008. The genesis of this program occurred during a periodic review of the information systems curriculum at The University of Tampa. This review was conducted jointly with the university's Management Information Systems (MIS) Advisory Board and MIS faculty. On a campus where females constituted about 60 percent of the university enrollment, the gender demographics of the MIS program reflected national trends with only about 20 percent of the enrolled MIS majors being female. Furthermore, the university's honors program, the brightest students in the university, is about 70 percent female yet only 17% of these women are majoring in STEM fields. The honors students, in the opinion of the faculty, are the students most likely to succeed in the information systems major. The members of advisory board, which included several female information technology leaders, were asked for input on the issue with the goal of increasing MIS enrollments by high performing female students.

The idea to partner on this problem was an extension of an existing academic-industry partnership. The University of Tampa had partnered with The Tampa Bay Technology Leadership Association (TBTLA) for almost a decade. TBTLA is an organization initially established during the dot-com bust to address a vacuum in the Tampa Bay Information Technology community. TBTLA is a not-for-profit organization whose membership is limited to current or former technology executives. Its two-fold mission is to apply the collective experience of its members to real world technology management problems, and to mentor undergraduate and graduate students and other professionals who are anticipating careers in Information and Technology Management (TBTLA, 2009). TBTLA membership includes a mix of both male and female IT executives who have alternated leadership positions since its inception. As an extension of its mentoring mission several executives who were also MIS Advisory Board members stepped up to work on the female enrollment issue.

Leveraging this existing academic-industry partnership, several female IT executives proposed an organization to work with students in both K-12 and undergraduate programs. This organization planned to host periodic events with females to promote math and technology careers. Recruited to this organization's team were recent female honors graduates in the management information systems, math and mathematical programming majors. The events were to be planned and executed jointly by the organization, which included both male and female academics and executives. Playing off the math, science and technology education theme, this new program was called Getting Everyone To Study Math and Related Technologies or GETSMART.

This organization had a number of strengths. It consisted of executives and academics who had worked together for a number of years. The members had very highly developed organizational and leadership skills. The demographics of the team included multiple female industry executives as well as multiple female academics. The female executives had worked in IT and leadership positions for a number of years, providing several role models of long term career success. The addition of recent female graduates of the math and information systems programs, who were also honors program students, provided strong role models who could relate their recent university experience to the students. This also created an informal network for recent female graduates and female IT professionals.

Together, an idea was developed to create a junior advisory board composed of female middle and high school students. These students would participate in round-table discussions and their feedback would be used to assess how the GETSMART program can address the national issue of low female participation in computer science and information systems. The goal of this initiative was to create an innovative educational and mentoring program that encouraged women to pursue studies and careers in math, science and technology.

The GETSMART's mission is to assist young women in comprehending the value of math, science and Information Technology today and in their future. The program strives to increase the numbers of young women who select math, science and IT as further education or a career field by broadening their understanding of the importance of the discipline and the awareness that math, science and technology are interlaced into many career areas. The GETSMART targets middle and high school students because they are in formative years that are important in vocational orientation and career choice. Most college students, on the other hand, have already formed their career plans. This social structure takes advantage of the recursive relationship between social and structural factors proposed by Ahuja (2002) with the notion that positive influence provided by the GETSMART initiative will increase attitudes toward IT careers. This is a longer term objective not evaluated in this study.

GETSMART Participants

Adya and Kaiser (2005) suggest that studies conducted with middle school girls may be quite challenging, given the nature of the subjects. They point out the lack of richness that survey data alone provide and suggest the use of small focus groups conducted with peers and family. They suggest the use of multiple methods, especially when exploring gender stereotypes. Data for this study was collected from a group of 18 students using a series of anonymous survey questions collected electronically in a group setting, which were used as the basis for follow-up semi-structured group discussion. Parents attending the event observed from the rear of the room and completed a separate written survey.

All students in this study live in a high-technology area in the state of Florida, USA. This area has a high penetration of high speed internet access as well as a large number of information technology and related industries. The area has experienced decades of population growth with many people working in these high-tech industries moving to Florida from all over the world. This is a potential factor influencing the socio-cultural norms regarding information technology.

To recruit participants, the GETSMART Advisory Board contacted career guidance counselors at local middle and high schools and asked for nominations of one or two female participants for the program. Schools were asked to select good students who would be able to participate in a round-table discussion and express their opinions, but who were unsure of career choice rather than ones who had selected or had a strong predisposition for math or technology careers. The first event was held for 18 middle school students on a university campus. Another event was held for high school students. Several businesses donated supplies, manpower, and financial support for the meetings.

Both students and their parents were invited to participate. The parents were offered a brief survey at the beginning of GETSMART session and also provided consent for recording the discussion and surveying the students. Female IT executives and academics were present at the event to serve as role models. The discussion during a two and a half hour session focused on stereotypes associated with technology and science careers, advantages and disadvantages of those careers as well as student perceptions of math and science classes in their curriculum. It also included a series of questions to assess the degree to which the girls perceived STEM careers as possibilities. The questions were

Table 1. GETSMART activities to encourage interest in STEM

Career Choice Factors	GETSMART Activities to Influence Career Choice Factors
Social Expectations	Round-table discussion
Role Models	Several female academics and practitioners served as role models during sessions Introduced role models through images in IT magazines
Computer Training	Interactive hands-on session
Computer Attitudes	Round-table discussion Interactive hands-on session

pilot tested by a small sample, revised, and then given to the middle school group.

The semi-structured discussion questions were designed to elicit girls' views on STEM careers as well as how others (teachers, parents and their peers) influenced their views of STEM. Open-ended questions were also included which focused on what major they were currently considering, whether or not they participated in competitions with other students and/or engaged in public speaking as well as their perceptions of people who do participate. These questions were designed to elicit their views of events where competition was important to success and measured the extent to which the women were competitive with other students, both male and female. At the end of the session, the participants were asked whether they would like to continue to participate in future GETSMART sessions and the results were overwhelmingly positive.

GETSMART Session Activities

Klawe, Whitney, & Simard (2009) describe a number of activities specifically geared to K-12 girls to address the dearth of females in computer science:

- appeal to girls and their influencers,
- expose girls to positive role models,
- dispel computing-career myths,
- provide accurate information to influencers,
- provide girls with age-appropriate activities,
- summer programs,
- use the positive social aspect of computing as a motivator,
- college Computer Science department engagement of middle school girls, and
- encourage outreach implemented by companies.

Following these recommendations, the GETSMART program was designed to expose girls to applications of STEM, dispel math and technology career myths, showcase role models and their lives and careers, and introduce fun technology related activities. Table 1 summarizes the GETSMART session activities and maps them to career choice factors.

GETSMART SESSION FINDINGS

Round table discussion was facilitated by female IT executives, academics and recent math and IT graduates. The purpose of the discussion was to identify girls' perceptions of STEM careers and their experiences with science, technology and math classes in school. Table 2 presents the survey results of the middle school girls during the GETSMART session.

When the data for middle school participants was analyzed, some interesting results emerged. More than half of the participants stated they liked STEM and excelled at it. The remaining participants liked it and/or excelled at it somewhat and none of them indicated a total dislike. 94 percent

Table 2. Middle school GETSMART participants perceptions of STEM

Question	Middle School Responses n=18					
	Yes		No		Somewhat	
	#	%	#	%	#	%
Do you like STEM (science / technology/engineering/ math)	10	56%	0	0%	8	44%
Do you excel at STEM	10	56%	0	0%	8	44%
Have you ever had a teacher tell you that you can't or shouldn't pursue STEM	1	6%	16	88%	1	6% No response
Have you ever had a teacher encourage you to pursue STEM	17	94%	1	6%		N/A
Do you think STEM is mainly for boys	0	0%	18	100%	0	0%
If you do like math, would you consider it as a profession	3	17%	0	0%	15	83%
Who is better at STEM (Equal/Girls/Boys)	13	72% Equal	3	17% Girls	2	11% Boys
Do you spend more time concerned with fashion/celebrities/ socializing than with schoolwork	10	56%	8	44%		N/A
Do you ever think of math as a game where there are certain rules that must be followed in order to win the "game"?	5	29%	7	38%	6	33%
If you do like STEM are you afraid to admit it to your friends?	0	0%	16	89%	2	11%
Do you purposely "do poorly" at STEM because you don't want your friends to think you are "un-cool" or "nerdy"?	0	0%	18	100%	0	0%
If you knew you could make a lot of money in the STEM field, would you want to go into the field?	10	56%	3	17%	5	28%
Do you ever say something like "I don't know when I would use math in the "real world", so why should I bother to learn it?	4	22%	8	44%	6	33%

	#	%	#	%	#	%	#	%
If you score poorly on a test, who do you tend to blame (Yourself/ Teacher /Parents/Other)?	14	78% Self	1	6% Teacher	1	6% Parents	2	10% other

were encouraged by a teacher to pursue STEM and 100 percent felt that it was not mainly for boys. This indicates that, at least at the middle school level, the girls were not turned off to STEM. In fact, 100 percent would consider it as a career, with 17 percent in strong agreement. They also do not see it as a "boys only" career with 72 percent viewing STEM performance equally and only 11 percent believing that boys are better at STEM. Regarding stereotypes and perception by others, 100 percent were not afraid to admit to their friends that they like STEM. Neither did they report purposely performing "poorly" on STEM assignments lest their friends think of them as "uncool" or "nerdy". Thus, at the middle school level, girls are still excited about STEM. They have not developed negative views toward it and their views have not been negatively influenced by society or their peers. This finding is contrary to reports from the 1980s and 90s on the bias shown females by their teachers about their abilities in math and science (Klawe & Levenson, 1995). The students in this study have not developed a poor self-efficacy toward math or technology, a major advancement from earlier studies. This finding also appears to be contrary to the proposition of Ahuja (2002) that social expectations will negatively affect attitudes toward computers and their use as well as IT as choice of career in women.

Table 3. Quotes from middle school GETSMART participants coded by career choice factors

Career Choice Factors	*Quotes from discussion with Middle School Girls*[1]
Differences in Aptitude	"It's not like guys have genes that make them good at math. I personally don't think that boys or girls are better than each other."
Attribution of Responsibility	"All you really have to blame is yourself because you're the one that had the material, if it was a pop-quiz I could understand. If you don't study, that's your own problem. If you don't have time to study, you have to make time to study."
Societal Stereotypes	"People think men are into technology." "Tech and stuff is geeky guy thing."
Peer Pressure	"I'm kind of a nerd, geek, because I do my homework during lunch and I study, I try to get good grades. But when you're caught talking to a nerd or geek, they start making fun of you: 'Oh my gosh! Are you one of them?'"
Subjective Norm	"I think it's the environment you grow up in and people encourage you to do it and that makes you better at it. Because if you have encouragement that makes you want to do it, and that makes you even better at it." "My teacher tells me something I'm good at, and another teacher says something completely different. I like both of their ideas but I can't think of a way to incorporate them."
Usefulness of Math and Science	"I think there is some math in our curriculum that we shouldn't use because we are not going to use it in real life." "I thought that you really don't know when you're going to use it unless you actually chose your profession. You don't know what you're going to do in this grade, you don't know what you're going to do, it's nice they actually teach it to you and you know it." "When you think of all the jobs that are out there, mostly everything has to do with math and science, or something that just like you think it's just pointless but when you get right down to it, it has everything in it." "I think, sometimes if you get frustrated with math then you tend to think that you're never going to use it. Or if it is this obscure thing that you can't see practical use for, then you don't understand it much."
Experience in Math and Science Classes	"The teacher will show it in a way how people don't understand it, if the teacher is very harsh, kids will be so afraid to ask, I know one of my friends had that problem." "Teachers need to teach it, even if you're not going to use it. They have to teach it anyway, because of other people. There are so many career choices; all of them have some kind of math. Some kids are more interested in other things, just because they don't like it doesn't mean that it has to go away." "There are some things where they [teachers] drag on too much, it's like they are going to drill it into kids' heads, but that's not the stuff you're going to use. Sometimes they'll skip over the stuff that you think might be important."
Variety of Career Options	"I personally have no idea what I'm going to do when I get older, because I have a lot of interests. …I have personal interests, and my mom was telling me you don't have to major in something, you can do it as a hobby and work on something else that you like, so you can do both." "I think because there are so many other things out there right now, do I want to do math, social studies, teacher, archeologist, all this stuff. So many possibilities."

Girls' comments made during the discussion of STEM careers and societal stereotypes provide additional insight into their thinking and experiences. Discussion comments were analyzed and coded by the career choice factors that they addressed (Table 3). Many participating girls confirmed that in society both technology and science are perceived more as masculine careers, yet the girls themselves did not believe that was the case. While survey results showed that the girls are not afraid to excel in math and science, one of the girls commented on how it was unpopular to be a "nerd or geek" and these students were made fun of. So, peer pressure is an issue and a desire to fit in may affect girl's choice of classes and interests. The middle school girls are still in their formative years and their opinions can easily be swayed. One of the ways to improve the girls' opinions of science and technology is to affect their subjective norm, because the importance of role models and environment was apparent in girls' comments. The girls look up to their teachers and other authority and mentor figures in their lives and need encouragement to explore and pursue their interests. Since this study was conducted in a high technology region the difference in the socio-cultural norm reflected by the students appear to be consistent with the finding that gender differences are not universal but are different by region (Trauth, Quesnberry, & Yeo, 2008).

Most girls believed that math, science and technology were important and necessary in many professions. Yet they also felt that there was a mismatch between what they learn in the classroom and their understanding of how it may be useful for them in the future. While some of them expressed frustration that some material in their school's curriculum was not applicable in their lives, most of them realized that they may not yet know how to use what they learn, but should learn it nevertheless. The girls indicated that their experience with math and science teachers was important in forming their perceptions of STEM and their interest and desire to learn more about this field.

Middle school girls are undecided about their career choice. Several of them acknowledged that they have many diverse interests and are still exploring what they really want to do with their lives. Their comments also highlighted the role of teachers as mentors in encouraging them to pursue a particular area of study or particular career. Parent's opinions are important for girls and parents have a chance to help girls with career orientation. This finding contradicts the speculation by Adya and Kaiser (2005) that "teachers and counselors will continue to steer girls toward traditional careers." An explanatory factor could again be the socio-cultural influence of the high-technology region (Trauth, Quesnberry, & Yeo, 2008).

Parent surveys, also collected during the event, provided additional insight into middle school girls interests and experience with STEM (Table 4). 87 percent of parents reported that girls talk to them about math, science and technology teachers and 71 percent of parents indicated that their daughters complain about STEM classes. Poor experiences with science, technology and math classes in middle school may translate into diminished interest in high school and in college. It is important for parents to help their daughters appreciate the importance of math and science and all parents indicated that they encourage their child's interest in STEM. Yet teachers are also important in promoting girls interest in STEM. Improving girls' experiences in the classroom and teaching them the value of STEM can increase their appreciation of teachers and those classes. Almost all parents, 93 percent, stated that their daughters had a special interest in math or science. It shows that at middle school level girls are still highly interested in those areas, and they need to be exposed to opportunities to promote their interest. While one may assume that parents would always support and encourage their child to pursue a career in STEM, the family support of females toward IT careers has been found to vary significantly across cultures and regions (Trauth, Quesnberry, & Haung, 2008).

Future Plans for GETSMART

The GETSMART program is not a one-time event. After the inaugural workshop with middle school students another workshop was held for high school females in Spring 2009, and a follow-

Table 4. Parents perceptions of their children's interest and experience with STEM

Question	Middle School Responses n=15[2]					
	Yes		No		I don't know	
	#	%	#	%	#	%
Is there an adult in your child's life who has a career in science, math or technology?	11	73%	4	27%		N/A
Do you encourage your child's interest in science, math and technology?	15	100%	0	0%		N/A
Do you think a career in math, science or technology may be interesting to your child?	13	87%	0	0%	2	13%
Does your child complain about her math, science, or IT classes?	10	71%	4	29%	0	0%
Does your child express special interest in math or science classes?	13	93%	1	7%	0	0%
Does your child talk to you about teachers in her math, science, or IT classes?	12	86%	1	7%	1	7%

up workshop with the middle school students is planned for Fall 2009. The GETSMART program itself has evolved. The second session included computer hands-on-activities for the participants to provide interactive training and improve computer attitudes. The GETSMART program will continue to be held twice a year in the Fall and in the Spring. Plans include expanding the number of girls participating, offering more computer-supported hands-on activities, involving additional female IT professionals and college students to serve as role models, and seeking alliances with other professional groups. Aggregated survey and interview results collected during the workshops are communicated back to the industry executives and program sponsors to provide feedback and underscore the importance of role models and mentors for the girls.

Evaluating GETSMART

While it is too early to judge the effects of the GETSMART program, so far the participants indicated that they enjoyed the program and almost all middle school girls indicated their intent to participate in a follow up program a year later. A more formal questionnaire was offered to the high school participants at the end of their program to evaluate the different components of the program. Results indicate that the hands-on activities were a huge success and the GETSMART leaders will continue incorporating and expanding the offering of hands-on activities. 78 percent of participants indicated that seeing female role models was very important to them. Mentoring young girls and exposing them to opportunities in STEM is the major goal of the GETSMART. 67 percent of the girls plan on recommending the GETSMART to their friends, which could also serve as an indication of program success.

RECOMMENDATIONS FOR ORGANIZING SIMILAR PROGRAMS

To successfully impact the socio-cultural norms affecting recruitment of females to IT careers programs such as GETSMART need to be developed and mature so they can grow to reach a large number of students and role models. The complexity of the issues calls for a long-term approach to the issue of female participation in IT careers. Even though the participating university is in a high technology area with favorable socio-cultural attitudes, at least as reported by this group of students, female enrollments in the technology-related majors remains low. This chapter has discussed an approach for developing social networks for informing female IT executives and professionals of the need for their participation as role models.

The formation of an academic-industry partnership which includes female executive role models is a key attribute to the success of such a program. Girls need to be exposed to role models, specifically those in the technical areas from both industry and academia. Linkages with colleges and universities in which college-aged women majoring in STEM careers can mentor middle and high-school girls is also important. Leveraging existing academic-industry partnerships can accelerate the organizational aspects of these programs and increase the potential for success by:

- providing leadership from professionals and academics from established organizations;
- utilizing the strong working relationships already established by the participating organizations;
- utilizing the resources and skills of individuals from multiple organizations, most of whom are members of their professional community in order to serve others;
- providing access to facilities, resources, and skills from the academic community;
- providing access to facilities, resources and skills from industry;
- demonstrating the utilization of STEM to students in a number of career options;
- providing an array of female mentors and role models from the ranks of professionals, academics, and college student and graduates;
- offering events over a sustained period of time, which may be hard to accomplish with an ad hoc organization.

It is also important to involve parents and the media so that the girls' viewpoints and opinions are heard by adults and publicized in the media for other girls to hear and encourage the interested students. Finally, a formal seminar in which the girls' opinions can be heard is important, as it reinforces that adults are interested in their viewpoints and encourages them to pursue STEM careers.

LIMITATIONS OF THE STUDY

While this study offers interesting insights into academic-industry partnerships to encourage female participation in STEM, there are several limitations. The GETSMART program so far was only offered to a small group of middle school and high school girls, yet the collection of qualitative data during the sessions provided a richer set of data than just a lone survey. Girls were selected by counselors from a variety of schools, but do not necessarily represent a random sample of the population of potential STEM students. The counselors were asked to recommend good students for a round-table discussion rather than STEM oriented students. Students provided responses to objective questions via anonymous electronic response device; however, their parents were in the room as observers, which could bias results. The recorded conversations and follow-up questions were overheard by parents, which could also bias the results. Additionally, the GETSMART participants come from schools located in Florida's "High-Technology" Corridor, a state-designated economic zone. Students are exposed to a population that has a fairly high percentage of technology workers, so results would be expected to be generalizable only to similar regions (Trauth, Quesnberry, & Haung, 2008). Finally, the GETSMART is a very young program and it is too early to judge its effectiveness. We are able to use the feedback from the original sessions to craft a better program for the future and plan to include a formal evaluation component at the end of subsequent workshops.

CONCLUSION

The GETSMART program is a model of academic-industry partnership where both sides offer voluntary community service with the goal of reducing and eliminating gender disparity in math, science and technology fields. This partnership specifically aims to work with the future workforce by focusing on middle and high school girls and promoting interest in math, science and technology. This mutually beneficial partnership will not only help students make more informed career choices, but also help IT college programs and the IT workplace become more diverse. The academic members of the partnership provide their expertise in teaching and facilitate educational sessions targeted to dispel myths about STEM and encourage girls to consider STEM as a possible career choice. Building on industry experience and academic research on career choice factors, the GETSMART program involves girls in activities that are recommended for working with this group. Industry leaders serve as mentors and role models by sharing their experiences in STEM careers.

REFERENCES

Adya, M., & Kaiser, K. M. (2005). Early Determinants of Women in the IT Workforce: a Model of Girls' Career Choices. *Information Technology & People*, *18*(3), 230–259. doi:10.1108/09593840510615860

Ahuja, M. K. (2002). Women in the Information Technology Profession: A Literature Review, Synthesis and Research Agenda. *European Journal of Information Systems*, *11*(1), 20–34. doi:10.1057/palgrave/ejis/3000417

Ajzen, I. (1991). The Theory of Planned Behavior. *Organizational Behavior and Human Decision Processes*, *50*, 179–211. doi:10.1016/0749-5978(91)90020-T

Bureau of Labor Statistics. (2007a). *Changes in men's and women's labor force participation rates.* Retrieved on March 15, 2009 from http://www.bls.gov/opub/ted/2007/jan/wk2/art03.htm

Bureau of Labor Statistics. (2007b). *Employed Persons by Detailed Occupation, Sex, Race, and Hispanic or Latino Ethnicity*. Retrieved on January 25, 2009 from http://www.bls.gov/cps/cpsaat11.pdf

Bureau of Labor Statistics. (n.d.). *Occupational Outlook Handbook (OOH), 2008-09 Edition.* Retrieved on March 15, 2009 from www.bls.gov/OCO

Ceci, S. J., Williams, W. M., & Barnett, S. M. (2009). Women's Underrepresentation in Science: Sociocultural and Biological Considerations. *Psychological Bulletin*, *135*(2), 218–261. doi:10.1037/a0014412

Costa, P. T., Terracciano, A., & McCrae, R. R. (2001). Gender Differences in Personality Traits Across Cultures: Robust and Surprising Findings. *Journal of Personality and Social Psychology*, *81*(2), 322–331. doi:10.1037/0022-3514.81.2.322

Fishbein, M., & Ajzen, I. (1975). Belief, attitude, intention, and behavior: An introduction to theory and research. Reading, MA: Addison-Wesley.

Hemenway, K. (1995). Human Nature and the Glass Ceiling in Industry. *Communications of the ACM*, *38*(1), 55–62. doi:10.1145/204865.204878

Hofstede, G. (1980). Culture's Consequences: International Differences in Work Related Values. London: Sage.

Johnson, R. D., & Stone, D., L. & Phillips, T. N. (2008). Relations Among Ethnicity, Gender, Beliefs, Attitudes, and Intention to Pursue a Career in Information Technology. *Journal of Applied Social Psychology*, *38*(4), 999–1022. doi:10.1111/j.1559-1816.2008.00336.x

Johnson, R. D., Veltri, N. F., & Hornik, S. (2008). Attributions of Responsibility Toward Computing Technology: The Role of Interface Social Cues and User Gender. *International Journal of Human-Computer Interaction*, *24*(6), 595–612. doi:10.1080/10447310802205784

Klawe, M., & Levenson, N. (1995). Women in Computing: Where are We Now? *Communications of the ACM*, *38*(1), 29–35. doi:10.1145/204865.204874

Klawe, M., Whitney, T., & Simard, C. (2009). Women in Computing - Take 2. *Communications of the ACM*, *52*(2), 68–76. doi:10.1145/1461928.1461947

Lazar, J., & Lidtke, D. (2002). Service-Learning Partnerships in the Information Systems Curriculum. In J. Lazar (Ed.), Managing IT/Community Partnerships in the 21st Century (pp. 1-16). Hershey, PA: Idea Group Publishing.

Lazar, J., & Norcio, A. F. (2002). Service-Research Partnerships: Research Projects that Help Bridge the Digital Divide. In J. Lazar (Ed.), Active Learning in Higher Education: A Model and Roadmap (pp. 271-289). Hershey, PA: Idea Group Publishing.

Marakas, G. M., Yi, M. Y., & Johnson, R. D. (1998). The multilevel and multifaceted character of computer self-efficacy: Toward clarification of the construct and an integrative framework for research. *Information Systems Research*, *9*(2), 126–163. doi:10.1287/isre.9.2.126

McKinney, V. R., Wilson, D. D., Brooks, N., O'Leary-Kelly, A., & Hardgrave, B. (2008). Women and Men in the IT Profession. *Communications of the ACM*, *51*(2), 81–84. doi:10.1145/1314215.1340919

MyPersonality. Info (2009). *Frequency of Personality Types*. Retrieved on April 1, 2009 from http://www.mypersonality.info/personality-types/population-gender/

NCWIT. (2007). NCWIT Scorecard 2007: A Report on the Status of Women in Information Technology: National Center for Women & Information Technology.

NCWIT. (2009), *Our Alliances*. National Center for Women & Information Technology. Retrieved on March 29, 2009http://www.ncwit.org/alliance.what.html

Orser, B., Riding, A., Dathan, M., & Stanley, J. (2007, November 2). *Gender Challenges of Women in the Canadian Advanced Technology Sector.* Paper presented at the Canadian Council for Small Business and Entrepreneurship Conference, Kelwona, B.C., Canada.

Papastergiou, M. (2008). Are Computer Science and Information Technology Still Masculine Fields? High School Students' Perceptions and Career Choices. *Computers & Education*, *51*, 594–608. doi:10.1016/j.compedu.2007.06.009

Rosenbloom, J. L., Ash, R. A., Dupont, B., & Coder, L. (2008). Why Are There So Few Women in Information Technology? Assessing the Role of Personality in Career Choices. *Journal of Economic Psychology*, *29*(4), 543–554. doi:10.1016/j.jocp.2007.09.005

Siegel, D. (2008). Building a Pipeline for Diversity Through Intersectoral Collaboration. *Higher Education*, *55*(5), 519–535. doi:10.1007/s10734-007-9072-6

Snow, C. P. (1995). Women in Computing. *Communications of the ACM*, *38*(1), 26–28. doi:10.1145/204865.204873

Tannen, D. (1990). You Just Don't Understand: Women and Men in Conversation. New York: Ballantine.

TBTLA. (2009). *Welcome*. Tampa Bay Technology Leadership Association. Retrieved on March 29, 2009 http://www.tb-tla.org/

Trauth, E. M., Quesenberry, J. L., & Huang, H. (2008). A Multicultural Analysis of Factors Influencing Career Choice for Women in the Information Technology Workforce. *Journal of Global Information Management*, *16*(4), 1–23.

Trauth, E. M., Quesenberry, J. L., & Yeo, B. (2008). Environmental Influences on Gender in the IT Workforce. *The Data Base for Advances in Information Systems*, *39*(1), 8–32.

Venkatesh, M., & Small, R. V. (2002). Active Learning in Higher Education: A Model and Roadmap. In J. Lazar (Ed.), Managing IT/Community Partnerships in the 21st Century (pp. 17-51). Hershey, PA: Idea Group Publishing.

Venkatesh, V., Morris, M. G., & Ackerman, P. L. (2000). A Longitudinal Field Investigation of Gender Differences in Individual Technology Adoption Decision-Making Processes. *Organizational Behavior and Human Decision Processes*, *83*, 33–60. doi:10.1006/obhd.2000.2896

Wood, W., & Rhodes, N. D. (1992). Sex Differences in Interaction Style in Task Groups. In C. Ridgeway (Ed.), Gender, Interaction and Inequality (pp. 97-121). New York: Springer-Verlag.

ENDNOTES

1 Quotes have been edited for grammar.

2 One parent only answered the first three questions, thus the last three questions only have a sample of 14.

Profile 3
Engineering Career Spotlight:
Fabrication and Characterization of Radio Frequency Plasma Polymerized Thin Films from Terpinen-4-ol for Electronic and Biomaterial Applications

© 2008 Queensland Government Office of Women. Used with permission.

Kateryna (Katia) Bazaka, Electronic Material Research Lab, School of Engineering, James Cook University, Townsville, Queensland Australia

I am originally from Ukraine, a beautiful country in Europe, where I spent my childhood and adolescence. Solving crosswords and playing various logical games was my parents' favorite activity, and so I developed a great liking for finding solutions to logical problems from a very young age. The Encyclopedia became one of my favourite books and my passion for reading was satisfied in my great parents' generous home library. At about 5 years of age, when my friends wanted to be movie stars, sailors or champions, I 'modestly' declared to my family that I am going to be a well-known Professor in Maths or Physics (I doubt

Copyright © 2010, IGI Global. Copying or distributing in print or electronic forms without written permission of IGI Global is prohibited.

that I knew then what it meant). At school, I was fascinated with science and philosophy. Wanting to pursue a degree in Engineering (Bachelor and later Master Degree in Physical and Biomedical Electronics) came as no surprise. The choice of the biomedical component of the degree was greatly influenced by the fondness my grandfather, a vet, had for medicine. He really wanted me to be a doctor and spent endless hours explaining how important medical/biological sciences are for the society. Studying at the university opened up my horizons as I discovered the truly multidisciplinary nature of the engineering profession. To further my knowledge in microbiology, molecular biology and biochemistry I did a few more subjects at James Cook University (Townsville). Having realised how important it is for the engineer to be able to manage their own affairs and other people, I went on to graduate with a Master in Business Administration at JCU.

Introduction

Katia is working on her PhD project entitled "Fabrication and characterization of RF plasma polymerized thin films from Terpinen-4-ol for electronic and biomaterial applications", which presents an innovative concept for fabricating novel polymer materials from a non-synthetic source. The radio frequency (RF) plasma polymerization Katia has adopted for material fabrication is an environmentally friendly, practically pollutant free low cost technique requiring only electromagnetic waves and no additional chemicals. Furthermore, the essential oil derived monomer (Terpinen-4-ol) employed in the project is an agricultural product predominantly produced in Australia, and is typically used in aromatherapy and medicinal applications.

Katia's PhD started in March 2008, however, she has already managed to successfully fabricate polymer thin films from Terpinen-4-ol and the preliminary characterizations have already been carried out and the results summarized and reported to the funding body (Rural Industry Research and Development Corporation RIRDC). Katia has also published one peer-reviewed paper and submitted several manuscripts to international journal that are currently under internal review. This is the very first time a polymer thin film has been fabricated from a non-synthetic based Terpinen-4-ol (derived from Tea Tree Oil). Terpinen-4-ol is the active component of Tea Tree Oil responsible for its anti-microbial properties, and it is hoped that this project will make use of these properties to have a significant impact on biomaterial research. Polymers produced from these methods and materials have the added benefit of being fabricated in an environmentally sustainable manner, with no harmful pollution or by-products.

Katia is the recipient of an Australian Postgraduate Award. Due to the innovative nature of the project, the Australian Institute of Nuclear Science and Engineering has also provided her with a top up scholarship and the use of their facilities to complete some components of the research. Katia received a 2009 Science and Innovation Award for Young People in Agriculture, Fisheries and Forestry. The Minister for the Department of Agriculture, Fisheries and Forestry awarded this prestigious prize in recognition of the potential benefits Katia's project holds for Rural Industries. She has also received a JCU grant to improve current fabrication facilities.

Project Background

Surface engineering is a sub-discipline of materials science which deals with the surface of solid matter, and is currently one of the most prominent areas for research and development. The combined global value of surface-engineered coatings applied to products exceeds $50 trillion per annum. Thin films and coating technology are the backbone of many electronic and biomedical products and new and advanced materials with tailored properties are essential for the growth of the electronic and biomedical industries.

Katia's project is focused on identifying non-conventional applications for essential oils and will contribute significantly to Australia's National Research Priorities (NRP). The project will utilize and merge two aspects:

1. Emerging technology applications for organic polymer electronics with high potential for strong growth and expansion, and
2. Increased supply and demand for Tea Tree Oil thereby value adding to the growing Tea Tree Oil industry in Australia. The development of advanced materials will underpin growth in many areas of industrial and economic activity in Australia, while enhancing its capacity in frontier technologies to power world-class industries of the future and build on Australia's strengths in research and innovation.

According to the RIRDC (Rural Industry Research and Development Corporation), Australia strong growth of essential oils market is anticipated, especially in the areas of Tea Tree, Lavender, Sandalwood in order to meet escalating demand for these commodities, both domestically and globally. Essential oils have been used commercially for many decades in food flavoring, fragrance and cosmetics, pharmaceuticals, and industrial solvents industries. Stimulating the growth of world-class Australian industries using innovative technologies developed from cutting-edge research is one of NRP.

PhD Research Aims

The aim of Katia's PhD project is to develop radio frequency glow-discharge plasma generated polymer thin films from non-synthetic sources and investigate the fundamental properties of these materials in order to identify potential applications. Some specific aims from this project include:

- The fabrication of environmentally friendly polymer materials from non-synthetic sources,
- The development of a novel application for an agricultural based non-synthetic organic material and hence to add value to an Australian product, thereby revolutionizing the Australian essential oil industry,
- The development of new polymer materials for electronic and biomaterial applications, and
- The evaluation of the fundamental physical and chemical properties of the new materials, which will contribute to the better understanding of the polymer thin films

The project brings together a number of traditional disciplinary areas of study (chemistry, physics, materials engineering, etc), and will build on the current knowledge available in each of those disciplines, and hence is likely to be beneficial to the wider scientific community and the Australian community in general.

Background

Surface engineering incorporates a broad array of technologies. *Plasma based technologies are amid the most rapidly growing in terms of industrial application as they are highly engineering-friendly, deliver 'green', practically pollutant free surface chemistry, and can be easily integrated into manufacturing processes.* Early applications of plasma polymers made use of the physical nature of these materials, the barrier properties in particular, and included electronics, metal coatings and membranes. With subsequent identification of plasma conditions capable of delivering chemically functionalized polymer films which retained some of the original chemical and structural properties of the monomer, the range of applications broadened. Katia's project is building upon this particular aspect of polymerization to develop bio-materials which retain the medicinal properties of the essential oils, in particular tea tree oil.

Tea tree oil is an essential oil distilled from the Australian plant *Melaleuca alternifolia* and has been used medicinally by Australian Aborigines for centuries. Currently, tea tree oil is predominantly used in the cosmetic, personal health, animal care and household products industries. Australia dominates the tea tree oil market contributing more than 80% of global oil production, with the majority of commercial production centered on the Atherton Tablelands of northern Queensland and coastal regions of northern New South Wales. With approximately 300 growers and 6,000 ha of cultivated *Melaleuca alternifolia*, the Australian Tea Tree industry has become a significant regional industry in Australia with a farm-gate value exceeding $20 million.

Furthermore, as the polymer films are expected to retain the antibacterial, antiviral, antifungal and anti-inflammatory properties of its essential oil based monomers, the films will have an added advantage of avoiding infections, thus achieving better biocompatibility within the body which may have significant applications in the biomedical industries. The development and manufacturing of coatings for implantable medical devices in order to enhance their biocompatibility had considerable research and development effort dedicated to it for more than 20 years. However, in spite of recent developments in biotechnology and material sciences, implantable bionic devices are still faced with a number of challenges, especially with regard to poor biocompatibility and rejection/infection.

Methodology

Monomers of non-synthetic Terpinen-4-ol obtained from three sources will be analyzed. Polymer thin films of thicknesses in the range of 100nm to 1000nm will be fabricated from the monomer using RF plasma polarization techniques. The volatile monomer vapour will be released into the "cold plasma" region where the plasma energy will be used to break the bonds of the monomer molecules and form polymer chains on a substrate (e.g. glass, Quartz, Teflon). During the plasma process, the activated monomer introduced in a gaseous phase under vacuum will undergo fragmentation, excitation, and ionization and rearrange and react to form a cross-linked polymer matrix deposited on the desired substrate. As a result, the physicochemical properties of a RF plasma polymer from a specific monomer are expected to be significantly different to those of a conventional polymer.

A variety of sophisticated techniques are employed to characterize the physicochemical, optical and electrical properties of RF plasma thin film polymers such as:

- Spectroscopic ellipsometry,
- Impedance spectroscopy,
- X-ray photoelectron spectroscopy (XPS),
- X-ray Reflectometry (XRR),
- Neutron reflectometry,
- Fourier transform infrared (FTIR) spectrometry,
- Mass spectrometry (time-of-flight technique),
- Current-voltage (I-V) measurements, and
- Atomic force microscopy (AFM).

Biocompatibility studies (*in vitro* and *in vivo*) will also be done to investigate the suitability of the RF plasma thin film polymers as a material in biomedical applications.

Katia already has expertise in a wide range of state-of-the-art analytical techniques such as AFM, ellipsometry and FTIR. All of the results will be analyzed to describe the fabrication process of organic polymer thin films with advanced properties set by the demands of the respective technologies for which they are intended in a standardized and reproducible method. Finally, a number of possible applications will be determined based on the properties of the developed polymers.

Results

Katia's PhD project is focused on the organic monomer Terpinen-4-ol, which is the most abundant (nearly 40%) component of tea tree oil and is thought to be largely responsible for the oil's antimicrobial activity. Non-synthetic Terpinen-4-ol has been successfully polymerized and some of the fundamental properties have already been described in a report to the RIRDC. The results were encouraging with further research needed to determine the degree to which the films retain the antimicrobial properties of the monomer.

It has been suggested that since Terpinen-4-ol has strong antibacterial, antifungal and antiviral properties, the polymer films produced from such a monomer may possess similar antimicrobial characteristics, making them an appropriate interface between the biological and the electronic systems for the successful integration of implantable medical devices in the human body.

The thin polymer films fabricated from Terpinen-4-ol also possess desirable optical and electrical properties and can therefore be applied in the traditional areas of electronics such as organic integrated circuits (IC), radio frequency identification devices (RFID) and smart transponder cards, organic light emitting diodes (OLEDs), displays, renewable energy systems, organic photovoltaic devices, solar cells and polymer batteries to name a few.

Anticipated Final Outcomes

Katia's research will have many significant contributions:

- *The development of an environmentally friendly polymer fabrication process* – a non-chemical based polymerization technique,
- *The development of an environmentally friendly polymer* - a novel application for an agricultural non-synthetic based organic material for organic semiconducting flexible electronics,
- *The development of a novel biomaterial* – a polymer coating with antibacterial, antifungal, antiviral and anti-inflammatory properties,
- *Increased knowledge* – novel polymerization techniques to polymerize essential oils and the understanding of its fundamental properties,
- *The development of novel applications for essential oils* – a new application for an existing product that will value-add to the Australian agricultural industry,
- *Collaboration* - The inter-disciplinary nature of the research will significantly contribute to traditional disciplinary areas of study (engineering, chemistry, physics, materials, etc) and is therefore of great benefit to the wider scientific community. Australian Nuclear Science and Technology Organization, RMIT and Swinburne Universities are already collaborating with the project and are expected to benefit from the knowledge gained.

The successful development of novel applications for essential oil products will add value to Australian agricultural produce, both locally (North Queensland being one of the primary areas of the Tea Tree Oil industry) and on the national level. The North Queensland Essential Oil Company is also a collaborator on this project, supplying Tea Tree Oil as an in-kind contribution. Novel polymers fabricated as a result of these studies are expected to possess advanced material properties that would make them a commercially viable and competitive alternative to existing materials within the electronics and biomaterials industries.

Chapter 4
Increasing the Numbers of Women in Science

Gwen White
Gateway Community and Technical College, USA

ABSTRACT

The fields of engineering and technology have traditionally been male dominated. The numbers of females in engineering and technology are traditionally less than their male counterparts. Within the past thirty years, women have had more opportunities to enter these fields. Previously stereotypes, misinformation and ignorance have shut women out of these important career choices. The short and long term goals are to eradicate stereotypes and make sure there is enough pertinent information available for women and girls about engineering and technology. There are a number of programs available that encourage females to enter the fields of engineering and technology; they just need to move up on the priority lists of corporations, guidance counselors and parents.

INTRODUCTION

Women are not entering the field of technology or engineering. They tend to see science and engineering as a male dominated field with no room for women. Women make up less than one quarter of the nation's scientist and engineers.

Women are held back due to stereotypes, lack of education, parental involvement and academic advising. These problems have continually contributed to the low numbers of females in engineering and technology.

This chapter will look at the overall placement of women and recommend ways to increase women's participation in technology and engineering. "The main hurdles in getting women to study engineering occur long before they reach college, and those that make it that far often have had a role model engineer like a brother or close family acquaintance to introduce them to engineering" (Wolcott, 2001).

DOI: 10.4018/978-1-61520-657-5.ch004

Copyright © 2010, IGI Global. Copying or distributing in print or electronic forms without written permission of IGI Global is prohibited.

BACKGROUND

Statistics in various countries shows that women for the most part are lagging behind males in engineering and technology. Countries include the United States, Canada, India, Japan and European nations. In the United States, women earned approximately 60% associate's degrees, 58% of all bachelor's degrees, and 59% of all master's degrees (Knapp & Kelly-Reid, 2006, p. 3). Overall high tech firms are lagging behind when it comes to women on the payroll in a STEM related field. Women only make up 25.6% of U.S. computer and math occupations. Only 8% of women with four year degrees are engineering managers (Anita Borg Institute, 2007).

There is expected growth in engineering and technology in the next four years. Management, scientific, and technical consulting services jobs according to the Employment Projections by Industry: 2006 to 2016 is expected to grow from 920,000 to 1,638,000. The same growth in Computer systems design and related services is expecting to increase from 1,278,000 to 1,767,000 by 2016 (U.S. Bureau of Labor Statistics, 2007).

The U.S. Department of Education's outlook for Engineering, Technology and Science Occupational Projections from 2002 – 2012 show that 1,500,000 positions in these fields are likely to be added in the near future. Only 26% of degree holders are women with a degree in Computer Science. "Women received about 38% of the computer science bachelor's degrees awarded in the United States in 1985, the peak year, but in 2003, the figure was only about 28%, according to the National Science Foundation "(Dean, 2007, p. 1).

Even though there is an expected increase in the number of technological positions, conversely the proportion of women available to work in these fields have decreased in the past 20 years in comparison to other fields like mathematics and science (U.S. Department of Education, 2005). However, only 26% of women hold professional positions in the IT industry in 2006. This number of females involved in technology is likely to decrease as the year 2012 approaches.

The National Center for Women in Technology (NCWIT, 2009) published statistics related to females in computer science. In 2008, 57% of women graduated in the Unites States with a four year degree. Of that 57% only 18% were computer science majors. This is a significant decrease from 1985 where 37% of female graduates were computer science majors. Statistics (Anita Borg Institute, 2007) in other countries show low number of computer science/engineering majors: India (8%), France (19%) and the United Kingdom (18.2%). In contrast, Eastern Europe (57%), China (59%), South Korea (46%) and Japan (66%) in 2000 have a higher representation of four year graduates. According to the Bureau of Labor Statistics, "demand for computer scientists in the United States will only increase in coming years," Dr. Cuny said. "If you look at the demographics of the country, if we are not going to get our new professionals from women and minorities and persons with disabilities, we are not going to have enough (Dean, 2007, p. 5)."

Climbing The Technical Ladder Study (2007) surveyed 1,795 men and women currently in the field of engineering or technology at various high tech companies in the Silicon Valley. The sample was chosen because it was representative of a globalized population. 34.2% of the sample was women. Women in senior management positions numbered less than 5%. There has been a decline in the number of women in technology and engineering fields as of 2005 (from 41% to 32% by 2005) (Simard, 2008). Based on the statistics, there are many of the same issues in other countries. The only bright spot is Eastern Europe where women have a greater portion of technical degrees than women in other countries.

Eastern Europe women in 1992 had 57% of all university degrees with at least 50% of these degrees in the engineering fields, 70% in natural sciences and 73% in mathematics and computer science (Anita Borg Institute, 2007).

ISSUES PREVENTING GIRLS FROM ENTERING ENGINEERING AND TECHNOLOGY

Unfortunately many girls and women have opted not to enter engineering or technology due to one or more of these obstacles. This chapter will focus on the main themes of parental involvement, educational opportunities and guidance counseling, work life balance and stereotypes.

Parental Involvement

Erickson (1982) revealed the stereotype that boys do better at math is still believed to be true by faculty and parents. "Parents provide a substantial amount of influence regarding their career choices. They pass down what they have been taught. In many cases, parents do not encourage their girls to pursue engineering or technology. This can be due to experiences or even stereotypes. Not that it is intentional but many times parents pass down these stereotypes to their children"(Erickson, 1982).

The background of the parent can influence the career direction of the child. When parents do not understand engineering and technology, they have a tendency to avoid encouraging females to enter these fields. In total contrast, the parent that has knowledge in engineering or technology can provide the student with proper guidance regarding engineering and technology.

When there is a lack of understanding, there is room for misperceptions. A study of incoming female freshmen showed that their career choice was based on information provided by their parents (Martz & Tetua, 2008).

Tenebaum and Leaper (2003) showed that parents' use of language with boys and girls differed. Most girls have been socialized using non-scientific language. The opposite was true for boys. Also girls were not introduced to many opportunities related to science. Careers like homemaking, social work and teaching were a few of the fields parents would choose for their daughters. When a mother believed science to be difficult for their children, the self-efficacy and interest was lower. When the opposite occurred, the child's self-efficacy and interest increased (Tenenbaum & Leaper, 2003).

Educational Opportunities and Guidance Counseling

"I still hear anecdotes about guidance counselors steering girls away from engineering, telling them they won't be able to do the math (Lewin, 2008, p. 1)."

The role of the guidance counselor or academic advisor is complicated. The counselor has to be able to provide fairness, compassion and help. Erickson and Schultz (1982) studied academic advisors in a two year college. Meetings between students and academic advisors were tape recorded. When the advisor and student connected through a common background, the session was very positive. However when the student was from a different background or genders the interaction was very different. It was less likely they could build a common background and rapport. These students who did not have common backgrounds with their advisor were not given the same benefit. Students who wanted to be engineers or computer scientist were encouraged to look at another career choice because their advisor was not familiar with STEM (Science, Technology, Engineering and Math) (Erickson, 1982).

A second reason why women are not guided towards engineering or technology is they perceive the field to be boring. The study "Lost in Translation: Gender in High School Computer Science" (Goode, 2004) revealed there are very few computer science/engineering learning opportunities at the high school level. Females have a tendency to be more insecure and do not pursue the field like their male counterparts. Girls fear that taking a computer science and engineering class can lower their GPA's (Goode, 2004).

A third reason (Barker, 2006) females did not to consider engineering or science was female students did not want to be stuck in front of a computer all day. Girls did not view information technology or engineering as a people oriented career (Barker, 2006).

Many high schools do not have computer or engineering based curriculums. Lately, there has been a focus on creating STEM (Science, Technology, Engineering and Math) related programs in high school. These new programs are designed to bring more students into the technological and engineering fields. However, enrollments of female students are still low.

Racial makeup became a factor whether or not a female would consider engineering or technology as a career choice. In areas where the racial makeup was more people of color, the opportunities for technology and engineering were lower than those who were in predominately white areas (Zarrett, 2006). The number of resources available for students to pursue engineering and technology as a career is minimal (Zarrett, 2006). When the student was a minority and female, the chance of them enrolling in science or math would be even less.

"Women are the canaries in the coal mine," Lenore Blum, a computer scientist at Carnegie Mellon University, told an audience at Harvard University in March, in a talk on this "crisis" in computer science. Factors driving women away will eventually drive men away as well (Dean, 2007).

Work Life Balance

Women who are married or have children find it difficult to enter the fields of engineering and technology due to commitments of family life and the extended workday that sometimes comes with these fields. Extended workdays are required for success. The women who cannot work late on a regular basis are perceiving barriers to their advancement (Simard, 2008).

"Where women work will be predominately white male with long hours and obsessive behavior" (Simard, 2008). The workload in technology and engineering fields can be a 24/7 schedule leaving little additional time for family. Between being on call to taking work home can make it harder for a female to be successful in these fields. The problem of making a choice between family and work drives many women out of the field. Many women have turned down a job or promotion due to the increased demands of the role.

Stereotypes

Simard (2008) discusses that the stereotypes of women are too family focused have resulted in women being unfairly steered towards positions that are stereotypically female. When women act assertively they are punished for their behavior. The male counterparts are considered acting normal – like a man should. Stereotyping has endured for a long time. Unfortunately it continues today. Men have seen women as not as smart or cannot perform mathematical operations. No matter their thoughts, it is untrue. Simard (2008) emphasizes these assumptions have been detrimental to many women's careers.

Even with equal qualifications and achievements to those of their male counterparts, women are perceived less favorably in terms of their ability and accomplishments. This stereotyping tends to be reflected in evaluations and promotions which places women at a disadvantage for advancement (Eagly & Carau, 2002).

In short, the cultural stereotypes contribute to an environment in IT (Information Technology) where women are consistently undervalued. These cultural stereotypes help support what is generally considered a masculine perspective, or culture, which dominates the organization and power structure throughout IT, related industries and academic communities (Kvasny & Lynette, 2004).

A quote from a mid-level technical male found in Climbing the Technical Ladder stated "The women I've worked with have been as good as or better than men. Maybe they had to be. I would like to see that problem solved. "I think that our technological leadership as a nation is very much at risk – if we can't expand beyond white males, we're in real trouble" (Simard, 2008, p. 10).

Foremost among these stereotypes is the assumption that women are better at human skills of "nurturing, emotional expressiveness and communication activities," whereas men are better at "instrumental and task-oriented assignments" (McCorduck & Ramsey, 2005).

In American society, for instance, the interactional identities of elementary school teacher, secretary, or nurse are expected to go with the master identity of being female whereas the interactional identities of surgeon, engineer, or manager are expected to go with the master identity of male. To the degree an interactional identity is strongly associated with some master identity, whether it be gender, race, or age, that interactional identity takes on some of the broader master identity features with which it is associated (Tracy, 2002).

In addition (Tracy, 2002) discusses relational identities that happen when people talk about certain subjects and interact with each other regarding specific qualities of each other. Things like equality, superiority or trust work on the psychological being of the female employee or student.

"Men resent women in the field and do little to make them feel comfortable. If there's a momentary change now with younger generations, women will soon face another reality. Approximately half of the engineering students in America today are foreign nationals and often their cultures do not share Western norms for professional women and women as equals. Management tracks may seem more "peaceful" to women compared with the brutal competition for success and notice among IT engineers, she says. This drives women away from the technology" (McCorduck & Ramsey, 2005, p. 4)

WAYS TO INCREASE EXPOSURE

Since there are not that many women going into technical fields, there needs to be focused activities specifically for females. Programs like mentoring and shadowing, changing parental perceptions, and modifying curriculum are good ways to increase the numbers of females in engineering and technology. Also there are a variety of programs cited that are working to increase the number of women in engineering and technology.

Mentoring and Shadowing

Mentoring is one of the most important ways to encourage females to enter engineering and technical fields. Reaching back into elementary schools around the fifth grade and sixth grades can help plant the seeds of technology and engineering, especially for girls. Something as simple as a female engineer participating in career day showing girls what she does on a daily basis along with demonstrating her actual impact on a project can go a long way with a young impressionable mind. When girls see someone who is actively making a difference, they are more likely to want to explore the field. (Inspiring girls now in technology education, 2009)

The mentoring relationship must continue into high school. These women and girls must be followed to ensure that they have an understanding of engineering and technology. "It may sound like segregation but it is designed to make females comfortable in a field that she might not have perceived as welcoming" (Joanna Goode, 2004).

Another mentoring program that would allow girls to experience math and science in a STEM program would be one that encourages females to enter a summer STEM program and incorporate fun experiments and field trips to companies that are actively pursuing females as a part of their engineering or technology teams. Girls are shown they should not be afraid of engineering and technology. The girls are introduced to women who are currently in the field. The matching of these individuals allows for a mentoring relationship and a shadowing opportunity during the summer. When girls are placed into a setting with current role models, they are more likely to consider engineering and technology as a career choice (Townsend, 2002).

Parental Education

Parents introduce students to career choices through their actions and also through other family members. Many times these relationships are positive and an aspiring role model might be found right inside the immediate family.

However, parents must be educated about the engineering and technology fields. Barker (2006) discovered the impact of group sessions that address concerns and misconceptions about the field. When parents hold the stereotype that boys do better in math than girls, it can be detrimental to their girl's development in engineering or technology. As stated earlier in the Wisconsin study, girls are doing better now in math and science. They have been given greater opportunities to showcase their talents.

Fortunately or unfortunately, according to Epstein (1982) and Shore (1996), parents are very protective of their children and want to see them succeed. Parents as a whole see their child as successful but at the same time they are guarded about the careers they choose. Parental comfort levels can guide the direction their child chooses as a career. This means parents will sometimes not steer their child to a career outside of their own comfort zone.

The NCWIT (2006) encourages the use of informational sessions with the parents and the female child to help them understand engineering and technology. In addition, parents could be asked to participate in science and technology programs along with their child so the parents can see exactly what is involved in the fields.

Parents should be encouraged to participate in career day, especially if their career is not in engineering or technology. Fathers who have a strong influence on their daughter(s) and their career choices have daughters that enjoy math (Experiments aim to ease effects of stereotypes, 2007). If the parent does not understand, provide them with information and support. However, this could be a difficult venture because many parents might not be willing to come forward revealing their views.

Misinformation leads to misunderstanding.. Schools must provide students with gender friendly information about engineering and technology. The information must be specific and show how the various careers contribute to society and also their earning potential for each specific career. At a later date, the guidance counselor or even the math or science teachers should follow up to answer any questions.

Shore (1996) believes parents have to be aware that they could be indirectly rewarding their children's behavior through the activities the children participate. Activities that are generally stereotypically "female" or "male" can influence a child's experience and enforce those stereotypical gender roles. Nothing wrong with these roles, it's just as important to make sure that children, especially females are involved in more than just "female" type activities. If possible, try to take away the gender bias in activities.

Parents should take more of an interest in science activities at schools or even after school programs. Encouragement is the best way to have their daughters become or remain interested in science or engineering. Also, if a parent is aware of a female scientist, take their daughter to meet them and let them shadow the female scientist for a day.

Understanding of Engineering and Technology

Academic Advising

Academic advisors face a major dilemma; Tracy (2002) emphasized to be a fair gatekeeper, treating all students similarly, it's best to be distanced, impersonal and informal style. This can be achieved by everyone. In contrast, to be a friendly helper is best enacted by using an interest-showing style that seeks to build rapport. The more an advisor uses the impersonal style to ensure fairness, the less he or she will succeed as a friendly helper. The more a person seeks to build a rapport and to be interpersonally friendly, the less likely he or she is to become a consistently fair gatekeeper. Thus the ideal of a "good" academic advisor involves a pair of contradictory beliefs (Tracy, 2002, p. 30)

Information provided in STEM related careers must be "unbiased and free from male references". Pronouns like he and his must be removed or she and her must be incorporated in the literature. Guidance counselors have to be non-biased and reach out to parents to increase the level of understanding of non-traditional careers (Erickson, 1982).

Dr. Lazowska and Dr. Blum, with colleagues at the University of California, Los Angeles, and Google, are working on materials that high school teachers can use to tell students about the challenges and opportunities of computer science. They are developing them for teachers of Math, Science and English because, as Dr. Lazowska put it, "many young women have opted out of the field before they even get to computer science" in high school (Dean, 2007).

During the middle school years, females should be encouraged to explore non-traditional programs. The middle school age group is very impressionable and is making decisions about what career they might want to follow.

A study "Examining the Gender Gap in IT by Race" found that females would consider IT as a field when they were either counseled by parents or guidance counselors. This guidance formed a positive attitude for the female student. When the female students saw other females in engineering or technology, it would encourage them to become a participant (Zarrett, 2006).

Seminars about the STEM fields should be incorporated into the curriculum as early as the eighth grade. The development of a career pathway is crucial for female students to understand what is needed to become involved in engineering or technology. One example of a career pathway program has been developed in Northern Kentucky at Gateway Community and Technical College. The community college has a relationship with many of the schools in the area. There are area technical centers located inside the high schools that provide potential students to the community college. They can receive full college credit for certain classes through a transfer agreement. The career pathway was mapped out from the ninth grade to a Bachelor's degree using these transferrable courses as the focal point. The pathway depicted what courses must be taken in either engineering or technology career fields along with salary information. The information is not gender biased. Over the past two years since the pathways were initiated there has been an increase in female enrollments in the STEM related fields at Gateway (Gateway Community and Technical College, 2007).

Defeating Stereotypes

In a recent study (Hyde, 2005) at the University of Wisconsin, found that females in high school on standardized tests are performing just as well as their male counterparts. Although boys in high school performed better than girls in math 20 years ago, the researchers found, that is no longer the case. The reason, they said, is simple: Girls used to take fewer advanced math courses than boys, but now they are taking just as many.

"Now that enrollment in advanced math courses is equalized, we don't see gender differences in test performance," said Marcia C. Linn of the University of California, Berkeley, a co-author of the study. "But people are surprised by these findings, which suggests to me that the stereotypes are still there" (Lewin, 2008).

Women who work in a male dominated environment have to overcome the stereotype that they have to be better than a man. Their work will show that they are just a good or a better performer than a man. However, before this stereotype is changed all employees should participate in diversity and cultural training. Many times people don't realize they are using a perceived stereotype to judge or evaluate a peer or employee. This type of training is essential to knock down stereotypes, especially gender stereotypes. Tolerance is the key along with fairness. Each person should be able to stand on their own free from stereotypes.

Organizations that actively recruit women for positions can change the corporate culture. Attending women only conferences for recruiting purposes is important too. Recruiting materials can be enhanced by adding female specific brochures or media. The new recruiting materials should emphasize a quality work life along with flexibility. These small changes will encourage girls and women to consider engineering and technology positions at that company.

Women that are outstanding in their fields should be spotlighted. Putting their efforts and achievements in the forefront will show others, especially male counterparts, that women can be equal or better in engineering and technology.

Changes in Educational Programs

Moving emphasis away from programming proficiency was a key to the success of programs Dr. Blum and her colleagues at Carnegie Mellon instituted to draw more women into computer science. "At one time, admission to the program depended on high overall achievement and programming experience. The criteria now, are high overall achievement and broad interests, diverse perspectives and whether applicants seem to have potential to be future leaders" (Dean, 2007).

The field of engineering and technology should be a priority in the school curriculum. Also more information should be provided at an earlier stage in a girl's education. The basic fundamentals of engineering, science and technology can be incorporated in the curriculum.

In addition, K-12 programs must be redefined to dispel gender bias. Programs like those from the U.S. Department of Education Gender Equity Panel (2001) are ideal and should be used as national models.

Teachers should have gender equity training and include this training as a part of their educational goals (Klein, 2005). Teachers must continually participate in training that helps them to recognize gender stereotypes and bias. Schools have to be held accountable for gender bias problems which negatively affect female students.

Klein (2005) suggested that national accrediting bodies need to develop criteria that are specific to eradicating gender bias. Programs in various schools must be evaluated to determine if gender bias in prevalent. If gender bias is realized, then educational programs for faculty, staff and students must be enacted to eradicate the bias.

Programs Geared Towards Women in Engineering and Technology

There are various programs that are focused on increasing the interest of women and girls in engineering and technology. These programs are primarily located in the United States but they are very similar to programs worldwide. There are additional resources listed at the end of this section with programs designed to help women enter and remain in engineering and technology fields.

Anita Borg Institute for Women and Technology

The Anita Borg Institute for Women and Technology (ABI) was founded in 1997 by renowned computer scientist Anita Borg, PhD (1949-2003). Initially known as the Institute for Women in Technology, IWT was renamed in 2003 to the Anita Borg Institute in order to honor Dr. Borg (Anita Borg Institute, 2007). The Anita Borg institute features five programs to help women who are interested in technology.

Women in Computer Science

An organization called Women in Computer Science @Brown runs the Artemis Project, which brings ninth-grade girls from schools in Providence, R.I., to the university campus for five weeks each summer. Its goal is to help the girls learn both concrete computer skills and abstract computer science concepts "in a positive and encouraging environment" (Dean, 2007).

Grace Hopper Celebration for Women and Computing

The Grace Hopper Celebration for Women and Computing gives corporations access to a pool of females to help enhance their recruiting efforts. It is an annual conference where women come together to share their expertise and skills with each other. The GHC Conference is the world's largest gathering of technical women in computing. Over the last decade and eight GHC Conferences more than 7400 women have attended and nearly 1,300 scholarships have been awarded to students. Presented with Association on Computing Machinery (ACM) the GHC Conferences offer opportunities for mentoring, motivation, networking, technical and career development (Anita Borg Institute, 2007).

TechLeaders

TechLeaders give women various opportunities for leadership development and skill development. TechLeaders workshop curriculum is grounded in research on the barriers faced by technical women and are led by world-class facilitators. The workshops strengthen and expand technical women's professional networks as well as provide participants with vital tools to better navigate all stages of their careers. "Working with organizational stakeholders, we find an increase all technologists' potential to create a positive impact on the world's women as we help the industry retain vital talent"(Anita Borg Institute, 2007).

Women of Vision

Women of Vision is a social networking area that gives women in technology inspiration and the opportunity to interact with others in the same fields.

Systers

Systers is another site that provides women with a support community where they can exchange ideas and knowledge while strengthening their leadership. Systers' is a forum for all women involved in the technical aspects of computing. The list has over 3,000 members in at least 54 countries around the world. Systers welcomes technical women of all ages and at any stage of their studies or careers to participate (Anita Borg Institute, 2007).

ABI Awards

The ABI Awards gives recognition to women in the technology fields by providing national recognition for their accomplishments. Additional information about these programs can be found at http://anitaborg.org/get-involved/participate/.

MentorNet

MentorNet is an online mentoring program designed to address the diversity concerns in engineering and sciences. The program has been a positive influence to the females involved. Founded in 1997, MentorNet provides highly motivated protégés from many of the world's top colleges and universities with positive, one-on-one, email-based mentoring relationships with mentors from industry, government, and higher education. In addition, the MentorNet Community provides opportunities to connect with others from around the world who are interested in diversifying engineering and science (MentorNet). "MentorNet is a means for creating and building relationships to a wonderful and important end: to level the playing field so that we can unleash all the talents in these crucial disciplines and enable it to flourish." - David Porush, PhD, President & CEO, MentorNet.

Black Data Processing Association

The Black Data Processing Association has always provided mentoring support to female minority students. BDPA (Black Data Processing Associates) was founded in 1975 by Earl A. Pace, Jr. and the late David Wimberly after the two met in Philadelphia to discuss their concerns about minorities in the data processing field. There was a lack of minorities in middle and upper management, low recruitment and poor preparation of minorities for these positions, and an overall lack of career mobility (Black Data Processing Association, 2009). Females that are considering the field of technology are embraced and surrounded by many females already involved in the field. Many local groups are creating female study only groups, mentoring relationships and training for female only students.

Girls in Science

The San Diego Zoo has a program which encourages girls ages 12 through 14 to explore science by participating in hands-on activities at the zoo. The girls are required to talk to scientist including researchers, animal behaviorist, veterinarians and zookeepers. Female scientists are introduced to the girls to help them feel comfortable. During the program the students have a chance to meet other female scientist in other fields. At the end of the program there is a special sleep over to emphasize that "girls can do it too!"

Sisters in Science

Sisters in Science was founded in 2001 to help girls gain a better understanding of science and mathematics. The program is based in Philadelphia, Pennsylvania. The target demographic is females in the fourth grade. "The multiple programs of Sisters in Science address the need for girls and boys in elementary, middle and high school to gain equitable access to science and mathematics education. Specifically, this need is based on the rising research-based public concern over the diversity gap in science, mathematics, and technology. Sisters in Science offers six innovative science programs designed to foster equity, access and inclusion in science, technology, engineering, and mathematics (STEM) education' (Sisters in Science). The program is designed to help girls with the national science standards for elementary schools. There are 20 weekly science experiments which help girls to become more interested in the sciences and change the stereotypes that girls cannot do science.

Women in Science and Engineering

Women in Science and Engineering (WISE) have a pre-college K-12 outreach program for girls who are interested in science and engineering. In addition to the K-12 program, there is a college level and faculty support program. The program is located at www.uicwise.org at the University of Chicago Illinois.

Women in Engineering and Outreach

Women in Engineering and Outreach at North Carolina State University is a program designed to recruit and retain young women who do not have engineering or technology role models in their lives. The program works with K-12 girls featuring hands on demonstrations of math and science. The program targets fourth and fifth grade girls because the critical time to gain their interest are earlier than originally thought (Wolcott, 2001).

Lore-El Center

The Lore-El Center for Women in Engineering and Science (MentorNet) at the Stevens Institute of Technology, Hoboken, New Jersey focuses on the K-12 curriculum to ensure females are exposed to engineering and technology. The program takes young women and encourages them to participate in a two week summer camp focused on engineering math and science. "Since 1978 we have found that 94% of the females who participated in the program major in science or engineering in college (Wolcott, 2001, p. 3)."

Additional Programs with Websites

IGNITE (Inspiring Girls Now in Technology Evolution) -http://www.ignite-us.org
TechDivas - http://www.techdivas.com/
Women in Technology International - http://www.techdivas.com/
Webgrrls - http://www.webgrrls.com/
Society of Women Engineers - http://www.swe.org/
Engineer Girl - http://www.engineergirl.org/
American Society of Mechanical Engineers/Minorities and Women - http://www.asme.org/Communities/Diversity/
Women Chemists Committee of the American Chemical Society - http://membership.acs.org/W/WCC/
Association for Women in Computing - http://www.awc-hq.org/home.html
Association for Women in Mathematics - http://www.awm-math.org/index.html
Association for Women in Science - http://www.awis.org/
Women in Aviation International - http://www.wai.org/ASPIRE – Girls in Engineerng – K-12 - http://aspire.swe.org/
GirlStart - http://www.girlstart.com/
AWSEM – Advocates for Women in Science, Engineering & Mathematics - http://www.saturdayacademy.org/GroupOptions/AWSEM/tabid/97/Default.aspx
Girls' E-mentoring in Science, Engineering & Technology (GEM-SET) - http://www.uic.edu/orgs/gem-set/index.htm
Techbridge - http://techbridgegirls.org/

CONCLUSION

Women are increasingly becoming more visible in engineering and technology. About three decades ago, women were steered in directions other than engineering or technology. Fields like teaching, social work or homemaker were the dominant occupations for a woman. Now things are changing for the better. Recent studies have shown that women are equal or better than males in math and science testing. Also the stereotypes that girls are not as good as boys in math are slowly being eradicated. Continuing with the education of parents, guidance counselors and girls themselves will help increase the numbers in the near future. Engineering and technology should not scare girls; it should welcome them with open arms. In the near future we should see an increase of females in engineering and technology.

REFERENCES

Anita Borg Institute. (2007). *History of website*. Retrieved February 12, 2009, from Anita Borg Institute: http://anitaborg.org/about/history/

Baker, D. P. (1988). School charter and parental management in West Germany. *Sociology of Education*, 255–265.

Barker, L., & Aspray, W. (2006). The state of research on girls and IT. In J. M. Aspray, Women and information technology: Research on under-representation. Cambridge: MIT Press.

Becker, H. J. (1982). Parent involvement: A survey of teacher-practices. *The Elementary School Journal*, *83*, 85–102. doi:10.1086/461297

Black Data Processing Association. (2009). *About BDPA*. Retrieved March 3, 2009, from Black Data Processing Association: http://www.bdpa.org/portal/index.php?option=com_content&task=view&id=16&Itemid=43

Crowley, K. C. (2001). Parents explain more often to boys than to girls during shared scientific thinking. *Psychological Science*, *12*, 258–261. doi:10.1111/1467-9280.00347

Dean, C. (2007, April 17). *Computer science takes steps to bring women to the fold*. Retrieved February 27, 2009, from The New York Times: http://www.nytimes.com/2007/04/17/science/17comp.html

Eagly, A., & Carau, S. (2002). Role congruity theory of prejudice toward female leaders. *Psychological Review*, *109*(3), 573–598. doi:10.1037/0033-295X.109.3.573

Epstein, J. L. (1982). Teachers' reported practices of parent involvement: problems and possibilities. *The Elementary School Journal*, *83*, 103–114. doi:10.1086/461298

Erickson, F. (1982). The counselor as gatekeeper. New York: Academic Press.

Experiments aim to ease effects of stereotypes. (2007).

Gateway Community and Technical College. (2007). *Student climate survey.: Gateway community and technical College*. Covington: Gateway Community and Technical College.

Henderson, A. (1981). Parent participation-student achievement: The evidence grows. Columbia: National Committee for Citizens in Education.

Hoover-Dempsey, K. V., Bassler, O. C., & Brissie, J. S. (1987). Parent involvement: Contributions of teacher efficacy, school socioeconomic status, and other school characteristics. *American Educational Research Journal*, *24*(3), 417–435.

Hyde, J. (2005, September). The gender similarities hypothesis. *The American Psychologist*, *60*(6), 581–592. doi:10.1037/0003-066X.60.6.581

Inspiring girls now in technology education. (2009). Retrieved June 30, 2009, from IGNITE: http://www.ignite-us.org/

Joanna Goode, R. E. (2004). Lost in translation: Gender in high school computer science. Representation of Women in Information Technology.

Klein, S. (2005). Handbook for achivening gender equality through education. Routledge.

Knapp, L. G., & Kelly-Reid, J. E. (2006, February). Postsecondary institutions in the United States: Fall 2003 and degrees and other awards conferred: 2. *Education Statistics Quarterly*.

Kvasny, A. H., & Lynette, T. (2004). Recruitment is never enough. *SIGMIS '04*.

Lewin, T. (2008, July 25). *Math scores show no gap for girls, study finds*. Retrieved 28 2009, February, from The New York Times: http://www.nytimes.com/2008/07/25/education/25math.html?_r=1

Martz, B., & Tetua, C. (2008). Students' perception of IS academic programs, IS careers, and outsourcing. *Journal of Education for Business*, *84*, 118–125. doi:10.3200/JOEB.84.2.118-125

McCorduck, N., & Ramsey, P. (2005). *Where are the women in information technology?* Boulder: Anita Borg Institute for Women and Technology. MentorNet. (n.d.). Retrieved March 1, 2009, from MentorNet: http://www.mentornet.net/

Shore, L. (1996). Developing gender equity at home: The influence of the exploratorium science at home project on parental attitudes, perceptions, and behaviors. St. Louis.

Simard, C. P. (2008). *Climbing the technical ladder: Obstacles and solutions for mid-level career women in technology.* Anita Borg Institute. Sisters in Science. (n.d.). *Sisters in science*. Retrieved March 13, 2009, from Sisters in Science, LLC: http://www.sistersinscience.org/

Tenenbaum, H., & Leaper, C. (2003). Parent-child conversations about science: The socialization of gender inequities? *Developmental Psychology*, *39*, 34–47. doi:10.1037/0012-1649.39.1.34

The National Center for Women in Information Technology. (2006). The future of IT. Boulder: The National Center for Women in Information Technology.

Townsend, G. C. (2002). People who make a difference: Mentors and role models. *Women and Computing*, *34*(2), 57–61.

Tracy, K. (2002). Everyday talk: Building and reflecting identities. New York: Guilford Press.

U.S. Bureau of Labor Statistics. (2007, November). *"Industry output and employment projections to 2016,'*. Retrieved March 21, 2009, from U.S. Bureau of Labor Statistics: http://www.census.gov/compendia/statab/tables/09s0601.pdf

U.S. Department of Education. (2005, March). *Postsecondary institutions in the United States: Fall 2003 degrees and other awards conferred.* Retrieved March 12, 2009, from U.S. Department of Education: http://nces.ed.gov/pubs2005/2005154.pdf

Viadero, D. (2007, October 24). *Experiments aim to ease effects of 'stereotype threat'*. Retrieved February 10, 2009, from Education Week: http://www.edweek.org/login.html?source=http://www.edweek.org/ew/articles/2007/10/24/09gap.h27.html&destination=http://www.edweek.org/ew/articles/2007/10/24/09gap.h27.html&levelId=2100

Wolcott, B. (2001, April). Role models needed. *Mechanical Engineering (New York, N.Y.)*, 1–9.

Zarrett, N., Malanchuk, O., Davis-Kean, P., & Eccles, J. (2006). Examining the gender gap in IT by race: Young adults' decisions to pursue an IT career. In W. Aspray, & J. McGrath, Women and information technology: Research on the reasons for under-representation. Cambridge: MIT Press.

ENDNOTES

1. Baker, J. 2003. A Report on the Study of Land-Sourced Pollutants and their Impacts on Water Quality in and Adjacent to the Great Barrier Reef.
2. The State of Queensland and Commonwealth of Australia. 2003. Reef Water Quality Protection Plan; for catchments adjacent to the Great Barrier Reef World Heritage Area, Queensland Department of Premier and Cabinet, Brisbane.
3. Haynes, D. and Michalek-Wagner, K. 2000. Water quality in the Great Barrier Reef World Heritage Area: Past perspectives, current issues and new research directions. Marine Pollution Bulletin 41(7-12): 428-434.

4. Bell, P.R.F. 1991. Status of eutrophication in the Great Barrier Reef lagoon. Marine Pollution Bulletin 23: 89-93.
5. Sellner, K.G. 1997. Physiology, ecology, and toxic properties of marine cyanobacteria blooms. Limnology and Oceanography 42(5): 1089-1104.
6. Carmichael, W.W. 2001. Health effects of toxin-producing cyanobacteria: "The CyanoHABs". Human and Ecological Risk Assessment 7(5): 1393-1407.
7. Lenes, J. M., Darrow, B. P., Cattrall, C., Heil, C. A., Callahan, M., Vargo, G. A., Byrne, R. H., Prospero, J. M., Bates, D. E., Fanning, K. A. and Walsh, J. J. 2001. Iron fertilization and the Trichodesmium response on the West Florida shelf. Limnology and Oceanography 46(6): 1261-1277.
8. Karl, D., Michaels, A., Bergman, B., Capone, D., Carpenter, E., Letelier, R., Lipschultz, F., Paerl, H., Sigman, D. and Stal, L. 2002. Dinitrogen fixation in the world's oceans. Biogeochemistry 57(1): 47-98.
9. McTainsh, G.H. 1989. Quarternary aeolian processes and sediments in the Australian region. Quarternary Science Reviews **8**: 235-253.
10. Access Economics Pty Limited. 2005. Measuring the economic and financial value of the Great Barrier Reef Marine Park. Great Barrier Reef Marine Park Authority Research Publication #84.
11. CSIRO. 2001. Climate Change Projections for Australia. Publ. Internet - http://www.cmar.csiro.au/e-print/open/projections2001.pdf.

Profile 4
Science Career Spotlight:
Phytoplankton Growth in Queensland Coastal Waters

© 2008 Queensland Government Office of Women. Used with permission.

Emily Shaw, Climate Change Research Centre, The University of New South Wales, Sydney NSW 2052, Australia, e.shaw@student.unsw.edu.au

I grew up in the Redlands in southern Moreton Bay, Queensland. On weekends we would go to North Stradbroke Island and go surfing and diving, which exposed me to the amazing diversity of marine life. My family also went camping on the Great Barrier Reef every year. Over the years it became apparent that there was much that we did not know about marine ecosystems and that we were exploiting them in an unsustainable manner. It was upsetting to go to places that I'd been in previous years and see that the number of fish had markedly declined or that a beautiful coral reef had turned into an algal bed. However, I had noticed that every year there seemed to be more and more humpback whales, leading me to believe that if we have the desire and the knowledge then it is possible to conserve marine species and ecosystems.

Copyright © 2010, IGI Global. Copying or distributing in print or electronic forms without written permission of IGI Global is prohibited.

I am currently completing a Bachelor of Science with Honours in Australian Environmental Studies at Griffith University and intend to embark on further research into the future with PhD studies and post-doctoral research. I am very passionate about conserving our natural environment, in particular the marine environment, and my research interests lie in this area. It was this passion and enthusiasm that led me to choose to undertake a Bachelor of Science majoring in marine biology upon graduating from school with an OP1, and that gave me the determination to achieve High Distinctions in all of my university courses and therefore graduate as the highest achieving student in my degree.

Introduction

Sound environmental management decisions can only come about through a thorough understanding of our environment, which requires good science. With this in mind, it has been my goal as an undergraduate to learn as much as I can about the environment so that in the future I will be capable of undertaking research that will hopefully influence the management of our environment, both in Queensland and throughout Australia. My high academic achievements have given me the opportunity to undertake research projects during my undergraduate degree. The first of my major research projects was completed over two summer vacation periods, where I was awarded a scholarship through the Australian Rivers Institute to research phytoplankton growth in Queensland coastal waters. As part of my Honours degree I am also currently researching the effects of climatic variability on phytoplankton growth in coastal ecosystems.

Both of these research projects focus on phytoplankton, which are unicellular floating plants that inhabit aquatic environments. Despite their small size, phytoplankton are extremely important. It is well known that sequestration of carbon dioxide by land plants is important in helping to combat global climate change. However, a largely unknown fact is that phytoplankton draw in more carbon dioxide through photosynthesis than all of the land plants on the planet put together. Therefore phytoplankton play an enormous role in regulating our global climate. Phytoplankton also form the basis of most marine food webs. Therefore, the dynamics of phytoplankton populations often affect species further up the food chain, from zooplankton to fish to whales, many of which have commercial importance for tourism and fisheries. Some phytoplankton species are also toxic and can form harmful algal blooms or 'red tides'. These harmful species can cause environmental damage, human health effects, the closure of beaches and other recreational areas, and huge economic losses from fish kills of commercially fished species. It is this importance of phytoplankton that makes it essential to have a better understanding of the factors that affect their growth and population dynamics.

Project 1: Impacts of Dust Deposition on Phytoplankton Dynamics

This project investigated the effects of dust deposition on the growth of phytoplankton, including algae and cyanobacteria, in the Great Barrier Reef and Moreton Bay. From this project I was able to find that the deposition of dust during dust storms can cause phytoplankton blooms in Queensland coastal waters. Although this phenomenon has been observed in other parts of the world, this is the first time in Australian waters that this process has been observed.

Background

There is much concern as to the effects of declining water quality in the Great Barrier Reef, which results from anthropogenic, or human induced, enrichment of nutrients[1,2]. Indeed, research has found that concentrations of nutrients have increased significantly over the past century with increases in urban, industrial and agricultural development within the Great Barrier Reef catchment areas[3]. Such increases in nutrients, known as eutrophication, promote algal blooms and blooms of other phytoplankton, such as cyanobacteria. Phytoplankton blooms can have a number of deleterious effects both to ecosystems as well as to human health, through the bioaccumulation of algal toxins in seafood and through direct recreational contact with toxic species[4,5,6].

In parts of the world, for example the North Atlantic Ocean, deposition of dust into the ocean has been shown to stimulate blooms of cyanobacteria by increasing nutrient levels, as dust contains high levels of nutrients[7,8]. However, in Australia, research on aeolian, or air transported, dust as a means of delivery of nutrients has been very limited in comparison to research on fluvial, or river-based, transport of nutrients[9]. Indeed the recent *Reef Water Quality Protection Plan* which aims to address the levels of land based pollutants, including nutrients, entering the Great Barrier Reef, only addresses fluvial sources[2]. It is important to gain a better understanding of all of the processes which lead to enhanced nutrients within coastal waters. This is particularly true given the environmental, recreational and economic value of Queensland coastal waters, where the Great Barrier Reef alone has a gross domestic product of AUS$5.9 billion *per annum*[10].

Methods

In my research project I reviewed the trends of cyanobacterial growth on the Great Barrier Reef, the factors that affect cyanobacterial growth, and the possible effects of aeolian dust deposition on cyanobacterial blooms in the Great Barrier Reef. In particular, the October 2002 dust storm, one of Australia's largest recorded dust storms, which traversed the east coast of Queensland and New South Wales, was examined to see if there were any phytoplankton responses to this dust event. This event was examined via the use of satellite data obtained from NASA.

Results and Significance

The results of this research showed that there were peaks in phytoplankton growth up to twice what would be expected at that time of year. However, during this time there was little or no rainfall, and therefore no significant fluvial supply of nutrients. This therefore supports the hypothesis that the phytoplankton blooms experienced in locations off the Queensland coast at this time were due to deposition of dust from the October 2002 dust storm. This is in agreement with studies from other parts of the world which indicate that dust events can, and have been responsible for phytoplankton blooms[7,8]. This is the first time that a phytoplankton response to dust deposition has been observed in Australian waters and is therefore a very significant and innovative finding.

Dust storms may be a particularly important source of nutrients for the ocean during the dry season because they occur at times when coastal waters would receive very little nutrients through riverine output. Also, it would be expected that the influence of aeolian dust deposition on phytoplankton growth in the Queensland coastal waters would increase in the future with global climate change. This is expected as it is predicted that drying on mainland Australia will increase, which is expected to increase the frequency of dust storms[11]. In fact, I was quite surprised when I could find no record of any prior research into the effects of dust deposition in Queensland coastal waters. The outcomes of my research, along with what has been found globally, indicate that this is an area that requires further research. I strongly believe that more research needs to be undertaken in this field, as the consequences of enhanced nutrients in the Great Barrier Reef and Moreton Bay could be severe in terms of adverse ecological, economic and human health impacts, as a result of an increase in harmful algal blooms.

Project 2: Impacts of Climatic Variability on Phytoplankton Dynamics

I am currently investigating the impacts of climatic variability on phytoplankton dynamics for my Honours dissertation. In particular I am investigating the effects of climatic variability on phytoplankton dynamics in coastal waters, where climatic variability interacts with other pressures, such as pollution and overfishing, to affect phytoplankton growth dynamics. Understanding how phytoplankton populations respond to climatic variability will be extremely important in predicting the influence of future global climate change. In particular, the response of phytoplankton to climate change may have serious implications for global fisheries, marine ecosystems and may feedback to further alter our global climate. This area of research is currently not well understood and has been identified a major focus area for the Integrated Marine Biogeochemistry and Ecosystem Research (IMBER), a decade-long international research project, that investigates how coastal environments will respond to climatic variability.

Already I have found a significant relationship between the El Niño Southern Oscillation and the presence of harmful algal blooms. I look forward to completing this project over the duration of this year and hopefully seeing more interesting results throughout the year.

Sharing of Information with the Community

From my first project I have already had a review of the potential effects of dust deposition on phytoplankton growth on the Great Barrier Reef published in the *Bulletin of the Australian Meteorological and Oceanographic Society* August 2007 edition. A paper presenting my findings entitled 'Impacts of aeolian dust deposition on phytoplankton dynamics in Queensland coastal waters' has also been published in the international journal *Marine and Freshwater Research.* I have also be presented these findings in an oral presentation at the Western Pacific Geophysics Meeting (conference web page: http://www.agu.org/meetings/wp08/), which is an international conference that was held in Cairns in July 2008.

I have also presented findings from my second project at the 13th International Conference on Harmful Algae, which was held in Hong Kong (conference web site: http://www.hab2008.hk/). Upon the completion of this project I hope to get my findings published in international scientific journals. Earlier this year, I gave a seminar on the topic of my second research project to other honours students, PhD students and staff at Griffith University. At university, I have also spoken to three groups of visiting high school students about the research projects that I have worked on, how I got to where I am now and on the options and benefits of studying science once they leave school. I hope that this has helped inspire and encourage younger students to study science when they leave school.

Finally, I believe that the research projects that I have worked on, or am currently working on, along with the research that I intend to complete in the future, will contribute to the preservation and sustainable use of our marine environment. Not only will this be of environmental importance but also of economic benefit, as our marine resources are economically so valuable. I also believe that our marine environment has intrinsic values and that it will benefit the community by preserving these environments for our future generations to enjoy, just as we have before them.

Section 3
Programs for College and University Students

Chapter 5
Understanding the Links between Mentoring and Self-Efficacy in the New Generation of Women STEM Scholars

Elizabeth Yost
University of Alabama at Birmingham, USA

Donna M. Handley
University of Alabama at Birmingham, USA

Shelia R. Cotten
University of Alabama at Birmingham, USA

Vicki Winstead
University of Alabama at Birmingham, USA

ABSTRACT

American colleges and universities are in need of innovative approaches to recruit and retain the upcoming generation of new faculty members. Specifically within the STEM (Science, Technology, Engineering, and Math) fields, there is an additional need to focus on meeting the needs of women in order to begin to address gender inequity within STEM. This chapter examines the impact of mentoring on self-efficacy for female graduate students and post doctoral fellows in STEM fields. Using data from a national study of selected U.S. academic institutions, recommendations are made in order to enhance mentoring practices that will reduce the barriers women face within STEM fields. Quality mentoring programs represent a viable way to enhance institutional change that may result in increased numbers of women in STEM fields.

DOI: 10.4018/978-1-61520-657-5.ch005

Copyright © 2010, IGI Global. Copying or distributing in print or electronic forms without written permission of IGI Global is prohibited.

INTRODUCTION

Within the last two decades, the academic science fields have progressed in positive ways that are changing the face of traditional science departments within academia. American colleges and universities have experienced considerable changes in the demographic makeup of their student population, as well as an increasing number of individuals that are obtaining graduate degrees (Zumeta & Raveling, 2002; Valian, 1999). However, these promising accomplishments have not culminated in an increase in professional scientists, and in particular women scientists, within academia. The U.S. is not alone in struggling with issues of gender inequity. Multiple studies throughout the European Union (EU) have addressed gender inequity in STEM academia; while Finland and Southern Europe are more equitable, the rest of the EU is similar to the US with less than 10% of women holding STEM academic tenure track positions (Dewandre, 2002).

In the last decade, a common trend for U.S. STEM graduates has been to pursue careers in private industry and research (Zumeta & Raveling, 2002; Stewart *et al.,* 2007). The majority of individuals who are choosing careers in academia have been and continue to be primarily male (Stewart *et al.,* 2007; Zumeta & Raveling, 2002). Thus, although a larger, more diverse group of students are enrolling in science courses at universities across the U.S., the challenge remains for institutions to continue to increase the number of graduate degrees, and the diversity of those graduating within the science fields. A critical component of this challenge involves finding effective ways of mentoring women that not only enhance their recruitment into faculty positions, but also encourage them to remain in STEM academia as a long-term career choice.

Scholars have provided information on both the successes and failures of recruiting women into STEM academic fields. From as early as their middle school years, teachers can make a difference in the career choices of girls by encouraging them to participate and excel in science classes (Dyer, 2004). Other successes are attributed to the recruiting efforts of colleges and universities (Stewart, 2007), and the general social and demographic shifts of an increasing number of women in the professional workplace (Barber, 1995). Areas for improvement, found in past empirical studies, reveal that in terms of total numbers of earned graduate degrees, women outnumber men in terms of earned master's degrees, but fall short in meeting parity in earning doctorate degrees overall (Freeman, 2004). However, the gender gap identified does have the potential to be eliminated, and has been addressed through government interventions in both the European Union and the United States.

A long-term view of enrollments within academic science would indicate that traditional recruitment and retention strategies have resulted in a positive increase in the number of women within the STEM fields. The number of women in science and engineering rose from 16% of women obtaining bachelor's degrees in 1960 to 40% in 1989 (Barber, 1995). Doctoral degrees earned by women reported for science and engineering was 6% in 1960 and 28% in 1989 (Barber, 1995).

Despite the overall increase in the number of women within academia and specifically STEM departments, data reveal that a disparity in male and female enrollment still exists within science and engineering. Males are found to outnumber females in some STEM areas by rates as high as three to one throughout the educational pipeline (Freeman, 2004). Trends in the most severe gender divides in STEM areas have not changed substantially over the past two decades. Those with the greatest gender divide in earned doctoral degrees include: computer science (17% female), engineering (16% female), and physical sciences (26% female) (Freeman, 2004).

As a result, much work remains in the effort to recruit and retain women within the academic STEM fields. This study seeks to evaluate ways in which the mentoring environment, and thus the self-efficacy of these women, can be supported within their academic environment which can improve both the institutional culture and the career success of women in STEM. Identifying models of institutional support provided to graduate students and post doctoral fellows can provide a positive impact on self-efficacy through formal and informal mentoring mechanisms. The first section of the chapter presents current literature in the field regarding the mentoring of women in STEM areas, which then leads us to discuss how existing policies are currently affecting women in STEM. Lastly, we suggest further issues that should be addressed by researchers in this area in the future.

BACKGROUND

Barriers to the Success of Women in STEM

Multiple factors affect whether individuals choose to enroll and persist in STEM fields. However, research over the past 30 years indicates specific factors that are particularly salient barriers for women in particular. One explanation for the disproportionate enrollment of men in the science fields may be attributed to the stereotypes of the fields of agriculture, engineering, business, the physical sciences, and computer science as "male" occupations; therefore women encounter few positive female role models in these areas. Instead, these occupations have negative connotations attributed to them, including an unattractive stereotype for scientists, a lack of female co-workers and mentors, and a hostile work environment (Rosser & Lane, 2002; Jacobs & Winslow, 2004; Mason & Goulden, 2004; Rosser, 2004; Winkler, 2000; Yoder, 1994; Mason & Ekman, 2007).

Other explanations that impact the recruitment of women into STEM are not specific to the academic sciences. There are institutional barriers within U.S. colleges and universities that affect women across all levels of academia. Female junior faculty note four critical barriers to their continuation in the academy: 1) family unfriendly policies (Rosser & Lane, 2002; Jacobs & Winslow, 2004; Mason & Goulden, 2004; Rosser, 2004), 2) harassment (Winkler, 2000; Rosser & Lane, 2002; Rosser, 2004), 3) stereotyping or tokenism (Rosser & Lane, 2002; Rosser, 2004; Yoder, 1994), and 4) lack of female mentors (Winkler, 2000; Rosser & Lane, 2002; Rosser, 2004). Distinct to women in STEM may be the persistent impact of these barriers throughout their educational path. The lack of family-friendly university policies may be interpreted as a message to women (as well as men) that it is not acceptable to incorporate a personal family life while focused on one's professional career. Examples of family friendly policies institutions could adopt include tenure clock stoppage policies that enable parents of either gender to request an extension of the tenure-earning period in the case of compelling personal circumstances. When these types of options are not available, the policies in place can significantly hinder young academicians as they progress in their careers. Both harassment and tokenism are most evident at the departmental or school/division level, and a lack of "open door" policies, either formal or informal (Rosser & Lane, 2002; Rosser, 2004), may also communicate to women that the organizational culture neglects the discussion of such issues. Further, women may interpret that reports of abuse are uncommon and could result in a punitive treatment for the one who reports the infraction. A lack of female mentors also has a tremendous impact on the continuation of women and their choice to continue with a career in academia (Winkler, 2000; Rosser & Lane, 2002; Rosser, 2004). This issue will be discussed in detail later in this chapter. Ultimately, these challenges may tarnish many women's professional and personal experiences early in their careers, which may impact both their choices to enter academia as a faculty member and the advice they may give to students considering an academic career path.

Evidence indicates that work must continue to assess and eliminate these barriers that still exist in order to help universities recruit and retain women, especially within STEM. The next section specifically examines difficulties for recruiting and retaining women into STEM academic career paths.

Institutional Barriers and Difficulties in Recruiting and Retaining Women within STEM Fields

Issues at each level of STEM academia should be studied to determine the full impact that gender disparities have on the choices women make to enter or not to enter STEM fields. If there are not adequate numbers of female professors or role models then there will be a trickledown effect that demonstrates an unfriendly environment to younger women within those areas. Therefore, in order to successfully recruit and retain women at each level of STEM academia, an overall view of current protocol is necessary. We note that these methods are valuable for the recruitment and retention of women across academia, although our focus is primarily on women within the STEM fields.

Changes must be made at both the university and department level in order to be successful in the recruitment and retention of both students and faculty of all genders (Trower, 2008; Assendelft, Gunther-Canada & Dolan, 2001; Gavel, 2000; Markert, 1996). As noted earlier, the traditional academic environment especially in the sciences is seen as "chilly" or "hostile", especially for women (Markert, 1996; Ferreira, 2002). An unsupportive work environment can have a disparate impact on both faculty and students, thus compromising the retention of both groups of individuals. Women also attribute a negative institutional climate to universities in which they experience "reverse discrimination" or "tokenism," because of their gender (Markert, 1996). If female students witness their female mentors experiencing these inappropriate behaviors, the result may include long-term damage to the potential career of the student who may reconsider an academic position, as well as the academic career of the faculty member. Thus, colleges and universities must implement policies and educate their faculty and administrators on providing a positive and supportive environment for all students and faculty members.

Studies also indicate a tremendous change in the characteristics and preferences of scholars that are in graduate school and who are currently planning to enter the academic job market within the coming years. They have been called "new scholars" because of the tremendous shift in preferences demonstrated when compared to traditional academic demands, especially in considering a job offer (Gavel, 2000). New scholars seek a balance between their work and personal lives, in order to enhance their quality of life while building their careers (Trower, 2003; Trower, 2007). Thus, contemporary strategies to recruit and retain the most collegial and productive men and women in academia are being revised in order to reflect the changing priorities of faculty. As a result, the institutional climate has had a significant effect not only on the new generation of academics entering the job market, but also on the role of women, especially in STEM fields.

The inability to balance career and personal life is a third challenge that may cause women to opt out of a career in academia. Academic women often exit their career path during childbearing years, signaling a "brain drain" in academia (Hewlett & Luce, 2005). The brain drain often leads to a disadvantage in their careers when women then choose to re-enter the workforce. This opting out and in is often referred to as an off-ramp and on-ramp system, which places women at a distinct disadvantage over their male counterparts.

While overcoming institutional barriers are often the focus of many initiatives, instructional supports are vital to the success of women in STEM fields (Trower, 2007). Institutional supports include; mentoring programs, clear and fair tenure policies, workload equity, stop-clock tenure policies, and adequate resources (Trower, 2007). As early female STEM faculty enter academia, the institutions must begin to address the inequality in these areas so that the work environment is more female friendly and thus encourages women to maintain their status in the pipeline. Further attention to mentoring as a basis of institutional support will be discussed later in this chapter.

As this section has illustrated, the challenges for colleges and universities in retaining women include decreasing institutional barriers, a focus on the availability of family friendly policies, and enhancing supportive work environments. Next, we turn to the importance of mentoring in these processes.

Role of Mentoring in STEM

Mentoring has been shown to be monumentally important in the formative years of training for any prospective profession (Fox & Fonseca, 2006). In graduate education, mentoring has been shown to be a vital component to the educational process and a key variable in predicting a student's overall experience (Rose, 2005; Phillips & Pugh, 2000; Roberts & Sprague, 1995). Most studies that involve mentoring in STEM focus on the impact of mentoring at each individual level of education, not mentoring throughout the educational pipeline and the impact continued mentoring has on women in STEM areas.

Self-efficacy is the belief an individual has "in their capability to produce desired effects by their own actions" (Bandura, 1994; pp. 2). Mentoring has been shown to increase self-efficacy in specific tasks. Pajares (1996; pp. 547) states that judgments about self-efficacy are "task and domain specific."He further asserts that " self-efficacy beliefs should be assessed at the optimal level of specificity that corresponds to the task being assessed and the domain of functioning being analyzed. " Therefore it is important that measures of self-efficacy are specific to a certain area such as belief that one can accomplish writing and defending a successful thesis, skills which are modeled or developed by effective mentoring.

Self-efficacy is a learned skill (Synder & Lopez, 2007) typically through mimicry or imitation. Having female mentors for other females in STEM provides a model of success, and thus students can imagine themselves managing a professional career that is often "similar" to that of their mentor. In contrast, if there are insufficient numbers of women in the field available to mentor young women, the environment can appear hostile to women and they may interpret the lack of female presence as an isolating factor which impacts their career plans. Because of a lack of prior studies, this study will look at the impact of mentoring on self-efficacy in a sample of female STEM graduate students and postdoctoral scholars. Our approach examines the importance of mentoring at each step in the later portions of the academic pipeline in order to assess the relationship between institutional support and self-efficacy. By using this approach to examine female STEM scholars, we hope to add to the studies that examine one aspect as to why there is such gender disparity in STEM academia.

Prior research has attempted to identify the specific types of mentoring that are best for graduate students, but have failed do to so because an ideal mentor is dependent on the individuals in the mentoring relationship (Rose, 2005). Because there is no mentor that works best for every student and every situation, directly studying mentoring can often show little by way of significant findings and direct results.

Although each mentoring relationship is unique, research shows that there are several considerations that enhance **successful mentoring**. Phillips-Jones (2003) notes that mentors and mentees should utilize specific core skills to enhance mentoring relationships. These include: actively listening, building trust, giving encouragement, and identifying goals and current reality of the situation. She also notes the importance of specific critical skills for mentors, such as instructing and developing capabilities (e.g., modeling effective behaviors, teaching new skills, assisting in finding resources, and monitoring behaviors), inspiring mentees to greatness, providing useful and constructive feedback, helping mentees manage risks, and enhancing visibility for mentees (Phillips-Jones, 2003). In order for mentoring to be successful, there must also be clear communication on the part of the mentor and the protégé as to what is needed. In identifying what works for both parties, specific individual factors must be taken into consideration. First, the age of the protégé is vital because each step of the academic ladder is substantially different. Mentoring at earlier ages has been shown to be a predictor for the future participation of women in some form of STEM discipline (Dyer, 2004). Mentoring at the junior faculty level has been shown to be of great importance for women in obtaining tenure (Stewart, 2007; Nielson *et al.,* 2001). A more in-depth look at the effects and outcomes of mentoring at each stage is needed before reaching an overall conclusion as to the effect.

While college is not the first point of mentoring in an educational pipeline, it is often the first point one receives a formal mentor. (A complete discussion of different types of mentoring follows this section.) The importance of the college mentor can be seen as the first official mentor who "knows the ropes" of academic STEM. Unlike primary education, mentoring becomes the strongest predictor of women entering the STEM academic path in college (Hassoun & Bana, 2001). Because women have often not been as active in science in earlier education, they are less likely to major in a STEM area in college (Smyth & McArdle, 2004). The differences in STEM undergraduate course enrollments are startling considering that women outnumber men in college (Wirt, 2004) and in most STEM areas with rates as high as 5 to 1 (except Life Sciences, i.e. Biology) (Dyer, 2004).

For STEM academic scientists or women in STEM areas, the lack of female mentors can be seen as signaling that women are not welcome or that women in these fields cannot achieve a family and career simultaneously or perhaps that women in general will not succeed in the field (Valian, 1999). Early educational experiences and mentors impact decisions later in the educational pipeline. A study of 1504 children in the U.S. supported the generalization that girls are taught to see STEM areas as masculine (Jones *et al.,* 2000). When young elementary children were asked to draw a scientist, less than 25% of respondents drew a woman (Jones *et al*., 2000). While women in college probably do not see STEM academic positions and fields as strictly male, as younger children do, the gender of a STEM academician is still most decidedly male (Jones, *et. al.*, 2000). These early childhood impressions of gender dominance in a field can often be subconsciously carried into career decisions.

Regardless of exposure to the STEM areas from a young age, mentoring in the graduate and postdoctoral years is of greater importance because it has been shown to be a direct indicator of degree completion (Rose, 2005; Tinto, 1993). Furthermore, it is one of the first times multiple mentors and types of mentoring are utilized. Mentoring is a key factor for keeping women in the academic pipeline and securing a position after the postdoctoral years. Having a supportive, well-connected mentor means more than most other aspects of being a postdoctoral scholar. Mentoring is an important factor in maintaining women in the pipeline because of ability of the mentor to help facilitate or encourage work-family balance. Most of the faculty who supervise postdoctoral scholars are senior STEM faculty, who are mostly male, reflecting the gender imbalance in higher positions in academia (Fox &Fonseca, 2006). Having an advisor/mentor of the same gender is optimal to many women because they are thought to be better able to understand the demands the job will have on their family life (Kulis, Sicotte,

& Collins, 2002). As a result, women are able to identify with women who as mentors are successful in their professional careers, as well as demonstrating that balancing a professional and personal life can be an option. Most research on the effects of mentoring and its impact on women has been assessed at the junior faculty level. These studies conclude that without proper mentoring and support the women are not integrated into the faculty (Kulis, Sicotte, & Collins, 2002) which could be seen as a factor for women leaving academia or not obtaining tenure and promotion. The following section will discuss the various types of mentoring and the impact they have on recruitment and retention of women in STEM as well as the effect mentoring has on self-efficacy.

Types of Mentoring

Mentoring comes in many forms and can have a tremendous impact upon the professional and personal careers of young professionals. Research indicates that mentoring relationships not only increase professional socialization, productivity, and time management skills, but they also enhance the overall goal orientation and career direction of junior colleagues (Poel, Arroyos-Jurado, & Coppola, 2006; Sands, Parson, & Duane, 1991). As a result, ensuring that our students, both male and female, are involved in productive mentoring relationships in order to maintain the quality of our students and future faculty members in the academy should be a goal of all colleges and universities.

While a clear relationship between mentoring and continued success in STEM areas has been shown, little work has been completed on types of mentoring and the specific benefits that each type. Is there a best type of mentoring? Yost (2008) studied female postdoctoral scholars and the types of mentoring they received. Through a qualitative study of the scholars, two types of mentoring emerged as important: formal and informal mentoring. Further compounding the effect of mentoring, Yost (2008) found that the gender of the mentor and protégé played a substantial role in the overall benefit the mentoring offered. The remainder of this section examines different types of mentoring relationships and the effect each type has on the mentees.

Formal Mentoring

Although individuals can benefit from a variety of types of mentoring, formal mentoring is more structured and involved than informal mentoring. Phillips-Jones (2003) notes that formal mentoring involves: specifically arranged partnerships which are time specific generally, the mentee receives most of the assistance, mentor partnerships are monitored and supported, and mentors and mentees may engage in specific activities for mentoring partners. Formal mentoring is also used to describe mentoring for work related skills. Formal mentoring occurs in an organized setting under specific predetermined situations; it involves specific directions and suggestions in relation to aspects of work such as, but not limited to, grant writing, laboratory skills attainment, and other work related issues (Yost, 2008). Formal mentors are typically those assigned by the institution or those for whom the scholar is assigned. These types of mentoring relationships are vital for basic skill set knowledge, but lack the mentoring needed for women to be able to successfully navigate the gendered terrain of STEM academia. Formal mentoring relationships are not socializing agents into the academic network.

Informal Mentoring

Informal mentoring occurs sporadically, naturally, and without ceremony. Informal mentoring deals more with how to balance work and family, life skills necessary to enter academia, and can include personal anecdotes to help protégés understand the finer points that ultimately aid in their success (Yost, 2008; Phillips-Jones, 2003; Inzer & Crawford, 2005). Informal mentoring is seen as more advantageous in terms of networking and necessary professional socialization aspects (Inzer & Crawford, 2005; Raggins, *et al.*, 1999). However, due to its amorphous nature, informal mentoring must be sought out by the protégé and often is not clearly defined which inhibits complex studies of these types of relationships. Informal mentoring in STEM education is socialization into the STEM community, which could be more advantageous to women who are already disadvantaged from the male-centric academic environment. However, women can be doubly disadvantaged if there are few or no female mentors to facilitate the socialization. Ultimately, this type of mentoring is the most beneficial to women in STEM academia but is difficult to obtain.

Regardless of the type of mentoring or the mentor, studies show that the relationship is vital to the continued success in a position (Yost, 2008; Phillips-Jones, 2003; Inzer & Crawford, 2005). Understanding the importance of mentoring and prior research, the aim of this paper is to understand how mentoring helps female graduate students and postdoctoral scholars continue in their educational path. What effect does this mentoring have on the continued success of these women? Enhancing self-efficacy can be a beneficial outcome of mentoring.

Mentoring and Self-Efficacy

As noted above, the impacts of mentoring can be many. One particular way that mentoring may have an impact on women in STEM is by enhancing self-efficacy (Freehill *et al.*, 2007). Bandura (1995: pp. 2) defined the term self-efficacy as "people's beliefs about their capabilities to produce designated levels of performance that exercise influence over events that affect their lives." Simply put, self-efficacy is a measure of whether or not a person has confidence in their ability to achieve what they propose they will achieve. These efficacy beliefs of individuals are strong predictors of diligence for specific tasks, and strength of perseverance in these tasks even in the face of strong adversity (Pajares, 1996). People with low self-efficacy may perceive themselves as unable to accomplish certain tasks, and may have an exaggerated perception of how difficult the tasks are, which creates avoidance of these tasks. High self-efficacy affects the types of goals an individual sets and the effort expended in reaching those goals (Maddux, 2002).

These are crucial criteria for success in an academic setting especially for women who face the dual challenge of being a female in male dominated fields, as well as the academic rigor of STEM disciplines. Increases in self-efficacy could be seen as a result of having a mentor who encourages and supports a graduate student's success or by an advisor who produces the opportunities for the graduate student. Having a mentor who has successively navigated through the academic process could serve to raise efficacy beliefs for the mentee. High self-efficacy beliefs create a resiliency to barriers and enhance problem solving, necessary attributes in academia, so that goals become achievable (Maddux, 2002). It can therefore be hypothesized that women who have high self-efficacy would be more effective in moving past institutional barriers often encountered in academia.

While a direct link between mentoring and self-efficacy in STEM female graduate students and postdoctoral fellows has not been studied specifically in this subgroup of academicians, research has been completed on academic self-efficacy at younger ages with less specificity towards academic subject. Much research focuses on the impact of self-efficacy at a young age on success later in life (Zeldin & Pajares, 2000; Bandura, 1986), but correlating self-efficacy with mentoring can provide an interesting outlook and perhaps begin to address another set of problems with drawing women into STEM academe. In STEM disciplines specifically, women are already disadvantaged due to low numbers of female role models, implying that women do not belong in their specific STEM area.

While mentoring might not be the only solution to resolve the leaky pipeline for STEM women, by increasing self-efficacy, effective mentoring does encourage women to continue in their academic pursuits (Trower, 2002). With higher self-efficacy, the women are more likely to believe it is possible for them to be successful in their academic career pursuits. Since self-efficacy is learned in part through mimicry (Bandura, 1995) having younger female mentors is thought to be the best possible mentor. The gender of the mentor has been shown to greatly affect the mentoring relationship (Paglis, Green & Bauer, 2006; Yost, 2008). Mentoring relationships have been shown to increase self-efficacy simply because of the presence of a role model (Freehill *et al.,* 2007). Therefore, having female role models could be a retention factor for women in STEM post baccalaureate programs.

In this study, we examine the role of various aspects of mentoring, including institutional support, on self-efficacy. Based on prior literature, we hypothesize that women who have been mentored during their graduate and undergraduate STEM careers will have higher self-efficacy than those who have not been mentored. It is also hypothesized that women who have more mentors, either formal or informal will have higher levels of self-efficacy. Furthermore, women who are mentored by women will have higher levels of self-efficacy. We also hypothesize that women who report higher levels of institutional support, as a proxy type of mentoring, will have higher levels of self-efficacy.

METHODOLOGY

Research Design and Samples

With funding from the UAB ADVANCE program, we conducted a study examining factors affecting recruitment and retention among female graduate students, postdoctoral fellows, and junior faculty in small STEM departments at ADVANCE funded institutions. Following Dillman's Tailored Design (Dillman, 2007), we developed two cross-sectional web-based surveys. One survey was tailored for graduate students and postdoctoral fellows and the second one was for junior faculty. Survey respondents could enter their name and email address at the completion of the survey for a chance to win one of six $25 debit cards.

Samples

We utilized a multi-stage sampling design to select samples of female graduate students, postdoctoral fellows, and junior faculty from universities that had received a National Science Foundation (NSF) ADVANCE Institutional Transformation Award as of 2006 (n=28). The first stage of the sampling design involved identifying all programs and departments that were included as Science, Technology, Engineering, and Mathematics (STEM) departments and programs according to NSF criteria. From the 28 universities, we identified all small[1] STEM departments and programs with PhD programs (n=73 departments and programs at 16 universities). Table 1 lists the universities in our sample. The second stage of the sampling design involved stratifying the sample by department/program type. Then, we randomly selected 50% of each of the departments/programs for inclusion in the sample; however, we included all the departments/programs in geology and geography in our sample as there were only a few from which to sample.

After selecting small STEM departments/programs, we gathered name and email information for all (1) graduate students, (2) postdoctoral fellows, and (3) junior faculty who were female from associated web pages and via contacting departments. Surveys were then distributed to each of the two groups: (1) graduate students and postdoctoral fellows and (2) junior faculty

Table 1. Schools included in sample

Universities Represented
Brown University
Case Western Reserve University
Columbia University
Georgia Institute of Technology
Iowa State
Kansas State
New Mexico State
Rensselaer University
Rice University
University of Alabama at Birmingham
University of Arizona
University of California Irvine
University of Colorado at Boulder
University of Illinois at Chicago
University of Montana
University of Wisconsin at Madison

Table 2. Sample demographics

Characteristics		**Grad. Student/ Post. Doc (GSPD)**
Race	White	76.4%
	African-American	5.0%
	Asian	8.2%
	Hispanic	3.2%
	Other	7.2%
Mean Age		29 years
US Citizen		81.9%
Currently Married/Partnered		43.0%
Total Responses (Response Rate)		368 (26.0%)

members. Response rates varied between the two groups: 60% of junior faculty responded to the survey while less than 30% of the graduate students and postdoctoral fellows responded (see Table 2).[2] We utilized the graduate student and postdoctoral fellow data for the results presented below.

Measures

The primary outcome examined is self-efficacy measured by the Douvan and Walker scale (1956). The Douvan and Walker scale is comprised of three questions assessing how often respondents report (1) feeling pretty sure their life will work out the way they want it to; (2) when they make plans ahead they usually get to carry out things the way they expect to; and (3) would they say they always finish things once they start them. Responses ranged from (1) Never to (5) All the time, with higher scores indicating higher self-efficacy (alpha=.638).

The main predictor variable is mentoring. We assess mentoring via several measures. We ask respondents to report whether they were formally mentored (e.g., an official assignment by their department or university) as an: (1) undergraduate; (2) graduate; and (3) postdoctoral fellow. Responses were (1) Yes and (0) No. Prior literature also suggests that the gender of the mentor is important. For each of these, we also ask the gender of the main mentor (1=Female). We also asked the number of (1) formal and (2) informal mentors that respondents had as (a) graduate students and (b) postdoctoral fellows.

We also utilized an institutional support scale, using Likert ratings, assessed through the following questions: please rank how supportive the following are in relation to your academic career: colleagues within your department/program, colleagues outside your department/program, professional academic association, supervisor, and mentor. This scale is part of a larger subscale of support. The measure looks only at institutional support an individual has in relation to their academic programs. Higher scores indicate greater levels of support (alpha =.646).

We include the following as control variables in our multivariate models: marital status (1=married); age (in years); race (1=white); and citizenship (1=U.S. citizen).

Table 3. Bivariate differences in self-efficacy by mentor status

		Mean of Yes (SD)	Mean of No (SD)	Significance
Self-Efficacy	Undergraduate Mentor	11.37 (1.43)	11.25 (1.79)	.590
	Graduate Mentor	11.29 (1.64)	11.36 (1.57)	.779
	Informal Mentor	11.35 (1.56)	11.13 (1.67)	.472
	Formal Mentor	11.31 (1.62)	11.38 (1.71)	.790

Analytical Design

We examine univariate characteristics to describe our graduate student and postdoctoral sample and utilize bivariate analysis to determine whether differences exist in self-efficacy according to the varying mentorship categories. Ordinary Least Squares Regression analyses are used to examine the relationship between self-efficacy and mentoring, controlling for other sociodemographic variables. Model one examines the effect of having an undergraduate mentor on self-efficacy. Model two focuses on the importance of having a graduate school mentor. Models three and four examine whether having formal versus informal mentors makes a difference in self-efficacy. Models five and six incorporate gender into these relationships to determine whether having a female mentor is related to higher self-efficacy. Finally, model 7 examines the impact of levels of institutional support on self-efficacy.

RESULTS

As shown in Table 2, the majority of the respondents were white (76%); approximately 8% of the respondents were Asian. The mean age was 29 years, 43% were married or partnered. Over 80% of respondents were U.S. citizens.

At the bivariate level, no significant differences existed in either mastery or self-efficacy by mentoring status (see Table 3). All groups had relatively high levels of self-efficacy.

Although we found no significant differences in self-efficacy at the bivariate level, we next regressed self-efficacy on mentoring while also controlling for sociodemographics (see Table 4). The presence of mentors is not related to self-efficacy. Institutional support is the only variable that shows any significant relation to self-efficacy. Females who perceive higher levels of institutional support are more likely to have higher levels of self-efficacy than those who perceive lower levels of institutional support. In this sample of respondents from ADVANCE institutions, presence or absence of mentors does not appear to be associated with self-efficacy, at least as we measured mentoring and self-efficacy in the current study.

DISCUSSION AND FUTURE RESEARCH DIRECTIONS

The results indicate both positive news and areas of improvement concerning the impact of mentoring on the careers of women graduate students and post doctoral fellows in STEM fields in the U.S. The women surveyed within the study do not appear to currently be negatively impacted as a result of the presence or absence of a mentor during their college and graduate school careers. We did not find statistically significant differences in the respondents' age, race, or marital status when assessing their self-efficacy levels and mentoring experiences. This finding, however, is good news for women in STEM fields, because the lower numbers of women currently serving as faculty members may not be as critical of a factor in

Table 4. Regression with self-efficacy

	Self Efficacy						
	Model 1 Beta (S.E.)	Model 2 Beta (S.E.)	Model 3 Beta (S.E.)	Model 4 Beta (S.E.)	Model 5 Beta (S.E.)	Model 6 Beta (S.E.)	Model 7 Beta (S.E)
Sociodemographics							
Age	.042 (.022)	.031 (.022)	.026 (.023)	.018 (.022)	-.074 (.035)	-.037 (.029)	.016 (.022)
Race	.027 (.307)	.031 (.307)	.019 (.314)	.004 (.302)	.062 (.398)	.025 (.353)	.003 (.303)
Marital Status	.101 (.247)	.012 (.247)	.010 (.252)	.043 (.242)	.063 (.315)	-.026 (.288)	.024 (.245)
Citizenship Status	.038 (.346)	.042 (.345)	.035 (.359)	.075 (.346)	.079 (.483)	.017 (.384)	.029 (.343)
Mentoring and Institutional Support Status							
Undergrad. Mentor	.051 (.229)						
Grad. Mentor		.018 (.267)					
Formal Mentor			-.017 (.299)				
Informal Mentor				.077 (.317)			
Undergrad Female Mentor					-.073 (.295)		
Graduate Female Mentor						-.052 (.282)	
Institutional Support							.193** (.023)
Adjusted R^2	-.019	-.021	-.023	-.011	-.037	-.027	.017
F Values	.235	.145	.097	.575	.279	.165	1.722

** p<.01

determining female students' success or careers in academia. The lack of significance in female compared to male mentors also suggests that the men within STEM careers are cognizant of all students they are interacting with and thus women are not at a disparate disadvantage. However, it is important to note that all respondents came from ADVANCE Institutional Transformation Award schools. This selected group of institutions and respondents has received National Science Foundation funding to aid in the empowerment of women in the STEM fields through the implementation of institutional policy change.

Beginning in 2001, the National Science Foundation (NSF) began to address issues of gender inequity in U.S. STEM programs. As of 2009, the NSF has funneled over 130 million dollars into programs aimed at transforming institutions and increasing the participation of women in the STEM workforce (NSF, 2009). Currently, thirty institutions have been awarded transformational awards. These awards are given to institutions to encourage comprehensive, system-wide changes that help women excel. The sample for this study was of women in small STEM departments at ADVANCE institutions that had received transformational awards. These institutions are already focused on helping women succeed and fostering a gender equitable environment. Perhaps these environments have already begun to foster change needed in helping mentor women, thus yielding higher self-efficacy among the respondents. More than 75% of respondents stated that they were formally mentored as a graduate student and almost half of all respondents stated they were mentored as an undergraduate student. Perhaps the measures utilized were not sufficient for monitoring the effects and quality of overall mentoring.

As part of the institutional transformation, ADVANCE funded schools have started many mentoring programs throughout all levels of the university (NSF, 2009). Women and administrators at these institutions are keenly aware of the positive relationship between mentoring and having greater numbers of women complete degrees and advance on to higher levels of education. Thus, it is perhaps not surprising that our measure of institutional support was the only factor related to self-efficacy for our sample of female graduate students and postdoctoral fellows. Multiple studies show that mentoring has a positive effect on self-efficacy (Freehill *et al.*, 2007; Paglis, Green & Bauer, 2006). Due to these effects, mentoring programs at ADVANCE transformational award institutions are seen as vital to the complete "transformation" of the learning environment in the effort to better gender parity. It will be important for future researchers to examine these processes at institutions that have received ADVANCE awards and those that have not. We suspect dramatic differences would be present depending upon ADVANCE status.

A second explanation as to why the outcomes of this analysis did not mirror those of previous studies also relates to sampling issues. The female STEM graduate students were advanced graduate students, defined as being in the end stages of their graduate careers. At this point in their career, the women are very close to finishing graduate degrees; ones who perceived they could not achieve this most likely have already leaked out of the pipeline. It may be a selection effect in that the respondents to our study have survived and perhaps thrived to this point in their training. This trend of high self-efficacy at later stages could be due to the proverbial light at the end of the tunnel. That is, individuals at this stage have almost completed their degree and they feel as if they will simply because they have persevered up to that point.

Women in this study also came from small U.S. STEM departments, defined as having twenty or fewer faculty present. Small departments may have fewer graduate students and would seem to typify better mentoring strategies because of the smaller student to faculty ratio. Students in small departments may have more chances to meet with professors and benefit from close relationships with their mentors and other colleagues in the department. Unfortunately the measures used to assess mentoring in the current study only assessed whether a mentor existed or not, not the quality of the relationship between mentor and mentee.

Another methodological question for the analysis comes from the self-efficacy measure itself. Bandura (1994) and Pajares (2002) state that specific measures of efficacy often work better than generalized measures of self-efficacy. The scale used in this analysis was that of general self-efficacy. Because self-efficacy is typically studied in children at younger ages, an age and subject specific scale was not located. Future research is needed that develops STEM specific self-efficacy scales. Furthermore, looking at the longitudinal effects of mentoring on self-efficacy of this specific demographic of women would be useful. While it is known that mentoring affects self-efficacy (Freehill *et al.*, 2007) and specifically self-efficacy in girls in science (Zeldin & Pajares, 2000), a longitudinal analysis of girls from young ages throughout their educational path would be useful, albeit time consuming and arduous. A better understanding of the interrelationships between mentoring, institutional support, and self-efficacy could be better determined with longitudinal data.

CONCLUSION

Perhaps the most poignant message of this study is that the individuals surveyed showed positive self-efficacy and high levels of mentorship. While most of our mentoring measures were not significantly associated with mentoring, the state of the institutions as ADVANCE institutions points to a possible transformation at these institutions that would lead to fewer significant findings. We are reminded of data that indicates the presence of a mentor does impact a student's productivity, time management skills, and networking within the academic profession (Poel, Arroyos-Jurado, & Coppola, 2006; Sands, Parson, & Duane, 1991). Thus, the survey limitations present indicate a need for further research in order to determine the exact relationship between quantity and quality of mentorship and effect on self-efficacy at both ADVANCE and non-ADVANCE institutions. It is likely that patterns would vary across the type of institution.

A longitudinal study that follows women from their graduate student days to their early faculty days could further explore the successes and failures of women, thus allowing for a more conclusive look at the extent to which mentoring impacts their careers. The findings of these same women several years into their first academic job may provide further indications of the importance of mentoring. A longitudinal study would allow for a clearer conclusion that these women are possibly not yet experiencing the benefits of a mentoring relationship, such as co-authoring on academic research, networking within the field at professional conferences, and advising in regard to professional and personal life balance issues. A closer look at other factors that impact the students' self-efficacy could be beneficial in determining the exact role a mentor plays (if any) in these relationships.

The need to diversify university departments within the STEM fields is strongly evident as historical trends indicate decreasing numbers of women choosing STEM academic careers. Mentoring is a vital component of the success of students entering academia. Further research is needed in two main areas. First, future studies on the impact of mentoring both on men and women at ADVANCE institutions that received transformational awards is needed to determine the extent to which the transformation affects both men and women. Secondly, focusing on women in both small and large STEM departments at ADVANCE institutions to see if department size is a mitigating factor in the effect of mentoring on self-efficacy would add to literature on effects of mentoring.

While mentoring continues to be an important factor in graduate education, perhaps not in terms of self-efficacy for this specific sample, but overall, greater importance and training should be supported in an effort to continue to strive towards gender parity in STEM areas. Providing positive role models as mentors and continuing to support women in STEM will only continue to address gender equity.

ACKNOWLEDGMENT

This project was funded by a grant to Drs. Cotten and Handley from the UAB ADVANCE Office for the Advancement of Women in Science and Engineering, sponsored by the National Science Foundation.

REFERENCES

Assendelft, L. V., Gunther-Canada, W., Dolan, J., Palmer, B., & Swers, M. (2003). Political Science in a Different Voice: Women Faculty Perspectives on the Status of Women in Political Science Departments in the South. *PS: Political Science and Politics*, *36*(2), 311–315.

Bandura, A. (1986). Social Foundations Of Thought and Action: A Social Cognitive Theory. Englewood Cliffs, NJ: Prentice- Hall.

Bandura, A. (1994). Self Efficacy. In V.S. Ramachaudran (Ed.), Encyclopedia of Human Behavior, (Vol. 4, pp.71-81). New York: Academic Press.

Bandura, A. (1995). Self-Efficacy in Changing Societies. New York: Cambridge University Press.

Barber, L. A. (1995). U.S. Women in Science and Engineering, 1960-1990: Progress Toward Equity? *The Journal of Higher Education*, *66*(21), 213–234. doi:10.2307/2943912

Berg, H. M., & Ferber, M. A. (1983). Men and Women Graduate Students: Who Succeeds and Why? *The Journal of Higher Education*, *54*(6), 629–648. doi:10.2307/1981934

Dewandre, N. (2002). Women in Science: European Strategies for Promoting Women in Science. *Science*, *295*(5553), 278–279. doi:10.1126/science.1063487

Dillman, D. (2007). Mail and Internet Surveys, The Tailored Design Method. Hoboken, NJ: John Wiley & Sons.

Douvan, E., & Walker, A. M. (1956). The sense of effectiveness in public affairs. Washington, DC: American Psychological Association, Psychological Monographs.

Dyer, S. K. (2004). Under the Microscope: A Decade of Gender Equity Projects in the Sciences. Washington, DC: American Association of University Women Educational Foundation.

Ferber, M., & Teiman, M. (1980). Are Women at a Disadvantage in Publishing Journal Articles? *Eastern Economic Journal*, *9*(1), 189–194.

Ferreira, M. (2002). The research lab: A chilly place for graduate women. *Journal of Women and Minorities in Science and Engineering*, *8*(85), 85–98.

Fox, M. F., & Fonseca, C. (2006). Gender and mentoring of faculty in science and engineering: individual and organizational factors. *International Journal of Learning and Change*, *1*(4), 460–483. doi:10.1504/IJLC.2006.013912

Freeman, C. E. (2004). Trends in Educational Equity of Girls and Women. Washington, DC: US Government Printing Office.

Frehill, L. M., DiFabio, N., Layne, P., Turner-Johnson, A., & Hood, S. (2007). Women in Engineering. *Society of Women Engineers*, *53*(3), 1–20.

Gavel, D. (2000, April 14). Study Reveals New Priorities for New Generation of Faculty. *Harvard Gazette*. Retrieved from http://www.news.harvard.edu/gazette/.

Hassoun, S., & Bana, S. (2001). *Practices for Recruiting and Retaining Graduate Women Students in Computer Science and Engineering*. Paper presented at the meeting of the International Conference on Microelectronic Systems Education, Las Vegas, NV.

Hewlett, S. A., Luce, C. B., & Servon, L. J. (2008, June). Stopping the exodus of women in science. *Harvard Business Review*. Retrieved from http://hbr.harvardbusiness.org

Hussar, W., & Bailey, T. (2008). The name assigned to the document by the author. This field may also contain sub-titles, series names, and report numbers.Projections of Education Statistics to 2017.Washington, DC: Publisher name and contact information, as provided by the publisher; updated only if notified by the publisher.National Center for Education Statistics.

Inzer, L. D., & Crawford, C. B. (2005). A Review of Formal and Informal Mentoring: Processes, Problems, and Design. *Association of Leadership Educators*, *4*(1), 31–50.

Jacobs, J. A., & Winslow, S. E. (2004). Overworked Faculty: Job Stresses and Family Demands. *The Annals of the American Academy of Political and Social Science*, *596*(1), 104–129. doi:10.1177/0002716204268185

Jones, M. G., Howe, A., & Rua, M. (2000). Gender Differences in Students' Experiences, Interests, and Attitudes toward Science and Scientists. *Science Education*, *84*(2), 180–192. doi:10.1002/(SICI)1098-237X(200003)84:2<180::AID-SCE3>3.0.CO;2-X

Kulis, S., Sicotte, D., & Collins, S. (2002). More Than a Pipeline Problem: Labor Supply Constraints and Gender Stratification Across Academic Science Disciplines. *Research in Higher Education*, *43*(6), 657–691. doi:10.1023/A:1020988531713

Maddux, J. (2002). Self-Efficacy: The Power of Believing You Can. In C. Snyder & S. Lopez (Ed.), Handbook of Positive Psychology (pp. 277-287). New York: Oxford University Press.

Markert, L. R. (1996). Gender Related to Success in Science and Technology. *Journal of Technology Studies*, *22*(9), 21–29.

Mason, M. A., & Ekman, E. M. (2007). Mothers on the Fast Track: How a New Generation can Balance Family and Careers. New York: Oxford University Press.

Mason, M. A., & Goulden, M. (2004). Marriage and Baby Blues: Redefining Gender Equity in the Academy. *The Annals of the American Academy of Political and Social Science*, *596*(1), 86–103. doi:10.1177/0002716204268744

McDowell, J. M., & Smith, J. K. (1992). The Effect of Gender Sorting on Propensity to Coauthor: Implications for Academic Promotion. *Economic Inquiry*, *30*(1), 68–82. doi:10.1111/j.1465-7295.1992.tb01536.x

National Science Foundation (NSF). (2009). *ADVANCE: Increasing the Participation and Advancement of Women in Academic Science and Engineering Careers (ADVANCE)*. Retrieved March 30, 2009 from http://www.nsf.gov/funding/pgm_summ.jsp?pims_id=5383

Nielson, T. R., Carlson, D. S., & Lankau, M. J. (2001). The Supportive Mentor as a Means of Reducing Work–Family Conflict. *Journal of Vocational Behavior*, *59*(3), 364–381. doi:10.1006/jvbe.2001.1806

Paglis, L., Green, S., & Bauer, T. (2006). Does Advisor Mentoring Add Value? A Longitudinal Study of Mentoring and Doctoral Student Outcomes. *Research in Higher Education*, *47*(4), 451–456. doi:10.1007/s11162-005-9003-2

Pajares, F. (1996). Self-Efficacy Beliefs in Academic Settings. *Review of Educational Research*, *66*(4), 543–578.

Pajares, F. (2002). *Overview of Social Cognitive Theory and Self-Efficacy*. Retrieved February 8, 2008 from http://www.emory.edu/EDUCATION/efp/eff.html

Pearlin, L., Lieberman, M., Menaghan, E., & Mullan, J. (1981). Mastery scale. Measures of personality and social psychological attitudes. San Diego, CA: Academic Press.

Phillips-Jones, L. (2003). The Mentor's Guide. Grass Valley, CA: Coalition of Counseling Centers (CCC)/The Mentoring Group.

Phillis, E., & Pugh, D. (2000). How to get a PhD: A Handbook for Students and their Supervisors (3rd ed.) Buckingham, UK: Open University Press.

Poel, E., Wolfe, E., Arroyos-Jurado, E., & Coppola, B. (2006). Women, Mentoring, and a Border University. *Journal of Public Affairs Education*, *12*(4), 501–513.

Ragins, B. R., Cotton, J. L., & Miller, J. S. (2000). Marginal mentoring: The effects of type of mentor, quality of relationship and program design on work and career attitudes. *Academy of Management Journal*, *43*(6), 1177–1194. doi:10.2307/1556344

Roberts, C. G., & Sprague, R. (1995). To Compete or Educate: Mentoring and the Research Climate. *Professional Ethics (Gainesville, Fla.)*, *8*(4), 1.

Rose, G. (2005). Group Differences in Graduate Students' Concepts of the Ideal Mentor. *Research in Higher Education*, *46*(1), 53–80. doi:10.1007/s11162-004-6289-4

Rosser, S. (2004). Using POWRE to ADVANCE: Institutional Barriers Identified by Women Scientists and Engineers. *NWSA Journal*, *16*(1), 50–78. doi:10.2979/NWS.2004.16.1.50

Rosser, S., & Lane, E. O. (2002). Key Barriers for Academic Institutions Seeking To Retain Female Scientists and Engineers: Family-Unfriendly Policies, Low Numbers, Stereotypes, and Harassment. *Journal of Women and Minorities in Science and Engineering*, *8*(2), 161–189.

Sands, R., Parson, L., & Duane, J. (1991). Faculty Mentoring Faculty in a Public University. *The Journal of Higher Education*, *62*(2), 174–193. doi:10.2307/1982144

Smyth, F. L., & McArdle, J. J. (2004). Ethnic and Gender Differences in Science Graduation at Selective Colleges with Implications for Admission Policy and College Choice. *Research in Higher Education*, *45*(4), 353–381. doi:10.1023/B:RIHE.0000027391.05986.79

Snyder, C., & Lopez, S. (2007). Positive Psychology: The Scientific and Practical Explorations of Human Strengths. Thousand Oaks, CA: Sage Publications.

Stewart, A., Malley, J., & Lavaque-Manty, D. (2007). Faculty Recruitment: mobilizing Science and Engineering Faculty. Ann Arbor, MI: University of Michigan Press.

Tierney, J. (2008, July 15). A New Frontier for Title IX Science. *NY Times*. Retrieved from http://www.nytimes.com

Tinto, V. (1993). Leaving College: Rethinking the Causes and Cures of Student Attrition. (2nd ed.) Chicago, IL: University of Chicago Press.

Trower, C. (2002, April 12). Women without Tenure, Part IV, Why it Matters, What to do? *Next Wave*. Retrieved January 19, 2008 from http://www.sciencecareers.sciencemag.org

Trower, C. (2003). Leveling the Field. [NERCHE]. *New England Resource Center for Higher Education*, *14*, 6.

Trower, C. (2007). *Collaborative On Academic Careers in Higher Education*. Paper Presented at Harvard University, Boston, MA. Retrieved from www.oced.org

Trower, C. (2008). The name assigned to the document by the author. This field may also contain sub-titles, series names, and report numbers. Amending Higher Education's Constitution. *The entity from which ERIC acquires the content, including journal, organization, and conference names, or by means of online submission from the author. Academe*, *94*(5), 16–18.

Valian, V. (1999). Why So Slow: The Advancement of Women. London: MIT Press.

Winkler, J. (2000). Faculty Reappointment, Tenure, and Promotion: Barriers for Women. *The Professional Geographer*, *52*(4), 737–750.

Wirt, J., Choy, S., Rooney, P., Provasnik, S., Sen, A., & Tobin, R. (2004). The Condition of Education 2004. NCES 2004-077. Washington, DC: US Department of Education.

Yoder, J. (1994). Looking Beyond Numbers: The Effects of Gender Status, Job Prestige, and Occupational Gender-Typing on Tokenism Processes. *Social Psychology Quarterly*, *57*(2), 150–159. doi:10.2307/2786708

Yost, E. (2008). *Why have all the women gone? Exploring Gender Differences in STEM Postdoctoral Education.* M.A. Thesis, University of Alabama at Birmingham, Birmingham.

Zeldin, A., & Pajares, F. (2000). Against the odds: Self-efficacy beliefs of women in mathematical, scientific, and technological careers. *American Educational Research Journal, 37*(1), 215–246.

Zumeta, W., & Raveling, J. (2002). The best and brightest for science: Is there a problem here? Commission on Professionals in Science and Technology, Washington, DC.

KEY TERMS AND DEFINITIONS

STEM: Science, Technology, Engineering, and Math

Gender: The social classification of sex

Self-Efficacy: The belief in one's ability to succeed and overcome obstacles

Mentoring: The relationship between a senior individual and an entry level individual, where the entry level individual is guided, advised, and supported in a working relationship by the senior individual.

Institutional Support: Policies put in the institutional structure to benefit employees

ENDNOTES

1 Utilizing the definition provided by the funding organization, we considered departments/programs with 20 or fewer tenure track/tenured faculty to be small.

2 Although the response rate for senior graduate students and post docs was lower than we would have liked, it is not uncommon for web based surveys of students (Dillman 2007).

Profile 5
Technology Career Spotlight:
QUT Bachelor of Corporate Systems Management

© 2008 Queensland Government Office of Women. Used with permission

Helen Vahdat, Brisbane, Queensland

I first became interested in ICT when I accept a junior role in a web portal team for the Department of Training. It was here I learnt how connected a community, local or global, can be by making use of technology. I became fascinated with the idea making services available and accessible to large communities. What I became passionate about was the creativity and innovation required to advance ICT services. I thoroughly enjoy listening to people's needs and requirements, and delivering them through a sophisticated system. ICT offers endless opportunity fro career progression and global diversity. During the last quarter of 2009 I will have the experience to travel the world implementing a information system for the global company I work for Ausenco. ICT is an amazing career choice filled with creativity, innovation, logic and global collaboration.

Copyright © 2010, IGI Global. Copying or distributing in print or electronic forms without written permission of IGI Global is prohibited.

Introduction

Helen Vahdat is a part-time ICT undergraduate student at QUT and works full time as a Business Analyst for Global Engineering Company Ausenco. Helen is a Smart Woman because she uses her formidable intelligence to apply IT solutions to empower others. Prior to Helen's move to Ausenco she used technology intelligently, creating efficiencies to contribute to the Department Education, Training and The Arts commitment to environmentally-sustainable practices.

In her previous position as Senior Project Officer in the Skills First team Helen demonstrated superior ICT innovation and research skills through her initiation and management of an ICT project to determine the feasibility of developing a web-based assessment application for recognition of prior learning.

This involved working in liaison with a business analyst to:

- Conduct an extensive scan of any similar products available in Australia and the international market.
- Conduct literature reviews, documenting publications and studies in regards to online assessment, VET trends, international market and influences.
- Lead consultation sessions with senior departmental management, steering committee members, key industry stakeholders, training staff in public and provide RTOs.
- Create and conduct a survey for TAFE teachers to gain insight to the tools and assessment methodology practiced within the Institutes.
- Identify and analyse the implications of policy documentation specifically in relation to Whole of Government ICT architecture.
- Research current and future training, assessment and associated issues impacting specific industries. For example – labour and skills shortage in construction and how technology can bring learning and assessment onto the construction sites to help alleviate current pressures.

This project is unique in its quest to produce a system so sophisticated all of the vocational training areas could utilise it and collectively changed the standard of RPL assessment. The proposed system offers several business efficiencies including the reduction of labour -intensive assessment and offering cost-effective assessment solutions. With the tight labour and skills market Helen's innovative ICT solution is an asset to the Department, the community and individuals trying to increase their income by having their skills recognise in a fast streamlined system.

Helen is strongly committed to providing a high level of client service to promote a public service that is responsive to community needs and innovative in identifying solutions. She applies knowledge from her tertiary studies in the field of Corporate Systems Management to her work and is constantly seeking ways to contribute to business improvement initiatives. She is to be commended for successfully meeting the demands of a senior role in the Department while pursuing her undergraduate studies. The Department has greatly benefited from the skills Helen has acquired through her studies at QUT.

Helen's recent nomination for a leadership award in the 2008 Staff Awards acknowledged her ability to manage innovative projects and paid tribute to her enthusiasm and passion for the Department and community.

In addition, Helen is a volunteer member for the 'Technology Takes You Anywhere' initiative by Education Queensland 'Girls in IT' project. She is responsible for the event marketing and participates on the committee as an advisory member.

A colleague recently commented on Helen's assistance:

I especially benefited from Helen's IT expertise as I returned to the workforce after taking a 5-year-break to look after my family. The IT advancements made in the time of my absence were overwhelming for me at first, but Helen constantly empowered me by showing me how to use technology to save time in my work life and to help me regain my work confidence. I appreciate the way Helen creates an inclusive work environment, sharing her expertise with others in the team to ensure maximum benefit

Helen is inspirational as a young women studying, achieving high grades and working full time applying her knowledge in ICT to better the community. Helen models the way young women can strive to achieve great outcomes and empowers women to embrace ICT in their working and personal lives.

Chapter 6
Go WEST – Supporting Women in Engineering, Science and Technology:
An Australian Higher Education Case Study

Jacquie McDonald
University of Southern Queensland, Australia

Birgit Loch
University of Southern Queensland, Australia

Aileen Cater-Steel
University of Southern Queensland, Australia

ABSTRACT

Australia appears to be lagging behind countries in North America and Europe regarding the participation of women in engineering, science and technology courses and careers. This chapter reports on a current project undertaken by a regional university to build a mentoring and support network among female Science, Engineering and Technology students, staff and industry professionals. As well as the context and history of the project, the chapter describes the activities undertaken and the challenges faced in making the project sustainable. Factors critical to the success of the project are identified and include securing funds and commitment from senior management, having a multi-disciplinary team with strong leadership and effective support, and using information technology to enhance personal networks and to promote activities.

INTRODUCTION

Women are traditionally under-represented in disciplines often summarized as STEM (Science, Technology, Engineering and Mathematics), or SET (Science, Engineering and Technology). Statistics (Australian Council of Engineering Deans' report, 2008;, Osborne, Rees, Bosch, Ebeling, Hermann, Hilden, 2000) indicate that this phenomenon is prevalent in developed countries around the world, commencing with lower female participation in high schools, continuing into tertiary education, and it is even more pronounced in professional life due

DOI: 10.4018/978-1-61520-657-5.ch006

Copyright © 2010, IGI Global. Copying or distributing in print or electronic forms without written permission of IGI Global is prohibited.

to the difficulties of combining work and family commitments with long working hours. In a university context, women tend to be disadvantaged in career paths with fewer women applying for promotion than men, and proportionally more women appointed at lower academic levels than men. However, research shows that women benefit from mentoring and networking, with positive results reported from programs that focus on women's needs (Mysyk, 2008). Research shows that mentoring during formative years of education is a predictor for the future participation of women in STEM study (Dyer, 2004). Also mentoring of junior female academics assists women to obtain tenure (Stewart, Malley & Lavaque-Manty, 2007).

The context for this case study is an Australian regional university: the University of Southern Queensland (USQ). USQ has both on-campus and external students, offering a flexible blend of distance and online education in the Faculties of Arts, Business, Education, Engineering and Surveying, and Sciences. USQ has approximately 25,000 student enrolments, with 75 percent of these students studying in external mode, including 30 percent international students. Many external students are mature age, with work and family commitments, studying part-time by completing one or two subjects a semester over four to six years of study. This flexibility provides access to many who would normally be unable to complete higher education study, but does create enormous pressure on students as they often study in isolation and need to balance study, work and family responsibilities.

So how could a networking and mentoring program be designed and implemented at a regional Australian university characterized by a high proportion of students enrolled at a distance, and with low female student and staff representation in SET? How would it fare in an environment where decision makers are predominantly male? This chapter provides a case study based on a project implemented at USQ. It describes the activities of the Go WEST project team (Go Women in Engineering, Science and Technology), an initiative funded through a university equity grant in 2007 to establish a university-wide, cross-disciplinary professional network to mentor and support female SET students and staff. The background of this chapter provides the context by describing the relevant policies and summarizing literature related to women in SET at Australian Federal and State Government levels. USQ's equity policies and practices are summarized. The case study methodology is then described followed by an account of the history and current activities of the Go WEST project. The challenges faced are outlined, followed by a discussion of possible solutions and recommendations. Future research directions are suggested.

This case study provides useful insights for other groups who are considering appropriate activities to address the imbalance of women in traditionally under-represented disciplines.

BACKGROUND

While many studies report that women are under-represented in SET areas in the developed world (Osborne et al., 2000), a comparison of four countries appears to indicate that female participation in Australia is at the lower end of that scale (Office for Women, 2006). For instance, where in 2005 the percentage of female professional engineers in the United States was 10 percent, in the United Kingdom 9 percent and in Canada 8 percent, it was only about 5 percent in Australia. The same paper reports that a quarter of the Bachelor in Information Technology (IT) graduates in Australia were female in 2003, compared to 28 percent across America in 2001-2002. In the UK, more than a quarter of postgraduate computer science degrees were completed by women and in Canada 28 percent of computer and information systems professionals were female in the year 2000. As noted in the following paragraph, not much improvement has been recorded.

At a national level, an Australian Council of Engineering Deans' report (ACED, 2008, p. 27) states that concern is expressed in all engineering schools about the low numbers of women in academic positions in engineering. In fact, the report provides total enrolment figures and proportions of female students in engineering awards between 1996 and 2006 (p. 34) which indicate that the percentage of female students in all engineering degrees has over these eleven years never exceeded 16 percent. The proportion of female first year students enrolled in undergraduate engineering degrees peaked in 2001 at 16 percent, but was less than 15 percent in 2008. The report comments that "Since women form the majority of all tertiary students (54.7% in 2006) their continuing gross under-representation in engineering is critical" (ACED, 2008, p. 61).

Another recent Australian Government report followed pathways of students from STEM high school courses to study and then to professional work and provides statistics on retention and influencing factors. These factors include parents' educational background, geographic location and career aspirations on the student's decision to enter a career in STEM (Anlezark, Lim, Semo, & Nguyen, 2008, p.6). The report concludes that "self-motivation is slightly stronger for males than females, with males less likely to be influenced by others and tending to rely more on their self-assessment of academic strength in STEM subjects than females", and that "it seems that careers advisors are perceived by young people as more influential in steering young people away from, rather than into, STEM careers" (Anlezark et al., 2008, p.6).

At the same time, compelling economic arguments are made for increasing the participation and retention of women in SET, for instance by authors from the UK (Greenfield, Peters, Lane, Rees, & Samuels, 2002). Some of these arguments are

- competitiveness (engaging the best people in SET),
- return on investment (retaining as many qualified people on the workforce as possible) and
- benefit to science (maximizing diversity).

This is not a recent phenomenon: in 1995, the WISENET Report (1995, p.6) demanded a greater recognition of the value of the different "perspectives, priorities and operating styles that women can bring to SET". An Australian State Government report states that

Attracting and retaining women in the labour market, especially SET industries, is an important strategy to address some of the skills shortages and ensure continued economic growth in Australia. Women's ongoing under-representation in SET education, training and industries will ultimately inhibit Australia's capacity to competitively participate in a global market (Office for Women, 2006, pp. 3-4).

There are currently a number of initiatives at national level in place in Australia. The following are two examples:

- The Equal Opportunity for Women in the Workplace Agency (EOWA) which reports to the Australian Government, awards *EOWA Employer of Choice* citations to women-friendly non-government organizations that recognise and advance their female workforce. This award gives public acknowledgment of their efforts in the area of equal opportunity for women (EOWA, 2009) in all areas of employment.
- Three industry associations support women in SET areas: Women in Engineering (WIE), a special interest group of the largest and most diverse engineering association in Australia, Engineers Australia; Women in Science Enquiry Network (WISENET), a national linking network with branches across the continent focusing on local

issues and with a newsletter-style journal publication; and Women in Technology (WIT), with the mission to create a sustainable network of members focused on the advancement of women in technology industries.

More locally, Queensland has been proactive in the support of women in SET. The State Government released a report on the issues hindering participation of women in SET (Clark, 2003). For example, based on statistical data from 2002, the female participation rate in education, training and professional employment in ICT was as low as 21 percent in Queensland in that year. Although 53 percent of women were studying science at university, they were poorly represented in the physical sciences, mathematics and engineering. In fact the report states that "women represent only 6 percent of the workforce in the latter field and occupy only 7 percent of senior positions in higher education in all fields of science, engineering and technology" (Clark, 2003, p.5). The report comprehensively documents barriers to female participation in SET, for instance:

The barriers to the participation of girls and women in SET are slow to break down, not because of their lack of ability in these fields, but because of entrenched attitudes in society of what work is appropriate work for girls. Many SET occupations also have a serious image problem with many girls seeing them as boring, male-dominated and lacking social value. The lack of appropriate up-to-date career information and visible female role models serves to re-inforce this perception. The content and teaching method of SET subjects too often fail to appeal to the interests and concerns of girls. The SET classroom and workplace can be hostile environments for girls and women and progress to higher levels in a SET career can be very difficult, particularly when women begin to juggle work and family responsibilities (Clark, 2003).

The report provides a good overview of existing successful programs and new initiatives and recommendations. Initiatives resulting from this report include the establishment of the *Smart Women – Smart State Awards* and the *Smart Women – Smart State SET Taskforce.* While the Awards were an annual event from 2004 to 2008 to recognize the achievements of women in SET in Queensland, the Taskforce provides advice to the State Government and to the Minister for Women and has developed the Science, Engineering and Technology Action Plan 2006-2009 (Office for Women). The Smart State Strategy commits to enhancing women's participation in SET through four priority areas for action. These relate to raising awareness and increasing exposure to opportunities in SET fields, fostering partnerships between Government, industry, educational institutions and communities to improve outcomes for women currently in or considering SET careers, promoting better pathways to transition, and monitoring and measuring progress.

A number of Australian universities have started projects to support or facilitate networking for women in under-represented STEM disciplines. For instance, focusing on the staff perspective is the WiSci (Women in Science) project at the University of Sydney, initiated by the Dean of the Faculty of Science when he realized that women were well represented among students and postdoctoral fellows, but under-represented among staff in senior academic roles. Women were also "over-represented in the lower levels of the academic hierarchy and tend not to apply for promotion in the same numbers as men" (Baz & Oxford, 2008). The WiSci project aims to identify obstacles that prevent women reaching senior positions in the sciences, to encourage more women in the sciences into senior academic roles, and to support more women in achieving satisfying and productive careers in science.

The Queensland Government responded to the decreased participation in STEM education and lack of skills by releasing a discussion paper

Table 1. Proportion of female academics in Faculties employed in ICT at USQ in SET roles in 2009

Faculty/ Discipline	Total number of full-time equivalent staff in SET roles	Number of full time equivalent female staff	Percentage of female staff
Faculty of Engineering & Surveying /Agricultural, Civil and Environmental	23	2.9	12.6%
Faculty of Engineering & Surveying /Electrical, Electronic and Computing	19	2	10.5%
Faculty of Engineering & Surveying /Mechanical and Mechatronic	17	1	5.9%
Surveying & Land Information	12	1.5	12.5%
Faculty of Engineering & Surveying Total	71	7.4	10.4%
Faculty of Sciences/ Department of Mathematics & Computing	21	4.5	21%
Faculty of Business/ School of Information Systems	18.5	4.5	24%

to examine "the issues Queensland must address to increase participation in science, technology, engineering and mathematics study and careers, from primary school to university and beyond" (Qld Government, 2007). While addressing STEM education in general, the paper points out that a significantly higher proportions of males than females undertake study in higher level mathematics, chemistry and physics at high school level. Women represented just over one-tenth of total STEM-related VET (vocational education and training) enrolments in 2006. About 60 percent of the state's domestic undergraduate university students were female, however less than 50 per cent of students in STEM related fields were women, and this included health-related fields with a higher proportion of women. The difference is even more pronounced in engineering and information technology, where women are representing 10 and 20 percent of enrolments respectively.

At USQ, the proportion of female academic and professional staff in SET follows the trend of low female participation. Table 1 shows percentages of female academics in faculties.

Furthermore, in the Division of ICT Services, only 22 percent of staff are women, with the majority of them employed at lower levels compared to males. Five of the 19 employees at manager level or above are female. Unpublished data on female academics in SET disciplines indicates a similar distribution at higher levels in the faculties.

Through the Equity Office, USQ annually offers the Women and Leadership Mentor Program open to all women employed at the university. USQ's Equity in Education Policy (USQ, 2008) states that the university is committed to proactively provide equitable education opportunities to disadvantaged groups such as women in non-traditional areas of study. Apart from these two initiatives, there had been no efforts to specifically support female staff or students in SET areas, which prompted the idea of starting the Go WEST project.

OUTLINE OF RESEARCH APPROACH

In this case study the authors followed a process of reflective practice and critical analysis as they reviewed documents relating to the Go West project. The data sources included drafts of project funding applications, the project proposal, ethics approval submission, meeting minutes, meeting action lists, media items, communication documents between project team members, project officer task lists and memos, invitations to events and web site documents, as well as participant responses to a call for ideas to address issues important to SET staff and students.

The review of data from project conception in September 2007 to June 2009, revealed that a range of processes and activities have been implemented successfully and have increased the profile of the project. The following section provides case study details of the Go WEST project, reflects on key aspects of the process and presents a context for reflection on issues and challenges for those planning similar projects to support women in SET.

THE GO WEST CASE STUDY

Establishing the Project

The Go Women in Engineering, Science and Technology project was envisaged after the project leader attended the Queensland Government's Smart Women/Smart State 2007 Award ceremony and noted that rural women and regional universities had an extremely low representation in the short listed candidates and none in the award winners. This indicated a double disadvantage and underrepresentation of rural and regional women in the Science, Engineering and Technology (SET) area. A subsequent university call for applications for USQ Equity projects, funded by the Australian Federal Government of up to $15,000AUD provided the impetus to plan a project to support women in SET at USQ.

The USQ *Equity in Education Policy* aims to proactively provide accessible and equitable higher education opportunities to the diverse USQ student constituency, in particular to target groups identified as disadvantaged in their access to and participation in higher education. The policy states that although USQ provides a wide range of programs and services that target all identified equity groups, specific areas of specialist teaching and support will be maintained and funded. Women in non-traditional areas of study were one of the disadvantaged groups identified in Equity and General Performance Indicators in Higher Education (Martin, 1994) and group profiles are currently used by the Department of Education, Employment and Workplace Relations (DEEWR) statistics to monitor sector-wide and institutional performance. The Australian Council of Engineering Deans' 2008 report about the low numbers of women in academic positions in engineering, and the proportion of female first year students enrolled in undergraduate engineering degrees (less than 15%) indicates that little has changed since Martin's 1994 study.

The first step in starting the project was via an email outlining the rationale and the focus of the project to potential team members. The identified team members were active advocates of increasing representation of women in SET, most had an active teaching role in a SET area and were representatives of a range of university Faculties and Departments. All women approached agreed to join the team, and through discussion and negotiation at several meetings, an application for equity funding was drafted and versions modified via circulated email drafts.

The Go WEST project received a USQ Equity Grant to implement the project during 2008. The project team aimed to establish links between USQ students and professionals in science, engineering and technology (SET) in rural and remote geographically isolated areas of Queensland, the Queensland Government's Smart Women/Smart State initiative and other high profile professional societies, and to build a mentoring and support network between industry, academia and schools via the hub of SET women at USQ.

The links with members of professional networks were to provide contacts for mentoring SET female students and academics to sustain and enhance their participation in traditionally male-dominated fields. Another goal was to establish networks among SET students to reduce attrition rates and difficulties afforded by gender and geographical isolation. The aim of the network was to embrace both students and staff at USQ to build cross discipline networks, professional and personal expertise.

The project application was deliberately aligned with strategic University goals and policy by addressing factors that contribute to the retention and progression of students - one of the recommendations of USQ's *Report of the Transition and Retention Working Party* (Taylor, 2006), and the USQ Australian University Quality Audit (AUQA) report (USQ, 2005) recommendation to develop an ongoing staff development program.

The budget proposal was also strategically focused, with two thirds of the funding for a project officer to be employed one day a week to work with the project team. This was considered a critical factor in sustaining the project, as team members, while committed to achieving the project goals, were doing so on top of existing workload. Workloads are an inhibiting factor for the sustainability of many equity initiatives as university staff are expected to undertake community service activities, but learning and teaching and research activities are usually where workload allocation and promotion opportunities reside.

Attracting suitably qualified female employees is an issue at USQ with women under-represented in a range of SET job roles. The project aimed to address the problem by supporting SET activities such as mentoring and recognising that younger employees should have opportunities to use USQ as a launching point for their career progression. In that context, the network is seeking opportunities to offer work experiences to female SET students.

The project planned to create links with Queensland Government's Smart Women – Smart State initiatives, industry professionals and societies such as Women in Technology or Women in Engineering. It also aimed to draw on these links to bring SET activities to rural and geographically isolated areas to support women/girls and increase USQ's profile. The project is a springboard to building better pathways for women and girls to transition between schooling, further education, training and employment (including career breaks) in SET industries.

Six strategies were outlined in the project proposal:

1. Establish a university wide network to build cross discipline links and professional and personal expertise to foster the success of academics and students in SET;
2. Provide a virtual support structure of professionals and academics to reduce existing isolation issues for SET students in rural and remote locations;
3. Identify female SET student concerns and priorities for action by network members;
4. Implement strategies to address student issues that may influence student retention and progression;
5. Link network members to Queensland Government Office of Women and industry initiatives; and
6. Establish mentoring relationships with SET students in rural and remote locations during critical times when female students are making decisions about their future studies and thus place USQ foremost in their minds as a competitive solution for higher education.

Implementing the Project

Once the project was approved the project team met several times to appoint a project officer, plan activities, organise budget processes, seek ethics approval and establish effective working processes.

The Go WEST project team met monthly to plan activities and address project issues such as building a database of names of SET staff and students, applying for ethics approval to contact staff and students, establishing a web presence and managing budget items. At the first team meeting, the project leader outlined her desire for horizontal leadership (Langham, 2009) and encouraged project members to decide where they wanted to take a leadership role. The project launch was the first public activity.

The USQ Chancellor agreed to launch the Go WEST project in April, 2008, along with a representative from the Queensland Government Office for Women. The launch was well attended by SET staff from USQ and provided an opportunity to communicate project goals, gain input from attendees and implement networking. The launch provided valuable visibility through media releases and personal links with Queensland Office for Women staff. This resulted in the project leader presenting to the Queensland Government Smart Women Smart State Taskforce and requesting financial support for a jointly funded scholarship for female Engineering students. While the Dean of Engineering and Surveying at USQ agreed to co-fund this scholarship for female engineering students and a draft scholarship proposal was sent to the Queensland Government Office for Women, this process proved very slow and was eventually overtaken by Queensland State Government elections, so was unsuccessful.

Project activities were planned based on issues identified by SET participants at the first group activity in May 2008, the "Implementing Ideas Workshop." At this session, attendees were encouraged to network by meeting two people from another discipline and introducing themselves to one new person. After a presentation of the project goals participants broke into groups to identify priority issues to provide a framework for project planning and implementation. The next activity was a Smart Women/Smart State award application workshop in June 2008, conducted by a project team member who was highly commended at the 2007 Awards. Project team members mentored applicants, including a USQ engineering student and an IT academic staff member who both consequently won awards in their categories. The award winners for both 2007 and 2008 are profiled on the Go WEST web site (Go WEST, 2009).

In August 2008, members of the project team met with the local Toowoomba Regional Council to discuss collaborative partnerships for mentoring and industry experience. This partnership has continued to develop over eighteen months, with Council engineering women attending USQ activities.

The Go WEST project team contributes at an institutional policy level, for example members were approached to provide data and advice to the Pro-Vice Chancellor, Social Justice and Equity, for the preparation of USQ's Social Justice five year strategic plan. Information related to women in science and engineering or women in the workforce at USQ in general, and several key future issues in the SET area that needed to be faced in the next five years were identified. Other initiatives included the profiling of the Go WEST project at USQ Open Day in the Faculty of Engineering and Surveying stand.

Another aim of the GO-West team was to undertake and publish research related to women in engineering, science and technology disciplines. This chapter and edited book are direct outcomes of the Go WEST project.

Sustaining the Project Activities

In recognition of the role of Go WEST in the SET arena, a member of the team was approached to participate in the Australian component of a large-scale project called *Practising Gender Equality in Science* (PRAGES) funded by the European Commission's seventh framework programme. Participation in the PRAGES project involved completing a questionnaire concerning the initiatives Go WEST employed to support women in leadership and decision-making positions. These responses were added to a database of 'best practice', contributed to by a number of countries in the Organisation for Economic Co-operation and Development (OECD), which will be a valuable resource for international knowledge sharing. The project was seeking examples of initiatives, programs or policies that promote the presence of women in senior and decision-making positions, particularly within the field of science and technology research, but also more generally. The project focus included an attempt to integrate the most important and relevant results derived from the studies and good practices relating to the struggle against vertical segregation in various professional, political and social areas, enhancing the understanding of the exclusion of women as being deeply linked to what may be called the lack of socialisation of gender in science. This last point is of particular interest and will be expanded in the following section detailing challenges faced by the project team.

The Go WEST project team started 2009 public activities with a stall at Orientation week to promote the project and a group activity in March for students and staff. This activity presented an overview of Go WEST goals and activities to date, followed by an engaging presentation by a recently graduated Engineering student who spoke about her study journey at USQ and recent transition to her career as a female graduate engineer. A group activity identified how Go WEST could support the needs of SET women in 2009 (Table 2). The suggested activities include social functions, professional development activities and networking with industry professionals, thus providing a framework for the planned 2009 activities.

The USQ learning management system, Moodle, is being used to provide access to an online community environment for both on-campus and external members. Activities include discussion forums, news updates, activities, resource sharing and listing of SET communities and societies. Online access is available to USQ SET staff and students identified in the established membership data base. The accompanying web site provides a public and visible presence for the Go WEST project (Go WEST, 2009).

The Go WEST team looks for opportunities to encourage female students and staff to achieve success in engineering, science and technology fields. The project provided financial support to enable a USQ final year female Engineering student to attend the International Institute of Women In Engineering (IIWE) conference in Paris in July 2009. As well, the nomination fee was provided for a female academic from the School of Information Systems to submit her project to the Queensland Australian Computer Society ICT awards in April 2009. The USQ project was awarded a certificate as 'highly commended' in the IT Service Management Category.

CHALLENGES

This section outlines the challenges faced by project team members as they worked to implement the goals of the project.

Institutional Restructuring and Funding

Go WEST received funding in the first instance to operate in the year 2008. As the university was undergoing major changes in that year, with staff reductions in academic and general sections, cuts to courses and programs and a period of low morale, the project was not able to achieve as much as had been planned. However, what had been achieved was seen as sufficient to warrant extended funding for 2009.

While the USQ Equity in Education Policy aims to proactively provide accessible and equitable higher education opportunities to the diverse USQ student constituency, in particular to target groups identified as disadvantaged, such as women in non-traditional areas of study, there is always a danger that projects such as Go WEST that are funded from external sources will be seen as 'an optional extra' and never be written into policy or budget for mainstream funding. The Australian Council of Engineering Deans (ACED, 2008, p. 61) noted that many university-based *Women in Engineering* programs had been initiated, but

Table 2. Activity responses from Meet & Greet March 2009 "What can Go WEST do for you?

Proposed events	
	Parties, e.g. Wine tasting event
	Opportunities to mix with other faculties and departments – professionally and socially
	More opportunities to meet with other women (especially for off campus women – social events etc …)
	Guest speakers – high profile women. Film local successful women for web site.
Industry related activities	
	Help to find work experience placements
	Mentoring – contacts with industry
Networking, mentoring, sharing practice	
	Share success stories; share practice in a safe environment
	Blog on Go-WEST site that anyone can post to about things they're doing
	Students mentoring students, share struggles and how to overcome on Go-WEST site, "ask an expert"
	Encouraging girls at school level to go into SET. Visit schools as role models and talk to Pre-Yr 10 girls to inspire
	Assistance in developing interactive area for small group external students (Go-WEST website etc); networking opportunities for external students
	Linking students doing maths/science
	Support – someone to talk to
	Personal stories
	Meeting on a regular basis
	Third year students being supportive of first year students
	Promote mentoring program: staff & staff; students & staff; students & students
	Nominate students & staff for awards
	More opportunities for general staff development (e.g. placements in different working environments) – varying environments to boost staff experience and morale
Professional Development workshops	
	Help find funding for women – prizes, scholarship, conferences, professional development.
	Equal employment and harassment training courses
	How do men/women work. Strategies for working in male-dominated environment
	Public speaking, confidence building, negotiation skills, research skills, statistical skills Professional writing, resume and interview preparation
	'Work skills' training
	Cross-faculty demonstrations e.g. showcasing how to use different equipment/methods
Provide funding to individual staff and students	
	Sponsorship of groups and individuals to meetings – give presentation to Go WEST on their return from meeting/conference
Promote Go-WEST	
	Promote information about events
	Recognize women on-campus. Media releases
	Widely distribute flyer for next meeting

over the past decade the number of initiatives has declined as funding was not sustained. Failure to continue funding for such programs is of major concern, as initiatives such as Go WEST attract and retain females in SET.

Sustainability

As the project team is 'volunteering' their time there is always conflicting pressure of mainstream teaching and research activities. While funding exists for a project officer, the core business and activities are maintained. However, if the funding expires at the end of 2010, the project team will have to decide how, and if, the project can proceed without project officer support.

Participation of SET Staff and Students

As three-quarters of USQ students are enrolled in external (distance) mode, attracting and including external students in activities presents a challenge. The web page and Moodle site provided access and information for external students, and external students are contacted when they attended on-campus residential schools. However, the low participation of external students is an ongoing concern. Workload and study commitments also prevent on-campus women from attending functions.

Corporate Stipulations

The launch of the project web site proved to be a difficult process. The team brainstormed a visually creative and interactive site to attract attention and promote the project. An academic staff member contributed to the process, however once the draft version was presented to the institutional web team, it had to be completely redesigned to meet institutional branding requirements. In this process the Go WEST logo was removed from the front screen, and several months of web presence lost. Organising access to staff and student contact details also proved difficult and time consuming, delaying contact with external students.

Feminine Way of Working is Undervalued

The involvement in the PRAGES project was of particular interest to the Go WEST project team members as one aspect of the PRAGES project is enhancing the understanding of the exclusion of women, which is seen as being deeply linked to what may be called the lack of socialisation of gender in science. While so called 'soft' skills, or 'feminine/female' ways of working can be devalued simply by giving them a 'soft' or 'feminine' label, women may always have to struggle for equity in leadership positions. USQ has been seeking to increase the percentage of women in Senior University positions for a number of years and to improve the percentage of female academics and students in SETS areas, but the existing male-dominated leadership situation can make the environment unfriendly to women. Addressing USQ's *Faculty of Engineering and Surveying Research Seminar*, Fortenberry (2009) stated that employers are now seeking graduates with 'professional' skills, as it is assumed that graduates will have the necessary discipline technical skills. Fortenberry noted that 'professional' skills were previously referred to as 'soft' skills, but are now an essential engineering skill.

CRITICAL SUCCESS FACTORS

The two main critical success factors for establishing and sustaining a project similar to Go WEST are people and funding and obviously these are interrelated.

The people will determine the success of the project. The Go WEST multi-disciplinary team included representatives from engineering, education, mathematics, information systems and ICT.

They are a group of determined women, able to work collaboratively and creatively to identify realistic, achievable goals, be flexible and sensitive to the time demands and priorities of other team members, and dedicated to achieving the project goals. The team members were able to be strategic in their approach to working within institutional processes and possessed the interpersonal contacts and skills to work with a diverse range of people to implement project activities. The Project Officer was an essential team member, able to successfully liaise across USQ's departments, staff, students and the team to implement the operational plans of the project. She was employed one day per week and was capable, demonstrated initiative, and used interpersonal skills and a range of community contacts to further project goals.

Funding was essential to employ the Project Officer and to support the activities outlined in the previous section. As the project members were committed to the project as a community service, there was no workload allocation for project activities. Funding included the initial competitive project application and an additional twelve months ongoing funding, which was essential to assure project sustainability. Ideally the funding will become part of mainstream budget allocation, and not be dependent on external equity funding.

Four further critical success factors are identified:

- Senior management support is essential, not only for securing funds but to raise the profile of the project within the organization. USQ is very fortunate to have a female Chancellor. The Chancellor took the role of project champion and attended the launch of the project. Senior management supported the project in terms of the proposed scholarship.
- Working within institutional processes insures the smooth implementation of project goals, for example, obtaining ethics approvals for accessing databases and contacting students, and also to link the project information to the USQ website.
- Use of IT is essential. In this project, team members used the USQ Learning Management System to provide a flexible portal for communication to/from staff and students. As previously mentioned, the project website enables promotion of the project as well as a central storage facility for photos and media reports. An important preliminary step involved selecting and extracting student contact details from corporate databases.
- Marketing is important. As well as letting staff and students know about planned activities, it is important to communicate success, by profiling individual and group awards. Media releases distributed to USQ News and the local newspaper also helped raise the profile of the project and ensure its success.

Recommendations

The authors support the Australian Council of Engineering Deans (ACED, 2008, p. 61) and *National Women in Engineering Committee* recommendation to promote the good employment market for engineers and believe these recommendations should be extended to include women in science and technology. Continued action to promote and attract women in SET, including redeveloping well-resourced dedicated programs such as *Women in Engineering*, with long term commitment from university academic leaders is required. The Women in Engineering committee also identified a number of curriculum changes to promote inclusion, for instance the implementation of inclusive curriculum for SET programs, reducing male stereotypes within the curriculum, and revitalising the best of the programs (ACED, 2008, pp. 106-107).

Future Directions

Go WEST has succeeded in increasing the visibility of USQ women in the SET areas through supporting women in their successful applications for Smart Women Smart State and other industry awards. It has also succeeded in promoting networking and lifting the profile of successful women in the disciplines of engineering, science and technology. As higher education institutions exist in a climate of accountability and tight budgets, in the future quantifiable measures could be identified and tracked. These include the retention rate of female students, career progression of graduates, and the promotion success rate of academic staff. This presumes that funding is available to provide the resources to collect, collate and publish such data.

In the future, the newly-formed links between project members and State, Federal and International lobbyists will be nurtured to identify opportunities to gain further funding and have an impact beyond the boundaries of USQ. Other activities of the Go WEST group could focus on is working with staff, students and employers in the areas of mentoring and work experience. Plans are in progress to work collaboratively with staff in the Student Services Division to ensure the sustainability of this process.

CONCLUSION

Evidence of the success of the Go WEST project is demonstrated by the fact that it has survived despite the many challenges encountered. This case study described a range of activities and initiatives designed to provide a mentoring and support network for staff and students in engineering, science and technology disciplines at an Australian regional university. The critical success factors identified in this case study may seem obvious in hindsight, but to other universities considering a similar project they may provide useful insights. For new projects, the strategies and activities described here could be adapted to fit with the organizational and legislative contexts of other institution.

POSTSCRIPT

In November 2009, the Go-WEST team won the Queensland Government's Our State, Our Women 'Public or Community Sector Award' and have secured funding from USQ to continue and expand the project activities throughout 2010.

ACKNOWLEDGMENT

The authors acknowledge the hard work and support of Sandy Sharman (Go WEST Project Officer), and the other executive members of the Go WEST team - Lyn Brodie, Senior Lecturer, Engineering & Surveying; Maggie Fryer, Principal Manager Service Delivery, Division of ICT Services; Petrea Redmond, Lecturer, Education; Dr Shirley Reushle, Senior Lecturer, Australian Digital Futures Institute; and also Shelly Grist, Lecturer in Information Systems for designing the web site. Support from senior management at USQ is also acknowledged, in particular the USQ Chancellor Ms Bobbie Brazil, Professor Peter Goodall (PVC Social Justice and Equity) for Equity funding in 2008-9 and Professor Frank Bullen (PVC Research) for supporting the scholarship bid and SET initiatives. We also acknowledge the support of the Queensland Government Office of Women and their staff: Karin Finsterle (Director); Peta Martin (Acting Senior Information and Education Officer) and Alexandra Winter (Principal Policy Officer).

REFERENCES

ACED. (2008). *Addressing the supply and quality of engineering graduates for the new century: Report on ALTC funded project*. Retrieved 12 May, 2009, from http://www.altc.edu.au/system/files/resources/Grants_DBIprojec_engineeringquality_project%20report_25march08.pdf

Anlezark, A., Lim, P., Semo, R., & Nguyen, N. (2008). *From STEM to leaf: Where are Australia's science, mathematics, engineering and technology (STEM) students heading?* Retrieved 12 May, 2009, from http://www.dest.gov.au/NR/rdonlyres/53AD328C-5620-4935-980A-FB8309F78F4F/24712/STEMreportOP03890100908.pdf

Baz, J., & Oxford, P. (2008). The University of Sydney's Women in Science (WiSci) project, *WISENET Journal*, 79. Retrieved 25 April, 2009, from http://www.wisenet-australia.org/issue79/Women%20in%20Science-WiSci.htm

Clark, L. (2003). *SET for Success in the Smart State: Increasing the participation of women in Science, Engineering and Technology in Queensland*. Retrieved 12 May, 2009, from http://www.learningplace.com.au/uploads/documents/store/resources/res_11094_setforsuccess.doc

Dyer, S. K. (2004). *Under the Microscope: A Decade of Gender Equity Projects in the Sciences*. American Association of University Women Educational Foundation. Retrieved 12 May, 2009, from http://www.aauw.org/research/microscope.cfm

EOWA. (2009). *Employer of Choice for Women*. Retrieved 12 May, 2009, from http://www.eowa.gov.au/EOWA_Employer_Of_Choice_For_Women.asp

Fortenberry, N. (2009). *Achieving the Engineer of 2020*. Paper presented at the Faculty of Engineering and Surveying Research Seminar.

Go WEST. (2009). *Go Women in Engineering Science and Technology: About Us*. Retrieved 09 May, 2009, from http://www.usq.edu.au/gowest/

Greenfield, S., Peters, J., Lane, N., Rees, T., & Samuels, G. (2002). SET Fair, A Report on Women in Science, Engineering and Technology from The Baroness Greenfield CBE to the Secretary of State for Trade and Industry, United Kingdom o. Document Number)

Langham, B. (2009). Horizontal Leadership. *Hay Group: People Torque,* May (3), 1.

Martin, L. (1994). *Equity and General Performance Indicators in Higher Education.*

Mysyk, F. (2008). Women Becoming Mentors: Reflection and Mentor Identity Formation as a Process of Lifelong Learning. *The International Journal of Diversity in Organisations. Communities and Nations, 8*(5), 207–218.

Office for Women. (2006). *Supporting Women's Participation in Emerging Industries: Science, Engineering and Technology.* Retrieved 12 May, 2009, from http://www.women.qld.gov.au/work-and-life/smart-state-strategy/documents/set-concept-paper.pdf

Office for Women. (2009). *Science, Engineering and Technology Action Plan 2006 – 2009, Smart Women - Smart State Strategy*. Retrieved 6 May, 2009, from http://www.women.qld.gov.au/work-and-life/smart-state-strategy/documents/set-action-plan.pdf

Osborne, M., Rees, T., Bosch, M., Ebeling, H., Hermann, C., Hilden, J., et al. (2000). *Science Policies in the European Union, Promoting excellence through mainstreaming gender equality*. Brussels, Belgium: European Commission. PRAGES. (n.d.). *PRActising Gender Equality in Science research project.* Retrieved 8 May, 2009, from http://www.retepariopportunita.it/defaultdesktop.aspx?page=2749

Qld Government. (2007). *Towards a 10-year plan for science, technology, engineering and mathematics (STEM) education and skills in Queensland.* Discussion paper released by the Queensland Government. Retrieved 12 May, 2009, from http://education.qld.gov.au/projects/stemplan/docs/stem-brochure.pdf.

Stewart, A. J., Malley, J. E., & Lavaque-Manty, D. (2007). Faculty Recruitment: mobilizing Science and Engineering Faculty. Ann Arbor, MI: University of Michigan Press.

Taylor, J. (2006). Report of the Transition and Retention Working Party. USQ.

USQ. (2005). *Quality Audit: Progress Report.* Retrieved 8 May, 2009, from http://www.usq.edu.au/resources/postauditprogressreport2005.doc

USQ. (2008). *Equity in Education Policy.* Retrieved 1 May, 2009, from http://www.usq.edu.au/resources/122.pdf

ADDITIONAL READING

Committee on Public Understanding of Engineering Messages. (2008) *Changing the conversation: Messages for improving the public understanding of engineering,* National Academy of Engineering, National Academy of Sciences, http://books.nap.edu/catalog/12187.html.

EdQuity Australia. www.adcet.edu.au/edequity

Girl, E. National Academy of Engineering, http://www.engineergirl.org/

National Science Foundation (NSF). 2009. "ADVANCE: Increasing the Participation and Advancement of Women in Academic Science and Engineering Careers. Retrieved 1 May, 2009, from http://www.nsf.gov/funding/pgm_summ.jsp?pims_id=5383.

Queensland Government Office of Women. a pictorial history of women's contributions to Queensland, 1859-2009, http://www.women.qld.gov.au/

Queensland Government Office of Women. Smart Women – Smart State Awards, http://www.women.qld.gov.au/work-and-life/smart-state-awards/.

Queensland Government Office of Women. Women in Hard Hats initiative. http://www.women.qld.gov.au/work-and-life/women-in-hard-hats/

Queensland Studies Authority. Equity Statement, http://www.qsa.qld.edu.au/downloads/about/qsa_equity_statement.pdf

Redmond, P. (2006). Increasing female participation in ICTs. QUICK. *Journal of the Computer Education Group of Queensland, 99,* 15–18. http://www.eprints.usq.edu.au/4479/1/Redmond_Quick_v99.pdf.

The National Academy of Engineering (NAE). http://www.nae.edu/nae/naehome.nsf.

USQ Equal Opportunity for Women in the Workplace Report 2005/2006. Retrieved 12 May 2009, from http://www.usq.edu.au/resources/eowa0506.pdf

WISENET. Women in Science Enquiry Network Inc, www.wisenet-australia.org

Women in Technology. www.wit.org.au

Profile 6
Technology Career Spotlight:
General Manager of Education Solutions at TechnologyOne

© 2008 Queensland Government Office of Women. Used with permission

Teresa Murphy, Brisbane

Teresa has more than 20 years experience in Information Technology covering scientific, commercial and government organisations. She has worked on business operational systems and data warehousing / management information projects. Teresa has an enviable blend of Business management, technical expertise, product, marketing and sales management. Teresa has been employed with TechnologyOne for the last 15 years and has spent a considerable part of this time working with the Student Management Product and Education Sector Vertical Market. Teresa has changed career paths a number of times over the years, starting work as a technician in a medical laboratory and eventually after working on a then leading edge automated computerised project with CSIRO ended up working in business software.. What she loves about working with business related process and problems is that there is always something new to consider and new ideas to be applied to improve the way people work.

Copyright © 2010, IGI Global. Copying or distributing in print or electronic forms without written permission of IGI Global is prohibited.

Background

Teresa Murphy is the General Manager of Education Solutions at TechnologyOne. TechnologyOne is one of Australia's most successful, publicly listed, multi-national software companies. Teresa commenced her employment with TechnologyOne in 1993 when she lead a team to develop the company's world class student management software. This project was initiated through a multi-million dollar, fixed time/ fixed price project in partnership with Curtin University. Teresa then commercialised the software application for sale to other higher education institutions. On the back of this initial project, Teresa has built TechnologyOne's Education Solutions division into a multi-national, $16 million dollar business, employing 65 staff and servicing more than 50% of universities and education institutions in Australia

Teresa holds a Bachelor of Science degree from the University of Queensland and a Post-Graduate Diploma in Computing from Macquarie University. Teresa has more that 25 years experience working in IT related roles and managerial positions for TechnologyOne, IBM, Telstra and various Queensland Government Departments. Teresa has been a member of the Griffith University ICT Industry advisory board for the last three years. The board provides input and reviews the curriculum being delivered or proposed by Griffith University in relation to ICT programs and courses. Teresa has also been involved for the past two years with a project being undertaken by James Cook University for the Australian Research Council. The project aims to better understand why young females are not participating in ICT related study at university level.

Teresa is a mother to three teenage children and enjoys pursuing a number of creative activities. She enjoys craft and has a passion for lace making, in particular needle lace, she is a long-term member of the Australian Lace Guild and is currently National Vice President, Proficiency Coordinator and Website Manager for the organisation. Teresa is also an avid tap dancer and volunteers in the production of dance concerts including making props and costumes (in particular dance head dresses), coordinating dress rehearsals and running backstage operations. Teresa believes participating in creative activities allows her to think outside the square when dealing with problems and issues related to running a highly successful business unit. Teresa also values highly the time she spends with her husband, children and parents.

Role and Responsibilities

As General Manager of the Education Solutions division at TechnologyOne, Teresa has total accountability for the success of the division, including financial performance, product development, sales success, marketing initiatives, customer support, and consulting services as well as all maintenance and upgrade services.

The Education Solutions division services more than 50% of Higher Education institutions across Australia. The division also services a growing number of institutions in New Zealand. Teresa is also currently planning to launch TechnologyOne's Education Solutions division in the United States within the next 12-18 months, so far Teresa has been involved in a number of trade shows and has met with a number of prospects in the region.

Teresa's role as General Manager, Education solutions involves but is not limited to:

- Providing input and influence into the overall direction of TechnologyOne's software offering to the education industry. Teresa is for example, currently working with the product manager for the company's new Customer Relationship Management product to identify and document the features and functions required for this product to be successful in the education sector.

- Providing high level business guidance to assist with the continuing evolution of the company's Application Development Framework (the architectural base for the company's entire product suite).
- Managing customer relationships through initiatives including the establishment of a Higher Education Strategic Advisory Committee, ongoing user group meetings, as well as facilitating the education solutions stream at TechnologyOne's Evolve conference staged for customers every two years.
- Managing and mentoring senior members of her team.
- Presenting regularly to the company's Executive team and board regarding the financial performance of the business unit, business strategies and market developments

Teresa also works closely with her management team to establish and implement the strategic direction for the Education Solutions division, which includes the following activities:

- Sales initiatives such as moving into new regions – working on, for example, the market evaluation and business case to expand TechnologyOne's reach into the USA.
- Consulting initiatives such as additional services to be offered, discounts and promotions.
- Product directions – the migration of the division's product range to a new technology platform and the addition of new modules including significant improvements around student communication facilities.
- Development of the company's Education Solutions product roadmap for the next three to five years.

Goals and Achievements

Since establishing the Education Solutions division of TechnologyOne, Teresa has built the business significantly, with customers throughout Australia and New Zealand. The Student Management software developed by Teresa and her team recently won a Consensus Award for innovation. The division was nominated by Microsoft and was one of nine award winners selected by an independent group of judges ahead of 56 other products. This places the business very well for its planned expansion into the US markets.

The long-term objective for TechnologyOne's Education Solutions division is to be recognised as a world leader in the provision of Enterprise Software Solutions to the education industry. The quality of the solution developed by Teresa and her team makes this goal very achievable. A key to building international success will be to listen to prospects, understand their unique challenges, watch and analyse developments in the education sector and maintaining a culture which delivers customer satisfaction. Essentially Teresa will be ensuring that all of the products in the solutions roadmap for education continue to evolve to meet the requirements of the industry.

Teresa's biggest achievement to date has been driving the creation of an internationally recognised, world-class software product. She is particularly proud to have been responsible for developing the TechnologyOne Student Management Solution which is the intellectual property of TechnologyOne, a company that was established in Australia and based in Queensland. Teresa has added substantial value to the business by creating a new product division which manages the company's central product for education in the enterprise suite of solutions.

Innovation and Value

The Student Management product developed by Teresa and the Education Solutions Division incorporates a unique Study Package architecture. It was the first software solution designed for teaching institutions which treats curriculum as a marketable product with a lifecycle therefore helping institutions optimise their revenue and focus resources appropriately. This architecture allows institutions to manage the introduction of new curriculum in a managed and controlled way with the retention of historical data integrity.

The unique Study Package architecture views each piece of curriculum as a building block. Meta data design philosophies provide the flexibility which allows an institution to invent different types of study packages for example courses, majors, minors, streams, sub modules and competencies; this allows for re-usability (for example, the same major can be used in many different courses but only has to be defined and created once.) This enables institutions to be more flexible in their offerings whilst still applying the various rules associated with studying, such as pre-requisites or choices. The other major advantage of this innovation is that creating new Study Packages and applying rules can be done on an ad-hoc basis without the need for costly code customisations.

The education industry is constantly evolving not only in the nature of how people study but also in what they study. New areas of research open up whole new branches of information which need to be passed on to the students. The ways students study and how they have the information delivered to them is also constantly evolving.

The Study Package architecture not only provides institution with mechanisms to manage this constant evolution and to keep track of what they did offer historically, but it also provides them with competitive advantage as they can respond very quickly to changes in education requirements. The architecture allows institutions to brand their offering, for example, some universities have courses whereas others have awards as a point of differentiation. The Study Package architecture subsequently allows for study plans to be built which are customisable to the individual student needs and result in better learning outcomes for the institutions clients (students).

Conclusion

Teresa Murphy has added substantial value to TechnologyOne by developing a product that has significantly contributed to the company's success. The Student Management product is the core product within the enterprise solution suite for higher education institutions. Teresa's contribution and commitment to developing and evolving solutions for education has added value not only to TechnologyOne but to the operations of the institutions and their clients. Teresa will continue to drive the Education Solutions division to meet TechnologyOne's objective of becoming the supplier of choice for enterprise solutions in higher education.

Section 4

Careers and Women at Work

Chapter 7
The Computer Games Industry:
Women's Experiences of Work Role in a Male Dominated Environment

Julie Prescott
The University of Liverpool, UK

Jan Bogg
The University of Liverpool, UK

ABSTRACT

This chapter focuses on the current position and experiences of women working within the computer games industry, the Information and Communication Technology (ICT) sector and the wider context of the Science, Engineering and Technology industry (SET). Global data collected as part of a larger quantitative study on women who are currently working in the male dominated computer games industry is reviewed, in relation to the long hour's culture associated with the sector. In addition, the lack of females, especially females with families in the gaming industry, skills shortages, work life balance and flexible working will be discussed in the games industry context. The research discussed will be related to the question of attracting and retaining women, in the games development workforce of the future. The issues discussed will be of relevance to employers, professional bodies, policy makers and researchers of the games industry and the wider ICT and SET industries. Recommendations from the findings and future research directions are provided.

INTRODUCTION

There is a need to recruit and retain more women as women are leaving the ICT industry in large numbers (Burns et al, 2007). This is despite the adoption of equal opportunities policies and various campaigns worldwide to attract women into ICT. Information, communication and technology (ICT) occupations represents a classic example of occupational gendered segregation and includes both vertical and horizontal segregation. The sector has been active in increasing women's participation through a number of government strategies with the aim of increasing the appeal and the image of the sector to young females. One initiative in the United Kingdom (UK) is a government scheme '*Computer Club for Girls*' that aims to combat the image problem of computers by emphasizing the

DOI: 10.4018/978-1-61520-657-5.ch007

Copyright © 2010, IGI Global. Copying or distributing in print or electronic forms without written permission of IGI Global is prohibited.

fun aspect. The aim of this is to maintain interest, in order that they may begin to consider a career within the industry. The games industry itself has also been active in recent years, in promoting the appeal and accessibility of a career in games to females. In 2009, 'Gamasutra' an online publication named the top 20 game writers (Ruberg, 2009), two women made this list (at number 15 and 19). Interestingly, one of the women Susan O'Connor was reported to be "*recognized as one of the most original and influential game writers currently working in the industry.*" Yet this woman appeared at number 15 in the list!

At a recent UK conference (2008 Women in Games), a number of these recent initiatives were discussed. Recent initiatives since 2007 have included:

- Microsoft 'Digigirls' workshop - 200 female participants aged 12-15 with focus on ICT and games.
- The University of Teesside, UK 'Girls and gadgets' conference - Females aged 13-16 with focus on the games industry and technology.
- University of Denver, USA 'P4Games' summer camp - Females with focus on game development

The widening women's work in ICT (Valenduc et al, 2004) project conducted in 2002-2004, in seven European countries (Austria, Belgium, France, Italy, Ireland, Portugal and the UK) found a number of factors contributing to the lack of women's participation in ICT. Findings from interviews and case studies with women working in ICT professions indicated that issues included the long hour's culture associated with the sector, the lack of females, especially females with families. In addition, there was a perception of the industry as masculine, 'geeky' and unsocial. Many of these factors have also been associated with the computer games industry as reasons as to why women are a minority within the industry (Haines, 2004, IDGA, 2004). The aim of the current chapter is to discuss these issues further and include findings from a larger qualitative study of over 450 women currently working internationally in the games industry (Prescott, forthcoming). The chapter aims to provide a brief overview of women's position within the wider ICT and SET industries, an overview of the games industries within the current economic climate and look at the position of women within the industries current workforce. The chapter will also discuss issues of work life balance; long hours and flexible working with the aim of providing the industry with an increased understanding and knowledge of how it may attract and retain a more diverse workforce in the future. The chapter contributes to the paucity of literature surrounding women working in the games industry in the UK and internationally as well as to provide a deeper understanding of women working under the umbrella of the ICT sector.

BACKGROUND

Women Working in the Science Engineering and Technology Industries (SET)

Data from the UK, 2007 Labour Force Survey indicates that women comprise almost a fifth of all SET workers (19% women and 81% men), the percentage of total women in SET occupations is the same as in 2002. In terms of proportional growth, overall, the number of SET women workers has increased by 12%, between 2002 and 2007 (representing 59,017 additional women in SET occupations). However, men in SET occupations also increased in the same period by 8.5% (representing 199,000 additional men in 2007). Women SET managers have increased by 2% since 2002 (14% in 2007).

In 2007 males represented; 86% of SET managers, 61% of science professionals, 95% engineering professionals, 86% of ICT professionals, 81% of building professionals, 78% of science and engineering technicians and 77% of IT service delivery occupations. The only SET occupation were women represent a higher percentage is research professionals with 51% and they are close to parity with their male counterparts in the teaching professions at 49%.

In 2007 (SET, 2009) across the EU countries, Switzerland had the highest percentage of female SET managers (43%) with 12 out of the country's 28 SET managers being female. The country with the highest percentage of female science professionals was Bulgaria (53%) and Cyprus is the country with the highest percentage of female ICT professionals (25%) however both countries have low a number of employees within the industry.

Data from UCAS (2007) shows that in 2007, for maths and computer science almost a quarter (24%) of degree applicants were women (21% in 1996) and that proportionally overall degree acceptances were 22% (23% in 1996). Women technology applicants and acceptances in the same period were 28%. Comparisons cannot be made between 1996 and 2007 for the technologies, as data categorization changed within the period.

Women's under representation in SET occupations is not only evident in the United Kingdom but also exists in most Western economies. In the European Union, Valenduc et al (2004) found that women represented 17% of all information technology professionals and 12% of engineers. Austria had the lowest representation of women IT professionals (14%) followed closely by Belgium and the UK (15% each). The highest representation was found in Ireland with almost one third (29%) female representation. The UK and Belgium had low representation of female engineers, at only 7% and Austria the highest level at 25%.

Figures from the USA are similar to those of the UK and Europe. This suggests a disparity exists globally. For example, in the life sciences women represent 45% of medical scientists, 47% of biological scientists, 34% of chemists and 22% of environmental scientists. For engineering occupations, women represent 17% of chemical engineers, 12% of civil engineers, 16% of computer hardware engineers and 8% of electrical engineers. For computer and math's occupations women represent just over a quarter (27%) of employees. (Labor Force statistics, CPS, 2007).

The under representation of women in the IT workforce is a worldwide problem. However, it may be a more predominant problem in Western countries such as the UK, USA, Canada, Australia and New Zealand. Research by Trauth (2002) and von Hellens and Nielsen (2001) has found that women from non western countries such as India and China have more positive attitudes towards computers and technology than their western counterparts. This appears to be particularly the case in India, where culturally families generally expect children to be successful in education. Boys and girls in this country are encouraged to strive for a career in either medicine or engineering (Adya and Kaiser, 2005). This cultural bias, may in part explain variations in the representation of women IT professionals and engineers in country comparisons.

The computer games industry is a relatively new industry of approximately three decades. The following two sections would like to focus more specifically on the games industry, in particular, looking at the current economic climate of the industry and women's position within the industries workforce.

The Computer Games Industries Current Climate

The games industry is one of the fastest growing sectors of the 21st century (Krotoski, 2004). More digital games are sold in the US and UK than books (Bryce and Rutter, 2001). In 2004, the industries worldwide worth stood at 20bln Euros for software and hardware (ISFE, 2004, see Krotoski, 2004). The estimated turnover of the UK computer games industry in 2008 was £625 million, having

Table 1. Game professionals UK salary by role and level, as of the 22nd January 2008

	Development Programmer	Artist	Designer	Producer
Junior	£18-25k	£18-21k	£18-21k	£18-27k
Middle	£20-35K	£20-27k	£20-28k	£28-40k
Senior	£35-50k	£27-35k	£25-35k	£35-60k
Lead	£40-60k	£38-55k	£38-55k	£45-80k

MCV (2008)

a direct contribution to UK GDP (Gross Domestic Product) of approximately £400 million (Oxford Economics, 2008). The UK industry supports an estimated 28,000 jobs, directly, indirectly and induced (Oxford Economics, 2008).

According to research and analysis of the US economy by IBISWorld, in 2008 the games industry (including development, production and retailing) was a $40 billion business. The Vancouver Film School (2009) suggests that in 2007 the worldwide games industries worth stood at 41.9 billion USD. According to the IBISWorld report, the industry is not expected to be hit by the current recession and it is expected to grow by 9.5% in 2009. The report also projects a yearly forecast thereafter of 10%, which is estimated to reach $63.2 billion by 2013 (IBISIWorld, 2008). A recent report by the Vancouver film school (VFS) in 2009, suggests the games industry is strong enough to survive the recession due to the growth in the industries software sales globally. The report suggests that in 2008 across the globe the games industry revenue figures in billions USD were; USA 21.3, Canada 2, Europe 17.9, Korea 1.7, China 7.9, Japan 6.3, Latin America 1 and India 0.2.

However findings in the UK may not be so optimistic. According to the Oxford Economics report (2008) despite the industries 10% per annum growth since 2004, the industries decline in the UK is a possibility. One of the major reasons the report suggests for this possible decline is due to the expansion of the industry in other regions such as Canada, China and Korea. The report goes on to suggest that the UK 's games development industry could be £350 million lower by the year 2013, with a loss of 10,000 jobs supported both directly and indirectly by the games industry.

According to the Oxford Economics report (2008) games development is part of the UK's 'knowledge industry' (p7). The high skills of the industries developers is reflected in the above average earnings of those employed in the industry, which the report suggests to be £30k per annum, compared to the UK average of £25k per annum. There is also evidence that a gender pay gap exists within the sector, in that women earn on average £7,126 per year less than men (Krotoski, 2004) and in America, according to IGDA (2004) women earn an average $9000 less a year than male counterparts.

In addition, research by Krotoski (2004) noted that less than half a percent of women in the UK games industry were in lead, director or management positions, whilst 1.2% of men were in such roles. Research by Haines (2004) found that almost a quarter (23%) of senior positions, within twenty UK games companies were filled by women. However, these were more often in managerial and marketing roles, rather than direct games development roles.

Average UK earnings in 2008 for games professionals ranged from 18k to 80k (see Table 1). Furthermore, in Canada according to the VFS (2009) report, game designers and artists earn $25,000 more than average Canadian salaries. The report went on to state that in 2007, salaries for game programmers and producers were double average earnings in Canada.

However, recent research by an online computer games magazine; reported very different findings, in that on average UK women in the games industry earn £2000 more than their male counterparts (MCV, 2009). The survey was based on 528 UK games industry professionals, from all sectors, including retail, publishing, HR, marketing and development. According to the MCV survey, women within the games industry were likely to be working in publishing, marketing, PR, retail and distribution services. MCV stated that only 6.9% of developmental roles were filled by women. This may explain the variation with other research.

A number of web 'blog' responses have been posted online in response to MCV's publication of salaries within the industry, a number of which were implying that the salary figures are untrue. It is interesting to note the response of a UK game developer and CEO, to a graduate who wrote a blog complaining about their junior role salary:

I'm the CEO of a UK developer. Starting salaries are as they are because it remains an attractive industry to join. Supply and demand. I'd set higher salaries if the right candidates weren't coming through.

Such statements reinforces the view point that the industry is an attractive or desirable industry to work in and therefore those that are involved in the developmental side of games are in it for the 'love of games', or a 'passion of games' as Consalvo (2008) refers to it, rather than purely for the money. A viewpoint the industry should take into account when looking at attracting and retaining a more diverse workforce.

Another interesting response from a different standpoint is the positive element of the research in highlighting salaries as opposed to keeping them secret:

Well done to MCV for doing this research, and making this data public. Developers in the UK are paid less than in the US, and there is a culture of 'salary secrecy' here that can undermine trust. We need to be more open about salaries - greater visibility.

The wider ICT industry, within the UK has been criticized for a gender pay gap (Tattersall, Keogh and Richardson, 2007), with one of the reasons put forward for this pay gap, due to what has been termed 'salary secrets'. An online blog also wrote of the industry having a culture of 'secret salaries'. Such practices are not equitable and do not assist in ensuring equity in the industry.

Furthermore, other research supports this viewpoint, for example the Directing Equality of Pay (DEPICT) in ICT (Keogh, Tattersall and Richardson, 2007) project ran from January 2005 to December 2006. The project was commissioned to explore the under representation of women in ICT, via the intended and unintended barriers within UK ICT pay systems. According to the DEPICT research, women in ICT earned 18% less than their male counterparts. The research found that due to the common practice of negotiated pay, bonuses and career advancement within ICT organizations, women were disadvantaged within the sector. Disadvantage was, in part attributed to women generally lacking the confidence and skills for negotiation. Keogh, Tattersall and Richardson (2007) went on to suggest a number of recommendations to assist in eradicating the pay gap, within the ICT sector. Recommendations focused on transparent pay systems, to eliminate 'salary secrets'. Promoting work life balance, to eradicate the 'long-hours' culture and encourage flexible working. Ensure transparent fair appraisals and remove salary negotiation. Value diversity, skills and potential and promote formal recruitment practices.

Women Working in the Computer Games Industry

In the UK women represent only 12% of the computer games industry workforce. This is a 4% increase from 2005 (Skillset, 2006). Similar percentages have found to exist in other countries, such as the USA, Canada and Australia (IGDA, 2005). In Canada recent figures suggest that women make up 20% of game industries workforce (VFS, 2009). An increase of 10% from 2007 and a 17% increase from 2004. The IGDA report found that the functions of writing, marketing/PR/sales and production have relatively healthy representation of females. However, male workers heavily dominate the majority of the core content creation roles. This supports earlier research conducted in the North West of England by Haines (2004), who found that the majority of women (73%) in the games industry work in managerial, administrative, marketing and PR roles. Haines research found that only 2% of programmers were female, 3% worked in audio, 5% were game designers, 8% production staff and 9% were artists. Women within the industry have little role in the content, interaction styles, character representation and the reward systems involved in games. This is crucial as it effects not only what is created, but how games are perceived by players (Flanagan, 2005).

Speaking at the 2007 Skillset Skills Week, Matthew Jeffery, Head of European Recruitment at Electronic Arts, noted that skills shortages were becoming more common in the gaming industry and measures had to be used to bring in 'more new talent'.

Games companies need to broaden out their recruiting scope and attract talent from other new industries and seduce more diverse groups into game teams, particularly women and ethnic minorities.

There have been a number of reasons suggested as to why there are so few women within the games industry and more specifically within a core creation role within the industry. For example, Fullerton, Fron, Pearce and Morie (2007) refer to the 'virtuous cycle' which is '*Making games that appeal to women and girls attracts more women to work on games, resulting in the creation of more games that appeal to women and girls*' (p141). Fulerton et al (2007) found that game design teams with more women in key design and production roles have produced products women enjoy. For example, the 'Sims' game is estimated to have a player base of approximately 50% females. The Sims development team has more women than average than other teams at Electronic Arts (EA) (Fullerton, Fron, Pearce and Morie, 2007).

Other research suggests that the portrayal of female avatars may be a significant contributing factor as to why the games industry may not appeal to women (Sheri Graner Ray, 2004). According to Graner Ray, if a company portrays female avatars in a demeaning way, potential applicants to that company may assume females will be treated in a demeaning way within the workplace. The sexual representation of females in games also contributes to the industry being viewed as for boys. Graner Ray goes on to say that 'booth babes', a term used to describe minimally dressed women at conference and shows and other sexually orientated themes, at professional games gatherings and conferences suggest the industry is for 'boys only' (p150).

There has been a paucity of research conducted on women who work in the games industry, with the exception of Haines (2004) and Krotoski (2004). Through a survey and interviews with female game workers, Haines (2004) found that women in the industry wanted encouragement and awareness for young girls to enter the industry and develop a career in gaming. Prescotts' (forthcoming) recent research confirms this viewpoint.

It is apparent that there are a number of issues which may impede women entering and remaining in the ICT and SET industries generally and the games industry specifically. However, other issues not necessarily specific to the gaming industry are relevant to consider in recruitment and retention. The next section will discuss work life balance, long hours and flexibility.

This chapter aims to evaluate the current position and experiences of women and issues of work life balance and flexible working, within the computer games industry and the wider context of the science, engineering and technology industry (SET). Research is discussed in the context of attracting and retaining women in the workforce of the future. Current findings from a larger quantitative study (Prescott, forthcoming) will provide the focus of the chapter in relation to these issues. The chapter will also add to the paucity of research on the games industries workforce, in particular women within the industry.

WORK LIFE BALANCE ISSUES

Work Life Balance, Long Hours and Flexible Working

Work life balance issues have become proliferate in the literature on women's careers and the career development of women. Women are still in the main responsible for the majority of domestic labour and childcare (Raskin, 2006; Bird and Schnurman-Cook, 2005; Wajcman and Martin; 2002; Blossfeld and Drobnic, 2001 and Simpson 1998). Even when women work from home, research has found that they tend to combine work with childcare (Noonan et al, 2007). The long hour's culture and a culture of presenteesim are important issues to consider when wanting to attract and retain women in the SET and ICT sector and the games industry specifically.

It has been suggested, that to show commitment to job and organisation there is a need for long hours and 'presenteesim' (Simpson, 1998). The long hours culture in the SET and ICT industries makes it more difficult for women and men with caring responsibilities (Gill, 2002; Perrons, 2003) and the industries have been accused of being modeled on men's rather than women's stereotypical lifestyles (Kelan, 2008). Working long hours is an issue throughout the ICT professions and is suggested as a major contributing factor for the low number of women within the sector (WWW-ICT report, European Commission, 2004). Research from western societies suggests that long hours are an issue across the sector. For example, Griffiths, Moore and Richardson (2007) through an online survey of 479 women in ICT across England found a strong long hours and presenteeism culture within the industry. Similarly, Diamond and Whitehouse (2007) through an examination of professional computing employment in Australia found a greater long hour's culture in the private sector and that part time work was a rarity in both sectors. They also found that men and women respond differently to workplace constraints. For example, females with children were more likely to move into areas where hours were more predictable. This suggests that in order to maintain a work life balance, women were choosing a career sacrifice.

As with the SET and ICT industries, the games industry has also been identified as having a long hours culture. This is mainly due to the project based nature of game development and long hours are especially prevalent at 'crunch time'. Crunch time within the games industry refers to when a game is due to be released, it is a notorious time for extremely long hours. In particular, the games industry has been viewed as a workplace which tends to favor young and unattached males, due to its long hours culture and the potential need to relocate (Deuze, Martin and Allen, 2007). Furthermore, the long hour's culture within the industry has been acknowledged by an IGDA report (2004). The reports suggests that three out of five developers work 46 hours or more in a normal working week and workers can spend anywhere from 65 to 80 plus hours per week at 'crunch time'.

Recent international research by Prescott (forthcoming), with over 450 women working in the gaming industry found that 77% of women felt that a long hour's culture was prevalent in their organization (a similar finding regardless of developmental, non-developmental or academic roles), supporting previous research alluding to long hours in the games industry (Haines, 2004; Deuze, Martin and Allen, 2007). Prescott (forthcoming) found that more than two-thirds (68%) of women reported working up to 45 hours a week. This reported average working hours could be considered reasonable and are for example within UK and European Union working time directives or regulations. However, a third of women (32%) reported working considerably more hours, with 22% working 46-55 hours and 10% more than 56 hours per week. It may be the case that those working longer hours were in the middle of a 'crunch time' period and were basing average hours on their current experience, as opposed to regular practice. How long such periods may last is variable and the impact on careers, work life balance and family issues is worthy of further exploration. This is important as Prescott (forthcoming) also noted that a significant proportion of women (44%) felt that the long hours worked impacted on their health and well being. In addition, 50% felt that hours worked had an impact on personal relationships.

Consalvo (2008) conducted in-depth qualitative interviews with 10 female game developers. The main findings of the research focused on 'crunch time' and long hours, with 'crunch time' in particular viewed as a significant challenge to women in the industry. Crunch time was thought to be ingrained in the work culture, with passion viewed as the main reason why they work the long hours. Consalvo suggests that women, like their male counterparts also have a passion for playing games. Consalvo viewed passion as problematic in that 'the ideal worker is constructed as someone possessing a 'passion for games', and that passion is used to help maintain work practices that may ultimately kill the passion' (p186). If this is the case, then for women with children or care responsibilities, there may be a passion for the work, but only limited hours. This may create conflict not only in work life balance, but also work based identity.

Prescott (forthcoming) noted that more than half of women (66%) reported that in their organization work life balance policies needed improvement and a quarter (26%) felt that work life balance was not considered important in their organizations. A lack of understanding or sympathy towards work life balance was also evident, with 63% of women feeling that colleagues were generally negative if they had to leave work for outside commitments. Importantly, the majority (85%) of women thought that work life balance was good for productivity and in order to recruit and retain women, employers must be willing to endorse and encourage initiatives such as organization cultural bias to long hours and encourage flexible working were possible.

Flexible Working

The Office for National Statistics (ONS, 2008), reports that women are more likely than men to work part-time, particularly if they have dependent children, 38% compared with 4% of men with dependent children. Despite women's increasing participation within the UK workforce, in 2008 almost half of women's jobs were part- time compared with approximately one in six of men's (ONS, 2008). Working part-time is often viewed as a solution in managing the coexisting roles of work and motherhood for many women with dependent children. However, combining work with other roles is a major disadvantage for women in the workplace (Straehley and Longo, 2006). Furthermore, research suggests that career development is structured around male experiences, but should include the diversity of women's career experiences gained in paid work and include domestic labour in the home (Tracey and Nicholl, 2007). Hughes and Parkes (2007) found that individuals who had control over their working hours suffered less work-family conflict, despite long hours. The authors suggest that employers should provide employees with some flexibility and control over work hours to reduce the negative impact that long hours can have.

Valenduc et al (2004), found that women with children were rare in the ICT sector and suggests that due to a lack of women with families in the industry, there is no pressure to change working practices or workload. Certainly findings from recent research would support this (Prescott, forthcoming), the majority of women were younger (35 or under) and without children, only 6% of women worked part time. However part time work, often meant full time hours, one woman noted that "*I'm paid for part time, 32 hours...but it always runs into full time*". Change is needed, as Griffiths, Moore and Richardson (2007), have suggested, instead of women and older workers adapting to the ICT culture, the ICT industry should change and broaden its appeal to recruit and retain a diverse workforce. Change in working practice such as promoting part-time or flexible working is required. In addition, ensuring that the long hours culture is no longer viewed as the status quo and removing the ingrained masculine culture may lead to improved workforce retention, and greater job desirability in the sector to women.

Another issue of importance to women in professional careers is relocation, this is especially when women have partners and or children to consider. It is often the case that women will relocate for their husbands, or partners careers, yet it is rarer for husbands or partners in a relationship to relocate for the woman (Bogg, 2007). This ability or willingness to relocate is considered a potential factor as to why the games industry is dominated by men and not as appealing to women (Deuze, Martin and Allen, 2007). Prescott (forthcoming) found that over a quarter of women (27%) stated that they would not relocate and 23% were unsure if they would relocate. This may be due to workforce loyalty, a fear of not obtaining similar game development work, what Consalvo (2008) refers to as a 'passion for games' (p186).

Relocation is currently an important issue within the games industry. In 2008, it was reported (Oxford Economics) that the UK games industry could face a downturn which would mean '*a permanent loss to the UK economy as the highly skilled and specialised game developers are most likely to be actively recruited by foreign companies as UK based opportunities disappear'* (p29). In addition, other countries successful in game development i.e. Canada, have recently been actively and successfully recruiting staff, from countries such as the UK (Oxford Economics, 2008). A decline in the UK industry is currently occurring and there is concern in the industry for job losses, but also a drive to retain the most talented developers (Frear, Develop magazine, February 25, 2009). This has positive implications for work life balance, as in order to recruit and retain the most talented employees, working practices will need reviewing.

Relocation will be a viable solution for some current UK game development employees, as opposed to the alternative of finding a new career. This could have a dramatic impact on the careers of many of the UK games industry employees, especially females and in particular those with families. Such a change in workforce demographics could lead to the feminization of the UK games development workforce, an event occurring in some other professions, such as medicine (Levinson and Lurie, 2004). Therefore, although country relocation may not be possible for many women with families or care responsibilities, the event may lead to an improvement in UK working practices, as the industry strives to recruit and retain a workforce that could be increasingly female and include more women who choose to have families. However, with the current UK and global economic downturn, its impact, both positive and negative on recruitment, retention, working practices and salaries is inevitable.

Recommendations

The current chapter has discussed a number of issues with the main focus being on work life balance. Major issues reported in the research in this area include the long hour's culture associated with the industry and the lack of flexible working practices. Current international research by Prescott (forthcoming) has highlighted that three quarters of women (77%) reported a long hour's culture in their organization and that a third (33%) were unhappy with their current work life balance. Women who are the primary carers, such as those with children, generally require predictability (mainly regular hours) in order to plan childcare. Improvement in working conditions may lead to women choosing to remain in the industry (for example after starting a family), a benefit of this would be that having a greater number of women in games development may also have the added benefit of enhancing the image of the games industry, as a potential career for young women.

Long hours and 'crunch time' are important issues within the games industry and reducing their impact on employees will provide benefits to the organization. For example, organizations could incorporate project management training for leads and managers, to yield greater effectiveness of project timelines and targets. Crunch time would be difficult to eradicate completely due to restricted budgets, last minute problems and deadlines. However, by effective project management training, employee work-life balance would be enhanced. It is suggested that the long hours associated with the role of games industry, become perceived as 'occasional', as they would often be in many professional roles, as opposed to 'regular'. Whilst individuals who have a passion for games and gaming enjoy their jobs, there is a growing recognition that achieving a work-life balance is of greater importance for many individuals.

Flexible working practices should be encouraged. This would not only improve the image of the industry as a 'family friendly' working environment, but would also assist in retaining more women, especially women with or considering having children. If more women with families are visible in the industry, this in turn, sends out positive messages that the industry is not just for the 'boys' and that women, and those with families can be successful and have a viable career in the games industry.

FUTURE RESEARCH DIRECTIONS

Due to the relatively recent emergence of the games industry, there are a number of possible directions for future research. Future research will not only inform recruitment, retention and progression in the industry, but also the ICT and SET sectors. Interesting questions include why are some women attracted to the industry and what makes them stay? Does the industry appeal to a certain type of women in particular?

Although research has started to look into these questions, i.e. Consalvo (2008), it would be beneficial to evaluate the career trajectories of men and women in the industry generally and also, how does gaming differ from other creative industries, such as film and television, in terms of the recruitment, progression and retention of women.

The issue of employees aging in the industry also requires attention, Prescott (forthcoming) reported that more than two-thirds (67%) of women in her study were under 35. Will employees choose to remain in a (often) perceived, 'young' industry? or will there be a constant turnover of employees (men and women)? As the industry ages, will the demographics of women within the industry change in terms of ages, children and care responsibilities? If more women within the industry have children, how will attitudes towards relocation and work life balance change? Are women in the industry not having children due to their careers, or are they postponing child bearing until their career established? In regards to flexible working, it would be interesting and extremely useful in terms of policy development and implementation to evaluate women who do work flexibly and the impact on their careers.

This chapter has highlighted a number of emerging themes, issues and questions for discussion that are of relevance to employers, professional bodies, policy makers and researchers. Further research will benefit the computer games industry and the wider ICT and SET sectors.

CONCLUSION

This chapter has focused on the current position and experiences of women working within the computer games industry, the ICT sector and the wider context of the science, engineering and technology industry (SET). Recent global data on women working in the computer games industry was discussed in relation to the long hour's culture associated with the sector, the lack of females, especially females with families, skills shortages, work life balance and flexible working.

The games industry not only needs to improve its appeal to a more diverse workforce due to the current skills shortage it also wants to improve its appeal to a more diverse workforce. The aim of a more diverse workforce within the industry has the potential of making games with more appeal to wider and more diverse audience. The current chapter has highlighted a number of areas were the games industry could significantly improve its appeal as a viable career to women. The image of the games industry like the wider ICT and SET industries is still very much 'boys work'. However, this image could change with a more diverse workforce and with an increasing number of female gamers worldwide, the industry in general may lose its 'for boys only', masculine image. The industry needs to widen its appeal through an increased awareness of the variety of roles and skills within the industry and highlighting the benefits and rewards of working in such a creative, competitive and growing industry.

Potential further research areas were suggested, as a method of contributing to the growing body of evidence on the experiences of women, working in male dominated professions. Understanding the experience of women working in the male dominated computer games industry has been highlighted as important in order to improve future recruitment and retention and the appeal of the industry generally.

REFERENCES

Adya, M., & Kaiser, K. M. (2005). Early determinants of women in the IT workforce: a model of girls' career choices. *Information Technology & People*, *18*(3), 230–259. doi:10.1108/09593840510615860

Bird, G. W., & Schnurman-Crook, A. (2005). Professional identity and coping behaviors in dual-career couples. *Family Relations*, *54*(1), 145–160. doi:10.1111/j.0197-6664.2005.00012.x

Blossfeld, H.-P., & Drobnic, S. (2001). Careers of couples in contemporary societies. Oxford, UK: Oxford University Press.

Bogg, J. (2007). Dr Jekyll and Ms. Hide: Where are the women in science? And what would attract them from other sectors? *Nature*, 447.

Bryce, J., & Rutter, J. (2003). The gendering of computer gaming: experience and space. In S. Fleming, & I. Jones (Eds.), Leisure cultures: investigations in sport, media and technology, (pp. 3-22). Eastbourne, UK: Leisure Studies Association.

Burns, B., Griffiths, M., Moore, K., & Richardson, H. (2007). Disappearing women: North West ICT final report. Manchester, UK: Salford University.

Consalvo, M. (2008). Crunched by passion: women game developers and workplace challenges. In H. Kafai, Denner & Sun (Eds.), Beyond barbie and mortal kombat: new perspectives on gender and gaming (pp. 177-192). Cambridge, MA: The MIT Press.

Current Population Survey (CPS). (2007). *US Labor force statistics*. Retrieved 20/02/09 from http://www.bls.gov/cps/wlf-databook2008.htm

Deuze, M., Martin, C.B., & Alen, C. (2007). The professional identity of game-workers. *Convergence: The international journal of research into new media technologies, 13*(4), 335-353.

Diamond, C., & Whitehouse, G. (2007). Gender, computing and the organization of working time: Public/private comparisons in the Australian context. *Information Communication and Society, 10*(3), 320–337. doi:10.1080/13691180701409879

European Commission. (2004). Towards a European Research Area: science, technology and innovation, Key figures 2004 Eurostat. Brussels, Belgium: Author.

European Commission. (2004). Widening Women's Work in Information and Communication Technology. WWW-ICT. Brussels, Belgium: Author.

Flanagan, M. (2005). Troubling 'games for girls': notes from the edge of game design. In *DiGRA 2005 Conference: Changing views-worlds in play*.

Frear, E. (2009, February 25th). The Career Crunch. *Develop Magazine*.

Fullerton, T., Fron, J., Pearce, C., & Morie, J. (2007). Getting girls into the games: Towards a 'virtuous cycle'. In H. Kafai, Denner & Sun, (Eds.), Beyond Barbie and Mortal Kombat: New perspectives on gender and computer games. Cambridge, MA: MIT.

Gill, R. (2002). Cool, creative and egalitarian? exploring gender in project-based new media work in Europe. *Information Communication and Society, 5*(1), 70–89. doi:10.1080/13691180110117668

Graner Ray, S. (2004). Gender inclusive game design: expanding the market. Hingham, MA: Charles River Media Inc.

Griffiths, M., Moore, K., & Richardson, H. (2007). Celebrating heterogeneity?: a survey of female ICT professionals in England. *Information Communication and Society, 10*(3), 338–357. doi:10.1080/13691180701409945

Haines, L. (2004, September). Why are there so few women in games? *Research for Media Training North West.*

Hughes, E. L., & Parkes, K. R. (2007). Work hours and well-being: the roles of work-time control and work-family interference. *Work and Stress, 21*(3), 264–278. doi:10.1080/02678370701667242

IBISWorld press release. (2008). The new American players: baby boomers and women take on video gaming. Retrieved 18/08/2008 from http://www.ibisworld.com/pressrelease/pressrelease.aspx?prid=133

IGDA. (2004). *Quality of life white paper*. International Game Developers Association.

IGDA. (2005). Game developers demographics: An exploration of workforce diversity. International Game Developers Association.

Kelan, E. K. (2008). Emotions in a rational profession: the gendering of skills in ICT work. *Gender, Work and Organization, 15*(1), 49–71.

Keogh, C., Tattersall, K., & Richardson, H. (2007). The directing equal pay in ICT (DEPICT) final report. European Social Fund and University of Salford.

Krotoski, A. (2004). *Chicks and joysticks: An exploration of women and gaming*. Entertainment and Leisure Software Publishers Association (ELSPA) White paper.

Levinson, W., & Lurie, N. (2004). When most doctors are women: what lies ahead? *Annals of Internal Medicine*, *21*, 471–474.

MCV. (2008). *Recruitment special: Industry salary survey.* Retrieved 26th January 2008 from http://www.mcvuk.com/news/29399/Industry-salary-survey

MCV. (2009). *Women earn more than men in the UK games industry.* Retrieved 03/03/2009 from http://www.mcvuk.com/news/32964/Women-earn-more-than-men-in-the-UK-games-industry

Newell, H., & Dopson, S. (1996). Muddle in the middle: organizational restructuring and middle management careers. *Personnel Review*, *25*(4), 4–20. doi:10.1108/00483489610123191

Noonan, M. C., Estes, S. B., & Glass, J. L. (2007). Do workplace flexibility policies influence time spent in domestic labour? *Journal of Family Issues*, *28*(2), 263–288. doi:10.1177/0192513X06292703

Office for National Statistics. (2007). Labour Force Survey: 2007. London: Office for National Statistics.

Office for National Statistics. (2008, September 26th). *More women in work but half in part time jobs*. London: Office for National Statistics.

Oxford Economics. (2008). The economic contribution of the UK games industry: final report. Oxford, UK: Oxford Economics.

Perrons, D. (2003). The new economy and the work-life balance: conceptual explorations and a case study of new media. *Gender, Work and Organization*, *10*(1), 65–93. doi:10.1111/1468-0432.00004

Prescott, J. (forthcoming). *Women in the computer games industry*. Unpublished Ph.D thesis, University of Liverpool, UK.

Raskin, P. M. (2006). Women, Work and Family: Three studies of roles and identity among working mothers. *The American Behavioral Scientist*, *49*(10), 1354–1381. doi:10.1177/0002764206286560

Retrieved 04/03/2009 from http://www.skillset.org/uploads/pdf/asset_9920.pdf?6

Ruberg, B. (2009 February 20th). *The Gamasutra 20: Top Game Writers*. Retrieved 27/02/2009 from www.gamasutra.com

SET. (2009). *UK resource centre for women in science, engineering and technology, requested dataset.* Retrieved from www.ukrc4setwomen.org.uk

Simpson, R. (1998). Presenteeism, power and organisational change: long hours as a career barrier and the impact on the working lives of women managers. British Journal of Management communication quarterly, 9(September), 37-50.

Skillset. (2006). *Census 2006: the results of the sixth census of the audio visual industries*.

Skillset. (2006). *Skillset: Workforce Survey 2006*. London: The Sector Skills Council for the Audio Visual Industries.

Skillset. (2007, October 10th). *Game boss calls for more ethnic minorities and women to combat skills shortages.* Press release. Retrieved 15/02/09 from http://www.skillset.org/skillset/press/releases/article_6286_1.asp

Straehley, C. J., & Longo, P. (2006). Family issues affecting women in medicine, particularly women surgeons. *American Journal of Surgery*, *192*, 695–698. doi:10.1016/j.amjsurg.2006.04.005

Tattersall, A., Keogh, C., & Richardson, H. (2007). The gender pay gap in the ICT industry. University of Salford, UK.

Tracey, C., & Nicholl, H. (2007). The multifaceted influence of gender in career progress in nursing. *Journal of Nursing Management, 15*(7), 677–682. doi:10.1111/j.1365-2934.2006.00677.x

Trauth, E. M. (2002). Odd girl out: an individual differences perspective on women in the IT profession. *Information Technology & People, 15*(2), 98–118. doi:10.1108/09593840210430552

UCAS. (2007). *Online datasets.* Retrieved 04/03/2009 from http://www.ucas.ac.uk/about_us/stat_services/stats_online/annual_datasets_to_download/

Valenduc, G., Vendramin, P., Guffens, C., Ponzellini, A.M., Lebano, A., D'Ouville L., et al. (2004, July). Widening Women's Work in Information and Communication Technologies, final synthesis report. *European Commission (IST-2001-34520).*

Vancouver Film School. (2009) The game industry: now and in the future. VFS Report.

Von Hellens, A. L., & Nielson, S. H. (2001). Australian women in IT. *Communications of the ACM, 44*(7), 46–52. doi:10.1145/379300.379310

Wajcman, J., & Martin, B. (2002). Narratives of Identity in Modern Management: The Corrosion of Gender Difference? *Social Compass, 36*(4), 985–1002.

ADDITIONAL READING

Women and Computer Games

Beasley, B., & Standley, T. C. (2002). Shirts vs. skins: clothing as an indicator of gender role stereotyping in video games. *Mass Communication & Society, 5*(3), 279–293. doi:10.1207/S15327825MCS0503_3

Bonanno, P., & Kommers, P. A. M. (2008). Exploring the influence of gender and gaming competence on attitudes towards using instructional games. *British Journal of Educational Technology, 39*(1), 97–109.

Clegg, S., & Trayhurn, D. (2000). Gender and computing: not the same old problem. *British Educational Research Journal, 26*(1), 75–89. doi:10.1080/014119200109525

Clegg, S., & Trayhurn, D. (2000). Gender and computing: not the same old problem. *British Educational Research Journal, 26*(1), 75–89. doi:10.1080/014119200109525

Crawford, G., & Gosling, V. (2005). Toys for boys? women's marginalization and participation as digital gamers. *Sociological Research Online, 10*(1), 1–13. doi:10.5153/sro.1024

Cunningham, S. (2000). Re-inventing the introductory computer graphics course: providing tools for a wider audience. *Computers & Graphics, 24*(2), 293–296. doi:10.1016/S0097-8493(99)00164-8

Dietz, T. L. (1998). An examination of violence and gender role portrayals in video games: implications for gender socialization and aggressive behaviour. *Sex Roles, 38*, 425–442. doi:10.1023/A:1018709905920

Divinch, J. (2008). "The divinich tapes: females representation in games across genres, consoles." www.gamasutra.com accessed 18/02/2009.

Heeter, C., Egidio, R., Mishra, P., Winn, B., & Winn, J. (2009). Alien Games: do girls prefer games designed by girls? *Games and Culture, 4*(1), 74–100. doi:10.1177/1555412008325481

Heeter, C., Egidio, R., Mishra, P., Winn, B., & Winn, J. (2009). Alien Games: do girls prefer games designed by girls? *Games and Culture, 4*(1), 74–100. doi:10.1177/1555412008325481

Kafai, Y. B., Heeter, C., Denner, J., & Sun, J. Y. (2008) Beyond Barbie and mortal kombat: new perspectives on gender and gaming. Massachusetts, London, The MIT Press

Mercier, E. M., Barron, B., & O'Connor, K. M. (2006). Images of self and others as computer users: the role of gender and experience. *Journal of Computer Assisted Learning*, *22*, 335–348. doi:10.1111/j.1365-2729.2006.00182.x

Natale, M. J. (2002). The effect of a male-orientated computer gaming culture on careers in the computer industry. *Computers & Society*, *32*(2), 24–31. doi:10.1145/566522.566526

Royse, P., Lee, J., Undrahbuyan, B., Hopson, M., & Consalvo, M. (2007). Woman and games: technologies of the gendered self. *New Media & Society*, *9*, 555–576. doi:10.1177/1461444807080322

WORK LIFE BALANCE ISSUES

Cunningham, M. (2007). Influences of women's employment on the gendered division of household labour over the life course: evidence from a 31-year panel study. *Journal of Family Issues*, *28*(3), 422–444. doi:10.1177/0192513X06295198

De Ruijter, E., & Van der Lippe, T. (2007). Effects of job features on domestic outsourcing as a strategy for combining paid and domestic work. *Work and Occupations*, *34*(2), 205–230. doi:10.1177/0730888406296510

Hamilton, E. A., Gordon, J. R., & Whelan-Barry, K. S. (2006). Understanding the work life conflict of never married women, without children. *Women in Management Review*, *21*(5), 393–415. doi:10.1108/09649420610676208

Keene, J. R., & Reynolds, J. R. (2005). The job costs of family demands: Gender differences in negative family-to-work spillover. *Journal of Family Issues*, *26*, 275–299. doi:10.1177/0192513X04270219

Liff, S., & Ward, K. (2001). Distorted views through the glass ceiling: The construction of women's understandings of promotion and senior management positions. *Gender, Work and Organization*, *8*(1), 19–36. doi:10.1111/1468-0432.00120

Rapoport, R., Bailyn, L., Fletcher, J. K., & Pruitt, B. H. (2002). Beyond work-family balance: Advancing gender equity and workplace performance. San Francisco, Jossey Bass.

Shapiro, M., Ingols, C., & Blake-Beard, S. (2008). Confronting career double blinds: implications for women, organization, and career practitioners. *Journal of Career Development*, *34*(3), 309–333. doi:10.1177/0894845307311250

Skillset (2006). Online datasets. http://www.skillset.org

UCAS. (2007). Online datasets. http://www.ucas.ac.uk/about_us/stat_services/stats_online

Wallace, J. E., & Young, M. C. (2008). Parenthood and productivity: a study of demands, resources and family-friendly firms. *Journal of Vocational Behavior*, *72*, 110–122. doi:10.1016/j.jvb.2007.11.002

Wharton, A. M., & Blair-Loy, M. (2006). Long work hours and family life: a cross-national study of employee' concerns. *Journal of Family Issues*, *27*, 415–436. doi:10.1177/0192513X05282985

Wickham, J., Collins, G., Greco, L., & Browne, J. (2008). Individualization and equality: women's careers and organizational form. *Organization*, *15*(2), 211–231. doi:10.1177/1350508407086581

KEY TERMS AND DEFINITIONS

Flexible Working: Practice that facilitates a work life balance

Games Industry: Computer or video games industry

Work Life Balance: The balance of work and other life or role commitments

Long Hours: Culture of long hours that may be used to demonstrate career commitment

Profile 7
Science Career Spotlight:
In Search of New Vaccine for Cattle Tick

© 2008 Queensland Government Office of Women. Used with permission

Ala Lew-Tabor, Brisbane, Australia

Associate Professor Ala Lew-Tabor is the youngest child of a post WW2 polish migrant family who had high expectations in regards to their children's education. Ala's interest in science and solving problems were evident at an early age leading her to a career in scientific research. She has published over 90 peer reviewed publications and conference abstracts. Ala currently leads a \$3.2m investment for the development of a novel anti-cattle tick vaccine utilising a genomic reverse vaccinology approach. Her research interests include the utilisation of genome-based tools to develop novel diagnostic assays and vaccines to better manage endemic livestock diseases and to prevent the potential introduction of exotic diseases. As an Acting Senior Principal research scientist with Queensland's Primary Industries & Fisheries and an Adjunct Associate Professor with Murdoch University's Centre for Comparative Genomics, Ala has utilised genomics or molecular biology to develop practical solutions for Australia's livestock industries. Ala was the winner of

Copyright © 2010, IGI Global. Copying or distributing in print or electronic forms without written permission of IGI Global is prohibited.

the 2008 Smart Women Smart State Award for Science in the Public Sector and in 2009 a finalist for the Women in Technology Biotechnology Achievement Awards.

Introduction

Pliny the Elder (23-79 A.D.) "Ill-favoured ticks... the foulest and nastiest creatures that be"

A/Professor Ala Lew-Tabor leads a team of 20 eminently renowned scientists and students (50% of whom are women) in the search for a new vaccine for the scourge of the cattle industry—the cattle tick – the single biggest health issue for Australian northern cattle.

Costing the world more than $US2.5 billion annually in lost milk and beef production and restriction of animal movement, the community, economic and environmental impacts of cattle tick diseases cannot be understated. A previous vaccine is no longer available and cattle dips are costly and environmentally insensitive. Using Queensland's smart science, Ala's team is well on its way to discovering a new, innovative cattle tick vaccine with a potential international market worth $US185 million annually—building on Queensland's strength and leadership in solving issues associated with global tropical agriculture.

This is the largest tick research program in Australia—both for the level of funding and team 'fire power'. Ala's initiative has provided the bridge between academia and the provision of practical solutions for the beef industry. Ala's ability to initiate and drive others towards common goals not only requires her to be innovative and smart ... but also inspiring.

The Ticks

Ticks (types of acarids) and the fatal pathogens the ticks carry are present in many parts of the world with *Rhipicephalus (Boophilus) microplus*, the cattle tick, being the most widely distributed species and economically important. Unfortunately the 'tasty' beef breeds such as the cute dairy types (British and European breeds), are also most susceptible to ticks—not surviving heavy infestations unless treated. To manage ticks, industry treats the cattle with pesticides (acaricides)—both harmful to the environment as well as the welfare of the animal. In addition, ticks have developed resistance to most current acaricides and there is a market imperative to reduce chemical residues in both cattle and the environment. The tick-disease complex is the most important affecting world-wide livestock production, leading to severe economic losses in milk and beef production and restriction in traffic of animals costing >$US2.5 billion annually [DeCastro 1977 Vet Parasitol.71:77; McCosker 1979 Recent Adv Acarol 2:45].

The 'Livestock Revolution' is a global food model that predicts the consumption of meat in developing countries will grow by 2.8% per year between early 1990s and 2020 [Delgado et al. 1999 Intl Food Policy Institute, FAO; Delgado et al. 2002 World Brahman Congress]. Australia is the world's number one beef exporter, supplying 23% of the world beef trade and is well placed to supply this predicted increased beef demand to the Asian markets. In 2003, domestic and export sales of beef exceeded $9.6 billion, with Queensland contributing to at least one third of this total ($3.2 billion).

Tick infested areas include the coastal regions and across northern Australia. The economic benefits from reduced input costs and increased productivity due to a reduction in parasites, improved animal welfare and increased market access (due to decreased chemical residues) has been estimated at $98 million annually. The previous tick vaccine (Tick*GARD*®*PLUS*) required three or four booster shots per year thus was subsequently poorly adopted and is now no longer manufactured commercially. Cattle ticks and the diseases they carry still pose the single biggest health issue for the northern beef cattle industry. Industry has identified that a tick vaccine with a 12 month duration of immunity and 90% efficacy is a top priority for Australia.

The Science

A new anti-tick vaccine with an annual boost will ultimately benefit Queensland rural communities, the environment (due to decreased chemical residue), Australian beef export and our economy. This research program, co-ordinated by A/Prof Lew-Tabor, has brought together a team of scientists who are employing modern biotechnology to innovatively identify putative anti-tick vaccine candidates. Using tools such as functional genomics, proteomics and bioinformatics, Ala's team is aiming to develop a single annual vaccination for strong immunity to cattle ticks. Once protective antigens are identified, QPI&F together with the Beef CRC will be securing a commercial partner to develop and market the product.

Research Highlights

- Current *in vitro* screening of 42 novel immunogenic vaccine candidates from the initial access to 13,643 tick genes from the USDA
- Discovery of tick-specific genes and gene pathways – novel for Arthropods to date

Highlights of A/Prof Lew-Tabor's Innovation to Establish this Program

- Support from industry – Meat & Livestock Australia (MLA); AgForce, Queensland Dairy Organisation, CTMQ (Committee for Tick Management Queensland)
- Support from commercial companies – Schering Plough, Pfizer
- Support from producers – regular updates for beef industry representatives and producers
- Securing collaboration with the USDA and access to tick genome resources for the research program
- USDA partner successfully funded Organisation for Economic Co-Operation and Development (OECD) fellowship to join the Australian team on an extended visit
- Securing critical researchers to the team e.g. Dr Rodriguez Valle (Cuban Gavac™ tick vaccine)
- Economic analysis by Prof Garry Griffiths – successful new tick vaccine worth $98 million p.a. to Queensland's economy

Environmental Benefit

By decreasing pesticide contamination, due to the minimisation of cattle dipping, there are a number of flow-over environmental benefits including:

- improved health and welfare standards
- cleaner, more sustainable environment
- improved competitiveness for international trade – e.g. organic beef industry.

Pesticides or acaricides are used to treat cattle by killing the ticks. With the use of an effective vaccine, the treatment of cattle with toxic chemicals (resulting in human occupational health and safety issues and stress to cattle) to manage tick loads will be minimised. Decreased use of pesticides will also minimise pre-slaughter 'Withholding Periods' required after the application of certain tick dipping chemicals.

A vaccine to effectively control ticks will decrease the stress and debilitation on cattle due to tick infestations thus improving welfare and possibly meat quality. This will contribute to Queensland's priority goal of developing foods (meat and milk) produced under safe and ethical environments and reduce the risk on non-tariff trade barriers being raised against Australian beef and dairy products—also supporting the organic meat and milk market. Through increased vaccine usage and more strategic use of chemicals to control ticks, chemical residues in milk and meat will decrease thus improving food safety within the value chain. The potential impact of decreased chemical usage on the environment and cattle is not trivial given the ≥11.3 million beef and dairy cattle population in Queensland.

Economic or Business Viability

The economic costs of $2.5 billion p.a. world-wide due to cattle tick include:

- direct effects of the tick on cattle: loss of condition, anaemia and deaths, susceptibility to drought, damage to hides, slow growth rate
- added costs to manage tick-borne diseases (babesiosis and anaplasmosis)
- risk of market being compromised by unacceptable chemical residues in meat or milk
- cost of controlling ticks, increased stock handling, costs of acaricides, cost of government regulations.

In Australia, the cost due to the cattle tick has been estimated at $175 million p.a. in 2005 [Playford, MLA AHW.054A]. On average, acaricides account for 11% of the costs, additional labour for 35% and production losses and animal deaths for 32%. A quarantine barrier was established in Queensland for the cattle tick to halt its westward spread and this barrier is maintained at an annual cost of around $3.3 million. A similar boundary exists between NSW and Queensland to prevent the spread of ticks to NSW. It has been predicted that global climate changes will affect pests and diseases in Australia and will have significant impacts on control costs and productivity. Various models have indicated a possible increase in costs of $18 million to $193 million per year [Sutherst 1996]. Arthropod pests, with their high reproductive rates, short generation times, efficient dispersal rates and ability to adapt rapidly, will respond quickly to climate change. The international beef export trade is worth $9.6 billion p.a. to Australia. An effective vaccine would reduce the need for artificial quarantine barriers, increase productivity in endemic areas (due to less tick damage and cattle dipping), and decrease the impacts of tick-borne diseases.

Estimated economic benefits from a new vaccine are $98 million p.a. for Australia with a further international market potential of $US100-185 million [Prof Griffiths, Tick vaccine business case]. Queensland supplies over 30% of Australia's cattle exports. Improved management of ticks will increase the value of these exports and by decreasing the costs involved with tick management will increase profits by $98 million for the state's beef and dairy industries ($33 million short-term benefit and $66 million medium to long-term benefit). Specific economic benefit estimates are available in South American reports where a version of the Tick*GARD*®*PLUS* vaccine called Gavac™ has been marketed i.e. ticks cost $US23 and $US12 per head of cattle in Cuba and Brazil (165 million cattle), respectively. If only 10% of Brazilian cattle were treated with an improved vaccine, with only a 10% saving in tick costs, the benefit would be $US20 million annually. These estimates aggregated across all cattle in the wet tropics of Africa and Asia and the rest of Central and South America would imply a benefit of a conservative $US100m annually.

Community Benefit

Research programs that will enhance important industries such as beef will address the broader issue of the global food shortage and benefit the community by the improved ability to produce high quality meat. Specific community benefit from the implementation of a new tick vaccine will arise from the resulting environmental and financial benefits for beef producers, government authorities and the consumer. Improved tick management will lead to fewer restrictions on the movement of cattle between regions which will ensure improvements in the sustainability and economic viability of beef producers (5231 specialists in Queensland) in rural communities, and the rural and regional providers (transport, abattoirs etc.) involved in the industry. As far as assisting with the global food shortage, a healthy cow can provide milk and feed a family—more people in developing countries die from starvation than malaria or AIDS—thus better management of ticks will improve cattle health and ultimately human health, particularly in developing countries. In Australia and developed countries, decreasing the reliance on chemical dipping through the implementation of a successful tick vaccine program would specifically:

- reduce the costs associated with maintaining regulatory controls and the costs of research and policy-making due to less need for dipping policies (costing Queensland Government $3.3 million p.a.)
- reduce the chemical treatment regimes and decrease the cattle withholding times prior to export. Currently acaricide treated cattle can be with-held from slaughter (for human consumption) or export for up to 42 days, as is legislated by the Australian Pesticides and Veterinary Medicines Authority
- provide clean green food diminishing the potential of chemically contaminated beef or milk and thus meeting consumer demands for clean organic beef and milk
- improve occupational health and safety for producers
- 'potentially' provide a solution which could be leveraged to develop a dog vaccine against the scrub tick.

Sharing Information with the Community

A/Prof Lew-Tabor has been committed to inspiring young female scientists, communicating with granting agencies, industry and across disciplines, demonstrated by her media releases, interviews, conferences, producer forum presentations, student supervision and mentoring. Specifically:

- she provides annual tick research updates to the Cattle Tick Management Queensland comprised of representatives from AgForce Cattle, Queensland Biosecurity Advisory Council, Australia Livestock and Property Agents Assoc., Australian Meat Industry Council, Livestock Transporters Assoc., Saleyards Assoc., Queensland Dairyfarmers Organisation and QPI&F Principal Policy Officers (Cattle Tick Control and Eradication)
- presenting invited seminars at a variety of forums including *Women in Technology* (AustBiotech satellite meeting, 2007), university lectures, and Australian Veterinary Association meetings
- 6 publications and 19 conference abstracts arising directly from this research alone
- student mentoring and supervision (Tick vaccine project PhD student Ms Piper recently won 'Early Career Researcher Award')
- inspiring high school students to study science and agriculture by presenting to Year 10 female students (ZONTA International 2006) and by preparing workshops for high school students e.g. *Designer genes* (CRC Fresh Futures, 2007). [NOTE: there is a current 'drought' of science agriculture students]

Summary Statement

A/Prof Lew-Tabor engaged key international and national researchers, secured external funding from a number of sources and developed novel research approaches. As evidenced by her publication rate and successful external funding record, she has **innovatively** applied her experience and knowledge in molecular biology and pathogens to develop solutions to improve the management of livestock diseases. Although her Smart Women Smart State nomination focussed on her tick vaccine project, she has been instrumental in developing novel molecular diagnostic tools for improved management of bovine reproductive diseases and tick fever e.g. recently she successfully co-led an Australian Research Council Linkage grant proposal to undertake genome sequencing to develop improved tools to diagnose bovine venereal disease. In effect, she has applied her science background, engaged universities, funding agencies and commercial companies to undertake research to ultimately benefit Australia's agriculture and export markets, and reduce the impact of harmful pesticides on the environment and the food chain. Her initiative has provided 'the bridge' between academia and the provision of practical solutions for the beef industry.

Chapter 8
"We Don't Have the Key to the Executive Washroom": Women's Perceptions and Experiences of Promotion in Academia

Jessica Guth
Bradford University Law School, UK

Fran Wright[1]
Bradford University Law School, UK

ABSTRACT

This chapter reports on a pilot study looking at the progression of academic women at one UK University. The chapter focuses on the promotions process and criteria as one important issue emerging from that research. Earlier research has shown that women are less likely to break into institutional networks which allow them to access information not only on formal and objective promotion criteria but also on hidden criteria and the way the 'academic game' is played. One result of this is that some academic women may have an inaccurate view of promotion criteria and processes. At the university studied by the authors, the Human Resources department has sought to make the promotion process more transparent and, officially at least, it no longer depends purely upon research achievements. However, these changes will not necessarily result in easier progression for women academics. The authors' study confirms that there is still a mismatch between what women think the criteria for promotion are, what the formal criteria are and how those criteria actually operate. Reliance on incomplete or inaccurate information about promotion criteria, coupled other factors, such as women's reluctance to promote themselves actively and traditional barriers to promotion such as caring responsibilities, puts women at a disadvantage when they attempt to progress into more senior positions within universities. Reform of promotions procedures needs to look beyond re-writing the substantive criteria for promotion and look to improving understanding of what is involved.

DOI: 10.4018/978-1-61520-657-5.ch008

Copyright © 2010, IGI Global. Copying or distributing in print or electronic forms without written permission of IGI Global is prohibited.

INTRODUCTION AND BACKGROUND

This chapter is based on the findings of a small scale research study carried out in 2008. The study considered the progression of women in higher education and focused on academic staff at the University of Bradford in the United Kingdom. The research outlined the European and national law relevant to this area before examining the university's own policies in detail. The majority of the study then focused on the lived experience of academics at the University exploring issues around promotion and progression, work life balance, mentoring and perception of policies amongst others through in-depth empirical work comprised of semi-structured interviews with male and female academics employed at the University (n=30). In addition to the interviews, some basic analysis was also made of statistics provided by the university. While the study did not specifically concentrate on the SET (science, engineering, and technology) disciplines, many of our respondents did in fact fall in that category. We interviewed academics in the engineering design and technology department as well as colleagues from life sciences, informatics, psychology and the school of health. Out of a total of 30 respondents, 17 were from SET disciplines while the others came from disciplines such as law, management, social sciences and languages. However, the analysis of our data showed no differences in responses by discipline.

This chapter focuses on one of the key themes emerging from the research: the promotions process in Higher Education. The emphasis is on promotion within and between the Lecturer and Senior Lecturer grades. This is partly because it was those promotions that most of our respondents talked about and also because these decisions are made internally, whereas assessors from other universities are involved with promotions to Reader and Professor. Promotions criteria in Higher Education are supposed to be transparent and clear to those employed in relevant institutions. The Higher Education Role Analysis (HERA) used to define roles within universities in the UK is supposed to allow "employers [] to ensure their pay and grading structures are designed to recognize the value of roles and ensure equal pay for work of equal value" (Educational Competencies Consortium Ltd 2007 (ECC)). A national role analysis was carried out, resulting in the development of national role profiles in UK Higher Education. These can be mapped onto a single national pay spine to ensure fairness and equality across the sector as a whole. The University of Bradford has adopted the single pay spine and the associated role descriptors. However this research casts doubts on the transparency of criteria and role profiles and the extent to which they are made explicit to university staff and/or applied consistently. In addition, as Deem and Morley (2006) note "although this methodology may deliver equal pay for equal work, [...] it may also restrict promotion opportunities since these often now depend on moving to a new job rather than upgrading an existing one" (p190).

The move to a single pay spine might also explain why we detected no disciplinary differences within our data. Academics working in SET disciplines are subject to exactly the same promotions criteria as those working in the social sciences and humanities and as such they have very similar views of those criteria and processes. Nonetheless the empirical work presented here takes as its focus those interviews conducted with colleagues in the SET disciplines.

The chapter first considers academics' perceptions of what the promotion process involves and their experiences in planning for and applying for promotion. It then turns to the university administration's own explanation of what the promotions process requires and involves. These are discussed, along with some "hidden" criteria for promotion. We then consider the implications of some significant differences we found between participants' and managers' understanding of the promotions process. We will conclude that it is

Table 1. Respondents by gender

	Male	Female
Number	5	25
Percent	16.7%	83.3%

unclear whether these differences have their roots in a failure to properly educate women academics about what they need to do in order to achieve promotion or whether the perception of women academics is correct but not accurately reflected in the university's formal criteria. Another factor is that this study was carried out at a time when the university had only just implemented some major changes to the procedures for promotion (including re-naming it as regrading) and this may have impacted on participants' knowledge of the process and criteria. Whichever of these explanations is correct, our findings suggest that attempts to reduce the imbalance between male and female academics at senior levels within the university are unlikely to succeed.

METHODS

The research strategy of this project employed a qualitative socio-legal methodology. It involved the analysis of law and policy from institutional to European level. This analysis was then complemented with in-depth empirical work comprising of semi-structured interviews with male and female academics employed at the University.

The empirical worksought to gain in depth knowledge and understanding of women academics in the University of Bradford. Qualitative semi-structured interviews were carried out with male and female academic staff at the University of Bradford at all levels of an academic career (n =30). The interviews covered areas such as

- Background Data
- Career trajectory
- Reasons for choosing to work in academia
- Advantages and Disadvantages of life as an academic
- Future plans
- Perceived barriers to career progression
- Views on promotions criteria and processes
- Public perception of women academics
- Institutional culture

In order to protect the confidentiality of our respondents we are unable to provide a full profile of our sample. However, the details in Table 1 and Table 2 may be of interest.

We did not find any striking differences according to seniority in the answers given and the gender differences were also minimal with male and female respondents identifying similar issues in women's progression.

Analysis of the empirical data was carried out using the software package Nvivo7. Interviews, which mainly took place in the staff's offices and ranged from 45 minutes to 2 hours, were recorded and the recordings were transcribed and entered. The data was split (or coded) thematically as the

Table 2. Respondents by seniority

	Numbers			Percentage		
	Total	Male	Female	Total	Male	Female
Lecturer or below	16	1	15	53.3%	3.3%	50%
Senior Lecturer or equivalent	3		3	10%	0%	10%
Reader/Professor/Senior position	11	4	7	36.7%	13.3%	23.3%

themes arose from the interview data and was then analysed in accordance with those themes.

As a qualitative study this work did not seek to make representative claims or present statistics or statistically significant data. The figures highlighting the lack of senior academic women are well rehearsed elsewhere. Instead this study aimed to understand the experience of academics at the University of Bradford and then situate that experience within the wider literature. It provides a detailed case study which highlights areas for future qualitative and quantitative research.

WOMEN ACADEMICS AT THE UNIVERSITY OF BRADFORD

The Equality Act of 2006 places a general duty on all public authorities in England and Wales, when carrying out their functions, to eliminate discrimination and harassment that is unlawful under the Equal Pay Act 1970 and the Sex Discrimination Act 1975; and to promote equality of opportunity between men and women. The university appears, on paper at least, to take this duty seriously. 'Confronting inequality: Celebrating diversity' is one of The University of Bradford's strap lines. The university's equality policies and schemes reflect a formal commitment to promoting equality. The policies examined as part of this study were found to be comprehensive and well thought out. The main gender equality policy and scheme is well supported by a number of other policies dealing with issues such as harassment and bullying as well as maternity, paternity and adoption leave and flexible working.

However, this does not translate into equal number of male and female academics in senior positions or indeed into a perception of equality within the institution (Deem and Morley 2006). Perhaps unsurprisingly, statistics we were provided with by the University show that the most marked gender differences arise at the more senior level. Out of the 444 members of academic staff below professorial grade 43% (189) are female. At professorial level, however, 76% are male. At Senior Lecturer level the proportion of men is 65%, and at Reader 88%. At Lecturer level however the genders are almost evenly split with men making up 51% of the total. The figures broadly match those of other studies (Forster 2000). Ackers and Oliver (2005) note that "[an] inverse relationship exists between the level of feminisation and seniority as women fail to progress in science careers at an equivalent rate to their male peers" (p3).

In spite of Fisher's proclamation 17 years ago that "[n]o organisation these days can afford to waste valuable brain-power simply because it is wearing a skirt (Fisher, 1992, p. 46) the figures suggests that women are less successful than men in achieving promotion, first to Senior Lecturer level and then to more senior grades (Forster 2000; AUT 2004). The university supplied us with the figures for the promotions round held in 2007. More men (87) applied for promotion than women (59) and men were more likely to be successful. 74% of men who applied were granted a promotion whereas only 66% of women were successful.

The main purpose of this study was, however, not to conduct detailed statistical analysis but rather to gain an insight into the experience of staff at the institution. While it must of course be recognised that there are disciplinary differences even within the SET disciplines and that different factors are likely to affect people differently as they progress through their life course and career trajectory, a number of common themes emerged.

In order to get a picture of the attributes and characteristics that women academics valued in their colleagues and the standards they were measuring themselves against, we asked respondents about their 'ideal' academic. It is clear that there was no single picture of an ideal academic, and different people placed their emphasis on different skills. We also asked respondents about their own decision making around promotion and about their experiences of applying for promotion. We got a strong sense that academics across the university felt daunted by the complexity of the promotions process; although information is available, our respondents did not find it easy to understand and to navigate. The difficulties were possibly increased because of the recent introduction of new promotions systems and criteria. We also found significant differences in respondents' perception of what the promotions criteria are – these varied not just from school to school but also within schools and departments. We detected an apparent mismatch between what people thought the criteria were, what the criteria actually were and how the criteria then operated on the ground.

The next sections in this chapter will explain in more detail what academics said about promotion in their interviews. We have dealt separately with views about what is needed in order to be promoted, and with what happened when our respondents actually tried to get promoted.

WHAT WOMEN TOLD US ABOUT PROMOTIONS CRITERIA

Researchers have found that lack of information about promotions procedures and criteria are not confined to new academics (Metcalf et al 2005). Our study also found confusion about the criteria for promotion. This confusion was not limited to junior academics but can be found across the seniority spectrum. Many respondents expressed anxiety about what exactly was expected of them. Our respondent Andrea for example commented "*No, I don't [know what I need to become a senior lecturer], I know that's where I want to be, but I don't know what I need to do to be there...*"'.

Many interviewees felt that there was a mismatch between the work the university expected them to carry out and the work that they were actually rewarded for in terms of promotion and progression. So while the university's focus was on teaching, widening participation and student engagement and these activities generally took up a substantial amount of academics' time, the reward and promotions structure was thought to be focused mainly on research activity. This focus on research in itself has been seen in previous studies to put women at a disadvantage (Forster 2000) but most academics still regard research outputs as the most important criteria for promotion. The quotations given below illustrate the point.

Researcher: What do you think it is in your discipline that gets people promoted, what's the key thing that has to be on the CV?

Interviewee: Undoubtedly, papers in peer reviewed, excellent, well-read journals. [Michelle, Lecturer, Female]

What does the institution reward? It rewards researchers who can bring in big research grants. It doesn't particularly reward teachers for good teaching [...] because if you want a promotion to senior lecturer then you have to have a research track record [Shannon, Lecturer, Female]

Debate about progression in academia is ongoing particularly in the context of recognising and rewarding teaching and learning (Parker 2008, Young 2006, Collins and Palmer 2007Gibbs and Habershaw 2002). This debate was not lost on our respondents and some mentioned that promotions criteria were beginning to change or that the emphasis or weighting of them was shifting; some thought towards teaching, some towards managerial skills and others to a more rounded approach:

So at Professorial level I've no doubt that we have to move from a focus on research. Now we have changed [...] some of the criteria, we've changed it so it's more about teaching excellence, it's about knowledge transfer, KTs, consultancy, it is wider [Daniel, Senior, Male]

However, even though a shift in emphasis and the availability of different career paths through academia was recognised at very senior level and welcomed at the lower levels of seniority, it appears that few believed that this shift in emphasis was actually being translated into practice. Three crucial issues emerged. Firstly, managers, most often academics themselves, do not necessarily have up to date and reliable information about how to apply promotions criteria in individual situations and are thus unable to give clear and constructive advice. Secondly, academics who are unsure of what exactly the criteria are and cannot get clear guidance from their managers, are unable to work strategically towards achieving those criteria or may believe they do not meet the criteria when in fact they do. Thirdly, as well as the criteria set down in HERA or those identified at departmental or school level, there are additional hidden criteria which operate in parallel with the formal criteria but which are far less transparent. These issues are explored further below but the quotation below highlights the problem women academics face:

There's a phenomenal repeat message from the females about how they went for promotion and they have the most research and you know, everything else they outperform guy X and guy X got the job and they asked for feedback and none of them got feedback. So I think the criteria might be different or the experience of the criteria might be different between the male academic and female [Nicole, Junior, Female].

What Women Told Us about their Experiences of Preparing for and Applying for Promotion

Researcher: How did you find the criteria?

I think it's still written in gobbledegook, I'm a plain speaker. I just find it really interesting that on the one hand the institution is saying that they want people to apply for a promotion, they advertise that this is going to happen. The information session run by the institution was a week before submission. So those 2 things do not go hand in hand. So they're saying one thing, when they actually mean something totally different. And that's my impression. If you really, really want to support people to do something you don't give them the information a week before, the amount of evidence that they want needs to go in, even down to when you've got to write your list of publications. We use Harvard referencing here, that's not Harvard referencing, so I had to redo it all. [Shannon, Lecturer, Female]

The personnel website of the University of Bradford devotes quite a significant amount of web space to matters related to promotions. Staff can access detailed information about the pay and grading structure, role descriptors and the promotions process and timetable. Nonetheless our respondents did not necessarily feel that they were well informed about the process and the applicable criteria, nor did they feel confident in going through it. There was a common perception that the promotions process was time consuming and involved complex paperwork. A number of our respondents actually felt unable to spare the time from the 'day job' to complete the required paperwork and those who did often found the process stressful. Sarah's comments encapsulate the feedback we received on the process: "*It was hard work, bureaucratic, lots of box ticking*". [Sarah, Lecturer, Female]

Michelle touches on a further issue which many respondents raised:

I don't want to want to be knocked back; I don't want to put a lot of time into something that I can see is going to be knocked back because I know I haven't got the quality publications recently [Michelle, Lecturer, Female]

She highlighted that she did not want to just 'have a go' at going for promotion but rather wanted to be sure that she met the criteria and would be successful when she applied. She did not want to go through being rejected or knocked back. Two related points emerge. In the view of some respondents, there was a gendered dimension to reluctance to submit a speculative application, and that men were less concerned about rejection and more comfortable with self-promotion (Bagilhole and Goode 2001). ' *I do think women are less willing to put themselves forward maybe because they're afraid of being hurt, you know'* [Stephanie, Lecturer, Female].

Further, those who were aware of the promotion criteria, for the most part felt it difficult to relate them to the work that they did or were frustrated with the application paperwork because it did not allow them to accurately portray what they felt was important. The criteria, said our respondents, are based on process and form and are *"a very traditional way of looking at it"*. [Elizabeth, Senior Lecturer, Female]. The application form used applied a restrictive view of the promotions criteria and thus did not allow applicants to portray *"a full enough or a round enough description of [themselves] and all the things [they] were doing"*. [Sarah, Lecturer, Female]

THE UNIVERSITY'S FORMAL PROMOTIONS/ REGRADING CRITERIA

The University has two distinct procedures for promotion, one applying to promotions within and between the Lecturer and Senior Lecturer roles and one applying to promotions to Reader and Professor. The focus of this section is on the former.

The University of Bradford has recently overhauled its promotions procedures. All staff have been placed on a single pay spine, over 260 role at the university evaluated, and new role descriptors introduced. This is part of a national initiative for pay modernisation. The new system was used for the first time in 2007. Most of the respondents interviewed for this research had not yet applied for promotion under the new system, but some had and others commented on it more generally. Given that one of the themes we detected in our interviews was lack of awareness of promotions procedures, the fact that many respondents failed to comment on the new process was instructive in itself.

The new procedure is initiated when the academic and their manager agree, at an annual Performance Review meeting, that the academic's "role" has undergone significant development, i.e., the academic has taken on additional responsibilities or developed new skills. (Regrading is not concerned with *potential* to perform at a higher level: unless that higher level has already been achieved, the position cannot be regraded.) The academic and manager then agree a new job description for the role. If the manager feels that the role has developed to the level of a higher grade, as defined by the HERA role descriptor, the new job description is forwarded to the Dean/Head of Planning Unit for approval and the manager submits an application to the School/Planning Unit Annual Promotions committee.

As can be seen, the academic's line manager has a central role in the promotions or regarding process. This is the person who carries out annual Performance Reviews, at which targets for further development are set. The line manager is also often the person who allocates administrative responsibilities. This is important because not all criteria for regrading can be satisfied simply by doing one's existing role well. If the role descriptor for a higher grade requires evidence of successfully completing a particular task, then the academic needs the opportunity to take on that task. However, in setting targets or planning an individual's future development, the manager is not concerned purely with the personal and career development of the academic: according to the document "Promotions Exercise 2008",

It is for management to determine how roles are to be developed in the context of strategic and operational requirements within Schools/Planning Units and across the institution. The potential development of any role, along with any associated development plans, should, therefore, be discussed in that context during the performance review (University of Bradford 2008 p1 emphasis added).

While formally there is no "rationing" of regrading, the operational needs of a particular school or planning unit are likely to have this result. If there is no need for more staff to carry out the higher level tasks required for the higher grade, then a staff member will not receive the opportunity to work at that higher grade and therefore will not be regraded.

The support of the manager is also required for any application to reach the School promotions committee: "All evidence submitted in support of a case must be verified by the appropriate line manager. No case for promotion will be considered unless the supporting evidence submitted has been verified" (University of Bradford 2008 p2). If the School committee approves the application, it is put forward by the School to the University committee. Self-submission to the School/Planning Unit committee is not possible but self-submission to the University committee is permitted where the School/Planning Unit committee does not approve the application.. The final decision on promotion is made by the University committee. The manager is therefore a gatekeeper for development and promotion opportunities.

Our findings echo those of Deem and Morley (2006) one of whose respondents articulates well what many respondents told us:

My experience of going through the promotion process... flushed out an awful lot of issues to do with gender... if you wanted to be promoted... you needed... managerial profile roles which were significant... the head of department decided who was going to get those roles... and that... operated as a way of blocking women from being promoted unofficially (Female academic, Cityscape) (p191).

Our respondents regarded publication of research as critical to promotion but it is not clear that the formal criteria in the role descriptors actually give research this central role. There are different HERA role descriptors that are intended to capture different academic career paths – that is, research focussed, administration focussed and all-rounders. However, all role descriptors are written in a very general way and are not discipline specific, let alone lecturer-specific. The result is language that seems strained as a description of what a lecturer or senior lecturer actually does. One of the elements in the HERA role descriptor for grade 9 (lecturer) and grade 10 (senior Lecturers) academics is "service delivery". At grade 9, the "[m]ain focus of role is responding to the needs of others and delivering an agreed service", with the academic "on occasions exploring the needs of students and others and adapting services to meet them." At grade 10, the focus changes to "being proactive in developing the service by approaching customers, exploring needs, developing new services" and the academic "will have responsibility and may be the lead role, for contributing to setting standards or determining the quality of the service to be provided". There are references to research, but there is no reference to *publication* of research at all. The closest that the grade 9 description comes to the perceived all-important requirement of publication, is that "Role holders will be conducting individual and/or collaborative research or scholarly projects and applying knowledge from projects to teaching/external activities." At grade 10, "role holders will be generating new approaches, developing new insights and contributing to the development of knowledge and practice in own field/educational strategies". If their focus is research, they will also be "principal investigator or collaborating on research projects and contributing to the development of research strategies". This language is far removed from the descriptions of what is required we heard from our respondents.

This suggests either that the respondents were out of date in their perception of what promotion involved or that the "official" criteria did not tell the whole story and that, in reality, research remained the main route to advancement, regardless of what the documentation stated. In either case, doubt is cast upon the transparency of the promotions process.

HIDDEN CRITERIA AND OTHER BARRIERS TO PROMOTION

As well as the formal criteria there are informal or hidden criteria which seem to be influence reward and promotions within the university. These criteria are subtle and relate to issues around expected workloads, value attributed to certain approaches to work and the importance of networking. We will consider these issues in turn.

Working Time and Work Life Balance

The long hours culture is a health and safety issue and is not consistent with a work life balance. It may also impact on progression: if conventions about how much work is required in order to advance within the university assume that long hours will be worked, employees who are either unwilling or unable to work long hours may be less likely to progress their careers.

Work load and the difficulties of achieving a satisfactory work life balance concerned our respondents, most of whom felt that it was impossible to carry out an academic job in a standard working week of around 40 hours.

I'm not sure it is possible, not if you want good research and good teaching. Ones got to give if you stick to 40 hours a week [Andrea, Lecturer, Female]

Studies looking at SET disciplines have long noted the high number of hours worked by scientists (Ackers 2003, Ackers and Oliver 2005) and a long hours culture was also prevalent within the institution studied. This was felt most keenly by those respondents with caring responsibilities. Respondents who managed to work a "standard" working week were few and far between. Moreover, many failed to take all the annual holidays or statutory holidays to which they were entitled. A number referred to coming to work when sick, in order to keep up with their workload. These findings are consistent with those of other studies. Ackers and Oliver (2005) make the point succinctly: "Scientists with children and mothers in particular, are less able to commit these kinds of hours and achieve an equivalent level of productivity" (p3). Caring responsibilities, either in relation to children, partners or other family members clearly do have an impact on how academics can manage their work and family life and as Forster (2000) points out "[t]he arrival of young children often coincides with the age when [scientists] are still expected to make an impact in their chosen fields through high quality research and a regular output of publications" (p319). Their interviews also confirmed "[t]he persistence of gender differences in the proportion of time spent on forms of unpaid/family work in the home restricts the ability of scientists with families, and women in particular, to devote a similar amount of time to their research"(Ackers and Oliver 2005 p3). While things are clearly better than they were in the past and caring responsibilities are being recognised more, it seems there is some way to go before academic institutions can be considered a family friendly place to work.

Gendered Approaches to Work

As various theorists have noted, there is a normative masculinist culture within academia that can disadvantage women (Wolffensberger, 1993). Forster (2000) talks of a 'deeply ingrained male view of performance' (p316) and Bagilhole (1993) notes that 'Universities are prime examples of Lipman-Blumen's (1976) ``homo-social'' institutions, being established and run by men. From this it follows that the rules pertaining to appointment are male driven and are evaluated according to male standards' (p 447). We were therefore interested in comments about a gendered approach to career planning

Many respondents agreed that it was too much of a generalisation to say that men viewed work and progression in a certain way while women did so in another. At the same time, many identified a gendered dimension in the way that the way academics approach their career:

I mean I don't want to make universalising statements about anything you know, but I think, I think men just tend to be more career focused and are having to think the next 2 or 3 steps ahead really, and I'm not saying that no women do that but I think women perhaps tend to be more invested in the job that they're doing at the moment and have perhaps more of a sense of responsibility. I know have a huge sense of responsibility towards students. [Karen, Lecturer, Female]

Some respondents commented that men were generally better at protecting their research time and were less likely to spend significant amounts of time devoted to pastoral care and the 'caring side' of teaching and learning (Martínez Alemán 2008). The quotation below gives an indication of the value assigned to these kinds of roles by the institution.

Women are much more open to doing that pastoral role, naturally, than men, they are perceived as sympathetic, etc, and again the loading there is not advantageous... [Julie, Senior, Female]

Given that the respondents also thought that research was the key to promotion, they were likely to regard this as an approach to work that helped men obtain promotion and hindered women's efforts.

When asking the questions about gendered approaches to work and promotions we were very aware that we were inviting respondents to generalise and draw out stereotypes and we were surprised to find that in many cases these stereotypes seemed to still be translating into practice. Overall respondents agreed that there were male and female characteristics and that we all had both to a greater or lesser extent.

I still think there is maybe a masculine style and a feminine style and it doesn't matter whether you are a man or a woman, it's whether you adopt those things and there is probably a sliding kind of scale [Crystal, Senior, Female]

The question of male and female characteristics and approaches to work often led on to a discussion about management styles and the value of having more women at the top level. This is relevant to the question of promotion in various ways: it provides an indication of whether women academics were motivated to progress to senior positions, what that motivation was, and what they thought had helped other women achieve promotion. Our respondents were overwhelmingly of the opinion that just having more women in more senior positions was not the answer; the key was having the right women there. There was a feeling that those women who had made it to the top level had to act in what was considered a 'male' way to get there.

I think in order to get on you have to take on that patriarchal role. The majority of the senior management are still men which has an impact

on who is employed.[...]. I think that you do have to prove yourself more than your equal male. You have to be better at being a man than they are. [Elizabeth, Senior Lecturer, Female]

The answers we received in response to this set of questions were interesting for a number of reasons. They highlight the importance of different approaches to working and the operation of stereotypes or perceptions of skills and approaches which may determine which roles are allocated to certain members of staff. These roles then in turn may impact on progression. The interviews also highlighted that there are different approaches to work and that these may be gendered to a certain extent. The value then assigned to these approaches is what impacts on progression (Martínez Alemán 2008; Lapointe Terosky et al 2008). Roles perceived as 'female' such as pastoral care, personal tutoring, representation on equality committees or heavy first year teaching loads were assigned less value for promotions purposes than roles perceived as 'male' such as research committee memberships, links with outside organisations/ industry or postgraduate teaching roles. It seems, the more prestigious the role, the more likely it is to be seen as and thus filled by a male. Or conversely, if a role is seen as female and mostly filled by women it will lack recognition and prestige. This is clearly problematic if promotion truly is based almost entirely upon research, but it is still problematic if it is wrongly thought that it does. Women academics who believe their involvement in teaching and pastoral care is irrelevant to promotion may not apply for promotion. Line managers who share the view that teaching and pastoral care are unimportant may not support promotion applications by women who spend a lot of their time on teaching and pastoral care. Members of promotions committees looking at applications for regrading may be less willing to accept strong performance of a pastoral or teaching-related administrative role as evidence of the competencies required. The pervasiveness of gender stereotypes therefore has the potential to influence the operation of promotions criteria quite independently of the content of those criteria themselves.

The Old Boys Networks – Alive and Well?

The importance of networks as a factor in career progression cannot really be overstated. It appears crucial for two reasons. Firstly networking increases the profile and visibility of the academic concerned both within the institution and further afield; "The main currency...is reputation" (Becher 1990 p. 52). Academics who are well connected within their institution can rely on support from colleagues and those with networks outside the institution 'look good' for the university. The university's formal promotions criteria do include criteria that are outward looking - for example, networking is mentioned in both grade 9 and 10 academic role descriptors - and our respondents showed some awareness of the importance of this. Rebecca makes the point below in relation to promotion to professorial level but similar views were expressed in relation to moving from lecturer to senior lecturer

I actually think that the thing that does you well for being promoted is being a known face and sitting on committees, networking. Of course you have to have enough research to be seen as valid but I actually think that the networking, [...] probably counts more than the research career, up to the chair level. [Rebecca, Senior, Female]

Secondly there is some evidence to suggest that many positions are awarded on the basis of informal networks even where departments go through the seemingly objective process of advertising and interviewing for a position (Gardiner et al 2007). Bagilhole and Goode (2001) claim that men and women rarely relate to each other as equals in professional circumstances and without access to networks and senior mentors, female academics have little chance of progressing in the same way men do. Networks provide access to information about promotions rounds, available positions and importantly to the way the academic game is played. It is common for academics to use job offers from other universities to fast-track internal promotions applications (retention report) and so developing networks outside an academic's own institution might be relevant to promotion even if there are no formal criteria relating to national or international reputation. Talking to 'the right people' at the right time can significantly increase the chances of a successful promotions application.

It does still so often come down to a kind of old boys network and sort of dodgy handshakes under toilet doors and stuff. ... there are cliques in the place and if you're in then you're ok, but if you're out then you're not [David, Senior, Male]

This culture of cliques or networks is seen to disadvantage women who are less likely to have access to these networks. Ackers and Oliver (2005) suggest that "To the extent that networks play an important role in the career progression in science, differential access to and ability to generate such networks is a factor shaping the career progression of scientists with family/personal responsibilities and of women in particular"(p8). Handley (1994) has argued that women often find themselves on the periphery in organisations including universities and continues "[g]iven this situation it is not surprising that women are often excluded from the informal network and hence may not have immediate access to relevant information or the decision making network within the organization" (p12). She further notes that "The woman who is not part of the informal network does not receive information vital to understanding organizational life "(p12). According to Thanacoody et al (2006) "White (2003) claims that Australian universities continue to be "boys clubs" and that the skills needed for a successful academic career is part of a socialisation process that some men and virtually no women participate in" (P539). It would appear from our research as well as other studies (Bagilhole and Goode 2001) that this is not a phenomenon peculiar to Australian universities. Our respondent Jamie sums this idea up succinctly in the quotation below from which the title of this chapter has been taken:

We don't have the key to the executive washroom, which is critical [Jamie, Senior Lecturer, Female]

The importance of networks lies not only in the visibility they give those within them or in the rather sinister idea of dodgy handshakes and deals made in 'executive washrooms' but also in the support and leadership they can provide. Many of our respondents talked about the importance of having support, strong and positive leadership, mentoring, role models and the management of staff. One of the perceived problems with the promotions criteria and process was the lack of transparency as discussed above and there was a sense that some of that could be overcome if the academic was well managed and supported. If the applicant cannot find out for herself what is required, a good mentor should assist and guide. Networks then are key to progression not only because there seems to be a hidden promotions criteria of 'being well connected' but also because networks allow academics to access information about procedures, processes and expectations that are not captured in any policy or formal document. It allows them to form a more accurate perception of what is expected to gain promotion regardless of what the formal criteria say. Women at all levels are not naturally part of these networks reinforcing what Lapointe Terosky and colleagues (2008) call the "Plexiglas room", making it hard for women to break into male dominated senior positions and forcing them to learn the 'academic game' by themselves.

While networks can provide some of these functions there is no guarantee that they will or that those excluded from them will have access to alternatives. In order to fill this "gap", many universities have introduced a more formalised mentoring scheme (Gardiner et al 2007). Mentoring was seen as having the potential to play a big part in helping staff make the most of their careers at all levels (Gardiner et al 2007; Thanacoody et al 2006, Joiner et al 2004, Heward et al 1997). Thanacoody et al (2006) in their study of Australian and Mauritian contexts suggest that "Senior male mentors act as the gate-keepers in both cultural contexts providing female academics with access to resources and networks that enable them to move up the university hierarchy" (p550).

The University of Bradford operates a mentoring system, and we asked respondents both about their experience of it and about more informal mentoring they may have received. As our respondent Karen as well as a variety of scholars (Gardiner et al 2007; Thanacoody et al 2006, Joiner et al 2004, Heward et al 1997) have noted "*If the university seriously wants to address the issue [of gender equality] then the mentoring process is key"* [Karen, Lecturer, Female] Gardiner et al (2007) note that emotional and personal support is important but "they are not sufficient in terms of a large investment in time and money in a well-run mentoring scheme. As such, positive career outcomes also need to be established" (P429) Most respondents felt that they were well supported emotionally but other types of support were not so forthcoming. Sylvia highlights the problem, saying "Oh, *my mentor was lovely but my mentor didn't have the right skills and lacked the expertise to support me* [Sylvia, Lecturer, Female]

The mentoring role could be much better used by women and our respondent Rachel considered why there is not the same mentoring network amongst women as there seems to be amongst men. In other words, if there is an 'executive washroom' for men why is there not an equivalent for women?

I think often getting to the top in a historically male dominated position, women have to go through things that change them in order to fit in and get promoted and get up there and by the time they get up there and they have that autonomy or a possibility, they've become changed, they've denied themselves. You kept hidden all the things in you that might help other women get there, in order to get there yourself and by the time you get there you're so changed and these things are so hidden, you don't think to do any backtracking to make it a little bit more appropriate for women to come up after you, which is not what the men do. The men do the old school thing, that's why the other men climb up, but when women get there they don't do it, so they don't because that's the very thing that they don't like about men, they don't do that because they are women, nothing is changed on the way up for the next woman who has to climb up that way. I can understand why women do it, but if we don't make changes, it's like rebuilding the wheel every time. Every woman who breaks through that ceiling, it's not like it gets a little bit weaker, it's like we are closing the door behind us and putting all the same locks back on it instead of using that influence to consider how we might make things different and more enabling down that chain [Rachel, Lecturer, Female]

The HERA role descriptors give no indication that research takes primacy over other skills in applications, and this might suggest that the problem of a research focus perceived by Forster (2000) has in fact been removed. However, the reality is more complex, because the HERA role descriptors do not capture the way in which the reward and promotion system is perceived by academics working in the institution. There is a prevalent perception that research is the key to promotion and that long hours and less of a focus on pastoral care and teaching and learning are therefore a necessity. Being in the right networks or "having the key to the executive washroom" were also seen as important in helping academics achieve promotion. In the next section of this chapter, we will look at the implications of this apparent mismatch between perceptions and reality for the efforts of women academics to achieve promotion.

PROMOTIONS CRITERIA AS A TOOL FOR PERSONAL CAREER PLANNING

The preceding sections highlight the problems faced by academics seeking promotion and draw attention to the fact that some of these problems are likely to affect women and men differently thus putting women at a disadvantage when seeking career progression. Confusion about what the criteria are and how they operate is however worrying for another reason. In order to undertake any meaningful career planning academics need to plan strategically, taking on those roles and responsibilities which will assist them in promotion while avoiding those that will hinder them. A lack of transparency or a misunderstanding of what is required to progress can therefore have disastrous consequences for academics. If for example a new lecturer with a particular strength and interest in teaching and learning believes that they can achieve promotion on the basis of excellence in that area with less emphasis on research activity, they are likely to focus their attention on teaching and learning matters. If the criteria in their department are then interpreted and applied in a way that emphasises research, the new lecturer is likely to have to spend a significant amount of time and energy changing the focus of their activities, if indeed this is possible or desirable on a personal level.

As the section above on networks has highlighted, women are likely to benefit less from mentoring and networks within the institutions which could help them to work strategically towards achieving both the formal and hidden criteria for promotion. In order for a promotions system to eliminate bias against particular approaches to an academic career, it has to have criteria which allow for career paths with different emphasis through academia and those criteria then have to be internalized by women academics; for them to be able to internalize the system, it has to be transparent and clear and unconflicting guidance from managers has to be given. Given the gate keeping role of line managers, it is not sufficient for academics themselves to fully understand the official criteria and their application as line managers may be interpreting them differently or be applying some of the hidden criteria discussed above. In other words managers themselves need to internalize the criteria and the extent to which line and senior managers have truly embraced the official transparent criteria which should allow for various pathways through an academic career remains questionable. In Bagilhole's (1993) words "…appraisers often reflect prevailing social expectations and attitudes which reinforce stereotypical expectations and assessments of women" (p. 435).

TRANSPARENCY AND FAIR PROCESS

An explicit goal of the new promotions process at the University of Bradford was to improve transparency. Lack of transparency in university systems has been criticised in earlier studies. Metcalf et al (2005) found that "The transparency of the procedures varied. It was clear that stated procedures and criteria were not universally followed and that staff had varying ideas on what these procedures and criteria were. …Many staff who had not yet considered promotion were unaware of the procedures and criteria. However, some other staff appeared not to know the actual procedure or criteria. A head of department in a new university thought this was, in part, the fault of the university, saying 'The criteria are not always clear" (p180-181). It was hoped that the adoption of HERA role descriptors, a system of performance reviews and greater scrutiny of what happened at promotions committees would result in greater transparency and fairness at Bradford: our study suggests that this may not yet have happened

In one respect, the Bradford procedures are fair and transparent: the role descriptors used for the regrading process are available to all through the university's intranet. This study did not interview line managers about how they decided whether to recommend a candidate for promotion, nor did it consider the decisions made by promotions committees at School or University level. It is not possible to say, therefore, whether line managers and committees themselves followed the formal criteria. However, some of our respondents doubted whether they did.

It's a bit disturbing that there are areas where there are quite staggering differences in practice, I'm not thinking particularly of personnel, but quite staggering differences in practice across the university schools. [Julie, Senior, Female]

Another respondent mentioned a committee making use of a department-specific checklist:

April: [The promotions panel] literally have bullet points and if you meet them all, they have to think of a really good excuse not to give you it [...] if I meet each of these bullet points then I have a really good chance [April, Lecturer, Female]

This checklist was shown to the panel member: it was not the HERA role descriptor and was not a document available on the Faculty's own website. The respondent then added that she herself only received a copy when she attended the promotions panel meeting.

But in any event, publication of job descriptors and transparency at the point when a decision is made about promotion are not the only requirements for a system that is genuinely fair or transparent. Promotion in academia is slow and many years may pass between each promotion: there are seven increment points in the Lecturer B band on the Bradford pay structure, and four further optional increments above that, suggesting it would not be unlikely for promotion from Lecturer B to Senior Lecturer to take seven or more years. Decisions made over a number of years are crucial to whether that promotion will occur. The belief, whether correct or not, that the true criteria differ from the formal criteria, adds to the complexity of individual academics' career planning.

Transparency and fairness therefore require an academic to have access to and a clear understanding of promotions criteria throughout their career, not just at the moment when a decision is made about whether they now satisfy them. A decision to reject an administrative post in favour of retaining research time, or a decision to focus on teaching rather than research, may not at the time be clearly a decision that is about promotion, but if those decisions are made without adequate information or support, then no amount of fairness in the University promotions committee will rescue the application. Many of our respondents lacked this kind of information; as indicated earlier in this chapter, they were not very familiar with the HERA job descriptors, did not know much about what promotion involved or required, and their picture of what they needed to do was often incomplete or at odds with the university's formal criteria. The authors themselves completed the university induction process in, respectively, 2006 and 2007: promotion was not covered during the two day induction. Whether the problem is that that the formal and informal criteria are different, or that there is widespread ignorance of the formal criteria, this points to a lack of transparency.

The other way in which our findings cast doubt upon both transparency and fairness is embedded in the whole idea of 'regrading' rather than promoting academics. As noted earlier, regrading is a retrospective process, and in order for an academic to be regarded they need to have already taken on new responsibilities or developed new skills, above those they possessed on appointment. The new promotions procedure is closely allied to an annual performance review system, but there are various ways in which this aspect of promotion lacks transparency and can result in unfairness. The university's guidance to staff preparing for annual reviews contains a number of things "to think about"; these do not include reviewing the role descriptor for the next role and considering what might be done to develop the skills not currently demonstrated. Academics are asked about their own perception of the previous year's performance, their objectives for the forthcoming year, and any training they need. There is no overlap in the language of performance review documents and those of role descriptors. As a result, the performance review system is a haphazard method of educating academics about what promotion requires, and assisting them to meet it, and yet it is through this system that decisions are made by line managers about whether to recommend a particular academic for promotion.

Another source of potential unfairness lies in the way in which line managers and Heads of Schools allocate administrative roles. Our respondents were clear that not all administrative jobs were equal: some were seen as the key to advancement, others as the "kiss of death". Given that regrading requires a lecturer to be performing the job of a senior lecturer, and that administrative responsibilities are central to this (at least, according to the job descriptors), lack of transparency in the allocation of these jobs is a precursor to a lack of fairness in the eventual promotions process itself. One respondent said about her appointment to a senior position in her School, that a senior manager just "put his arm around me" and said that she should be given the position. "*So in terms of getting that job, it wasn't fair and it wasn't open, but he wanted me urgently to play that role, to have some authority.*" [Amber Senior, Female]

Finally, our respondents often commented on their reluctance to put themselves forward, to take the risk of failure or rejection. This is not solely a problem at the time when a promotions application is made: an aversion to risk, will also mean that some women academics do not volunteer for the very tasks that they need to be doing in order to be promoted. This is also the point at which difficulty with taking on additional workload, because of caring responsibilities, may impact. The promotions committee may be gender blind, but gendered decisions both by women academics and their line managers filter out many cases for promotion before they ever reach those committees.

FUTURE RESEARCH DIRECTIONS

This chapter has attempted to highlight a further aspect of women's progression in academia by drawing attention to the way in which formal and hidden criteria operate in promotion in a UK institution and highlighting the impact the perception held by academics of criteria can have on their chances of gaining promotion. In order to fully understand how to address the continuing underrepresentation of women in senior positions in academia more work needs to be done on understanding how promotions criteria are perceived. The mismatch we detected between what the criteria are, what they are perceived to be and how they are applied needs to be understood fully so that it can be addressed. Our research indicates that the issues raised in this study are fundamental to promotion in academia. Other studies cited in this chapter would suggest that ours is not the only institution to face these issues but further research in this area would increase our understanding of how inequalities can be better addressed.

CONCLUSION

Our study reinforces the results from previous work on women in academia. It also however goes further by considering how academic staff perceive promotion and rewards processes in one institution in the UK.

We found a mismatch between what academics think the criteria for promotion are, what the criteria actually are and how the criteria then operated on the ground. This mismatch can have particularly serious consequences for women academics for a number of reasons. Research shows that women are less likely to break into institutional networks which allow them to access information not only on formal and objective promotions criteria but also on hidden criteria and the way the 'academic game' is played. This results in female academics having a perception of promotions criteria and processes which may actually not be accurate or which is formally accurate but which operates differently in their schools or departments. This, coupled with women's reluctance to promote themselves actively and traditional barriers to promotion such as caring responsibilities, puts women at a disadvantage when attempting to progress into more senior positions within universities.

The effect of the HERA role evaluations on the promotions process is to create a one-size-fits-all job description for academics. There are two ways in which this may impact on women in SET disciplines. The first is SET specific. The nature of an individual academic's job in reality may differ across academic disciplines: for instance, not all disciplines have a culture of collaborative or inter-disciplinary work. If concepts of the academic role that are not actually universal are embedded in role descriptions, those working in academic disciplines that do not conform may be disadvantaged. The second impact of the process is that it may potentially discriminate against women more generally. The difficulties our respondents encountered with balancing home responsibilities with work and their general risk-averse approach to promotion may mean that they are even less likely to pursue regrading. If taking on a major administrative role is a precondition of regrading, rather than something that follows promotion, women who are uncertain about their promotion prospects may conduct a personal risk assessment and choose not to take on that role, because the cost is too high when eventual regrading is so uncertain. It is not possible to conclude whether the move to a single spine will have these effects as yet, as the process is in its infancy, but it is something that should be studied in more detail in the future.

REFERENCES

Ackers, H. L. (2003). The Participation of Women Researchers in the TMR Marie Curie Fellowships. Luxembourg: European Communities.

Ackers, H. L., & Oliver, E. (2005). Gender, Mobility and Progression in Science Careers: MOBISC Summary Report. Leeds, UK: University of Leeds.

Association of University Teachers. (2004). *The Unequal Academy: UK academic staff 1995–96 to 2002–03*. London: Association of University Teachers. Retrieved from http://www.aut.org.uk/index.cfm?articleid=916

Bagilhole, B. (1993). Survivor in a male preserve: a study of British women academics' experiences and perceptions of discrimination in a UK University. *Higher Education*, *26*, 431–447. doi:10.1007/BF01383737

Bagilhole, B., & Goode, J. (2001). The Contradiction of the Myth of Individual Merit, and the Reality of a Patriarchal Support System in Academic Careers, A Feminist Investigation. *European Journal of Women's Studies*, *8*(2), 161–180. doi:10.1177/135050680100800203

Becher, A. (1990). Academic Tribes and Territories, Milton Keynes, UK: Open University Press.

Collins, R., & Palmer, A. (2007). *Perceptions of Rewarding Excellence in Teaching: Expectancy Theory Makes a Comeback?* Presented at the International policies and practices for academic enquiry International Colloquium, Southampton Solent University.

Deem, R., & Morley, L. (2006). Diversity in the Academy? Staff Perceptions of Equality Policies in Six Contemporary Higher Education Institutions. *Policy Futures in Education*, *4*(2). doi:10.2304/pfie.2006.4.2.185

Educational Competencies Consortium, Ltd. (2007). *About HERA*. Retrieved from http://www.hera.ac.uk/about/page.php?page=1

Fisher, A. (1992, September). When will women get to the top? *Fortune*, 44–56.

Forster, N. (2000). A case study of women academics' views on equal opportunities, career prospects and work-family conflicts in a British university. *Women in Management Review*, *15*(7), 316–330. doi:10.1108/09649420010378124

Gardiner, M., Tiggemann, M., Kearns, K., & Marshall, K. (2007). Show me the money! An empirical analysis of mentoring outcomes for women in academia. *Higher Education Research & Development*, *26*(4), 425–442. doi:10.1080/07294360701658633

Gibbs, G., & Habershaw, T. (2002). Recognising and Rewarding Good Teaching – a guide to good practice. Milton Keynes, UK: Centre for Higher Education Practice, Open University Press.

Handley, J. (1994). Women, Decision Making and Academia: An Unholy Alliance. *Women in Management Review*, *9*(3), 11–16. doi:10.1108/09649429410056290

Heward, C., & Taylor, P. (1994). Women at the Top in Higher Education. *Policy and Politics*, *20*(1), 111–121.

Heward, C., Taylor, P., & Vickers, R. (1997). Gender, Race and Career Success in the Academic Profession. *Journal of Further and Higher Education*, *21*(2), 205–218. doi:10.1080/0309877970210206

Joiner, T., Bartram, T., & Garretta, T. (2004). The effects of mentoring on perceived career success, commitment and turnover intentions. *The Journal of American Academy of Business*, *5*(1/2), 163–170.

Lapointe Terosky, A., Phifer, T., & Neumann, A. (2008). Shattering the Plexiglas: Continuing Challenges for Women Professors in Research Universities. In J. Glazer-Raymo (Ed.), Unfinished Agendas. Baltimore, MD: John Hopkins University Press.

Martínez Alemán, A. M. (2008). Faculty Productivity and the Gender Question. In J. Glazer-Raymo (Ed.), Unfinished Agendas. Baltimore, MD: John Hopkins University Press.

Metcalf, H., Rolfe, H., Stevens, P., & Weale, M. (2005). *Recruitment and Retention of Academic Staff in Higher Education*. Department for Education and Skills, Research Report RR658.

Parker, J. (2008). Comparing Research and Teaching in University Promotion Criteria. *Higher Education Quarterly*, *62*(3), 237–251. doi:10.1111/j.1468-2273.2008.00393.x

Thanacoody, P. R., Bartram, T., Barker, M., & Jacobs, K. (2006). Career progression among female academics, A comparative study of Australia and Mauritius. *Women in Management Review*, *21*(7), 536–553. doi:10.1108/09649420610692499

University of Bradford. (2008). Promotions Exercise 2008. Bradford, UK: University of Bradford.

White, K. (2003). Women and leadership in higher education in Australia. *Tertiary Education and Management*, *9*(1), 45–60. doi:10.1023/A:1022218403113

Wolffensberger, J. (1993). 'Science is truly a male world': The interconnectedness of knowledge, gender and power within university education. *Gender and Education, 5*, 37–54. doi:10.1080/0954025930050103

Young, P. (2006). Out of Balance: Lecturers' Perceptions of Differential Status and Rewards in Relation to Teaching and Research. *Teaching in Higher Education, 11*(2), 191–202. doi:10.1080/13562510500527727

ADDITIONAL READING

Bebbington, D. (2001). Women Scientists In Higher Education A Literature Review, Athena Project.

Glazer-Raymo, J. (1999), Shattering the Myths: Women in Academe, Baltimore, John Hopkins University Press.

Glazer-Raymo, J. (Ed.). (2008). Unfinished Agendas, Baltimore: John Hopkins University Press.

Guth, J., & Wright, F. (2008). *Women in the Higher Education Sector - Confronting the Issues for Academics at the University of Bradford* Project Report, Bradford: Bradford University Law School, available at www.brad.ac.uk/management/lawinbrief

Metcalf, H., Rolfe, H., Stevens, P., & Weale, M. (2005) *Recruitment and Retention of Academic Staff in Higher Education*. London: National Institute of Economic and Social Research & Department for Education and Skills. http://www.niesr.ac.uk/pubs/searchdetail.php?PublicationID=645

Morley, L. (2003). Quality and Power in Higher Education. Buckingham: Open University Press.

Morley, L., & Walsh, V. (Eds.). (1996). Breaking Boundaries: women in higher education, London: Taylor & Francis.

SETFair. (2002). *A report on women in science, engineering and technology from the Baroness Greenfield CBE to the Secretary of State for Trade and Industry.* www.setwomenresource.org.uk

ENDNOTE

1 We would like to thank our research participants who gave up their time to be interviewed and without whom this study would not have been possible. We have changed all their names and only given general details about their job titles to protect their identity. Thanks must also go to the Bradford University Human Resources Directorate who supported the study financially. Thank you also to the reviewers of this chapter for their insightful and constructive comments.

Profile 8
Technology Career Spotlight:
Developing a Dynamic Environmental Calibration Technology for Cost Effective Online Water Quality Monitoring

© 2008 Queensland Government Office of Women. Used with permission

Shoshana Fogelman, South East Queensland

Shoshana Fogelman is a research academic staff member at the Advanced Water Management Centre, The University of Queensland. She has over six years research experience in the area of developing intelligent on-line water quality monitoring sensors and tools for water and wastewater systems. Her PhD was undertaken at Griffith University, which focused on developing a new analytical method specifically designed for continuous and field based water monitoring applications. It was an interdisciplinary project that combined the areas of science and artificial intelligence. A range of intelligent water quality monitoring sensors for the water industry were developed, that were capable of dynamically responding in situ to their environments negating the need for calibration, reagents or strictly controlled measurement conditions. Shoshana previously worked as a Research Fellow on the Urban Water Security Research Alliance Project – Water Quality Information Management Program and was the Acting Program Coordinator for Water Quality and Diagnostics, at the Smart

Copyright © 2010, IGI Global. Copying or distributing in print or electronic forms without written permission of IGI Global is prohibited.

Water Research Facility. Shoshana is currently working as the Project Manager on the Anaerobic Digester Stability Sensor Project. She is currently the president of the Australian Water Association (AWA) Queensland Young Water Professionals special interest network and was recently elected to the AWA Queensland Branch Committee. Through her involvement in AWA, she actively promotes and supports the needs of the next generation of water professionals within Queensland, through her involvement on committees including branch, conferences and professional/technical workshops and other related activities. In addition she has published over 10 peer-reviewed journal and conference publications and has served on several international and national conference committees.

Introduction

I was born and grew up in Papua New Guinea, in the eastern highlands before moving to Australia with my family in 1985. I believe my experiences as a child with no television and living in a very small town, in an isolated, undeveloped area really developed my passion and love of the environment. In addition, as a child my father always promoted and exposed me to the science and technology area, so it seemed important to me to develop technology to better protect the environment. In 1999, I enrolled at Griffith University in a Bachelor of Science and met Dr Michael Blumenstein, who worked then as a lecturer in the School of Information and Communication Technology. Dr Blumenstein introduced me to this exciting field of artificial intelligence and pattern recognition, which was getting computers to "intelligently" think like the human brain. I thought this could really be applied to environmental field to better manage our water systems and consequently approached the university to do an interdisciplinary honours research project in the water quality area and my career blossomed from there.

I have currently worked on three projects aimed at developing more reliable and cost effective water quality monitoring technology for the water and wastewater industry.

Project 1: Developing a Dynamic Environmental Calibration Technology for Cost Effective Online Water Quality Monitoring

This innovative project develops a range of new online water quality monitoring sensors using a dynamic environmental calibration technology that can enable conventional sensors to intelligently respond insitu to dynamic matrix conditions, without the need for conventional calibration, reagents or sample pretreatment.

The project outcome pertaining to my PhD has been the development of a new generic technology that can enable a traditional laboratory based sensor to be converted into a direct insitu online sensor. The technology developed employs a combination of software and hardware to environmentally calibrate a traditional water quality monitoring sensor so that it is capable of autonomously and intelligently responding to its environment irrespective of sample matrix conditions. The applications of the technology have been successfully demonstrated for the online monitoring of Fluoride, Ammonia, Ammonium and pH.

The success of the project was based on the innovations that were created through a multidisciplinary project that combined the main areas of Science and Information Technology, in addition to a small engineering component. The environmental calibration technology incorporated the areas of analytical and environmental chemistry with ecology principles, combined with computationally intelligent techniques, such as artificial intelligence to instruct the sensors how to respond insitu to their environments. Once the selective sensors were environmentally configured using the technology, they were capable of intelligently responding to their environments, without ideal measurement conditions, calibration or reagents being required.

The project was designed to respond to the needs of the Australian water industry by developing simple to use, accurate, real time online water quality monitoring tools that are economically feasible to implement in rural as well as urban environments. This is so continuous information can be reliably and simply obtained by authorities to ensure the continued safe supply of potable freshwater to the Australian consumers. This is because currently, accurate and cost effective online monitoring of various water quality parameters has proven difficult to achieve, as direct sensor deployment often means sensors are unavoidably exposed to a wide range of measurement conditions. As the majority of current online monitoring technologies employed are direct adaptations of traditional laboratory based analytical methods, they are not originally designed for field based monitoring purposes. Hence they require frequent calibration, maintenance and large quantities of costly reagents to maintain their reliability in the field. The costs associated with maintaining the instruments based on this measurement principle has greatly reduced their wide spread application, especially for large scale environmental water quality monitoring in places like the European Union. Therefore technology that can improve the accuracy and economic costs associated with online monitoring such as maintenance and reagent consumption would be of great benefit to industry, government and research organisations.

An outcome from the intellectual property developed in my thesis was that it gained commercial attention, as I was approached in my final year of my PhD by an Australian Venture Capital Company. As I was successful in being offered a term sheet, this lead to negotiations with Griffith University to determine a potential venture capital investment. A paper describing the work has also been drafted to submit to Science.

Project 2: Western Corridor Recycled Water Project Urban Water Security Research Alliance

In 2008, through the development of my expertise in the area of intelligent water quality monitoring sensor design and tools, I received an appointment by invitation that was approved by the Vice Chancellor of Griffith University to work on a research project related to the Western Corridor Recycled Water Project. The project seeks to develop new online water quality monitoring technology and tools to enhance operational monitoring strategies in barriers 1 and 2. The technology and tools developed will be designed specifically for a closed loop purified recycled water system and are designed to assist operators and grid managers with the maintenance and management of barrier integrity.

Project 3: Anaerobic Digester Stability Sensor Project

In 2009, I commenced employment at The University of Queensland to begin work on the development of an on-line Anaerobic Digester Stability sensor that was part of an ARC linkage project in conjunction with Gold Coast Water and Brisbane Water. The sensor is designed to improve process stability of the treatment process and regulate renewable energy for the biogas waste stream, thereby decreasing green house gas emissions. The sensing system has been designed, built and we have installed two on-line monitoring systems and are currently validating and verifying the accuracy of the system.

Also during my PhD candidature and employment I have also been involved in industry and community services activities pertaining to my research that have aimed at promoting the area of water quality monitoring, in addition to encouraging and supporting younger people to pursue careers in the Queensland Water industry.

Community Benefit of Research

Through my PhD and work I have also been actively communicating the need for innovative, multidisciplinary research in the water industry, through my involvement in student supervision, international and national reviewing of research, conference organisation and through workshops and forums.

Student Supervision

In addition to my research I have engaged and mentored younger undergraduate students to get involved in water quality technology developments, through undertaking research projects which encompass the new concepts developed in my thesis. This can be demonstrated by my supervision of a third year project student, Wade Button, Chris Carney and Nicolas Brodu. Through their experiences, most of them have pursued a career in the water industry

International and Nationally Recognised Researcher

I have also been internationally and nationally recognised for my expertise in the field of sensor development and computational intelligence. This is evidenced through being a regularly invited to review research papers for science and IT journals and conferences in my respective research fields. In addition I have published a number of peer reviewed journal and conference papers in the area.

Founding Australian Water Association (AWA) Queensland Young Water Professionals (YWPs) Special Interest Group

To attract, support and retain YWPs in the Queensland Water Industry, I was a founding committee member who developed the first YWP Special Interest Group in Queensland in July 2005. This was developed after the success of the AWA Queensland YWPs Forum in 2004. Since its inception in 2005, the group has gained momentum, members and strong industry support. The technical and career develop sessions cover new and relevant issues that affecting YWPs working in Queensland.

I also served on the organising committee that first developed and launched the National Australian YWPs conference in Sydney in 2006. The aim of the biennial conference was to provide a national platform for YWPs working in the water and wastewater fields where they can engage, network and share their experiences and research outcomes in a relaxed environment. It also provided opportunities for YWPs to interact with senior industry professionals through mentoring and professional development activities. Since the launch of the 1st conference, the YWP group has gained in strength and numbers with active committees being developed in most states in Australia. We have developed our own logo, website and mentoring program and are continually gaining strong industry support.

In 2008, I Co-chaired the 2nd National Australian YWP Conference in Brisbane. Through my involvement with the committee, we highlighted the role that YWPs play in the water industry and encouraged a national insight into areas such as indirect potable reuse, desalination, sustainability, as well as new infrastructure and technology developments. Through my involvement as Co-Chair we were able to gain government support for the group with the Premier of Queensalnd Anna Bligh MP endorsing the conference. The conference was a great success bringing 150 of the brightest young minds to Queensland, representing all facets of the Australian Water Industry.

Developed National Mentoring Program for the Australian Water Industry

To ensure YWP have the skills necessary to become the next generation of leaders, I developed, launched and convened a national mentoring program for the Australian Conferences IWA/AWA 1st and 2nd National Australian YWPs Conference Water Industry.

This was to support YWPs in developing their managerial and technical skills, in addition to providing them with an opportunity to engage and build mutually beneficial relationships with senior industry and research professionals. Initially the mentoring program was developed in Queensland for the YWP group in mid 2005. However in late 2005, I was invited by the International Water Association (IWA) and AWA to expand the program nationally. In 2006, in conjunction with AWA members the program was launched nationally at the 1st YWP Conference in Sydney. Since then the program has gained strong industry and AWA support.

Chapter 9
Factors Contributing to the Success of Women Working in Science, Engineering and Technology (SET) Careers

David JF Maree
University of Pretoria, South Africa

Marinda Maree
University of Pretoria, South Africa

ABSTRACT

Women's under-representation in Science, Engineering and Technology (SET) careers is a problem in South Africa. This chapter discusses structural and individual barriers responsible for the under- representation of women in SET. Self-efficacy as a requirement for success in SET is discussed. The results are illustrated with a report on a study done with a sample of 29 women in successful SET careers. These women experienced some form of gender discrimination at some stages of their development from school to career. They also struggle with family and work balance and similar issues. However, the sample ascribed their success or the fact that they stay in a SET career mostly to personal characteristics and strong self-efficacy beliefs. Aspects such as a drive for achievement, strong goal orientation, passion for their work, determination and perseverance were identified. Strong self-efficacy beliefs which can be associated with resilience and cognitive hardiness came to the fore. These women believe that they can achieve their goals and they do to a large extent.

INTRODUCTION

The under-representation of women in Science, Engineering and Technology (SET) careers is a worldwide well-known phenomenon and presents many challenges that are not yet solved (Hodgson, 2000). South-Africa is no exception. Although more than 50% of students at tertiary educational level were women in 2001, data showed that women were still severely under represented in SET fields of study (National Advisory Council on Innovation (NACI), 2004). The purpose of this chapter is to discuss some of the barriers to women entering

DOI: 10.4018/978-1-61520-657-5.ch009

Copyright © 2010, IGI Global. Copying or distributing in print or electronic forms without written permission of IGI Global is prohibited.

science and difficulties accounting for women leaving science in order to find a clue to what makes women in SET careers successful. The ideas are illustrated with research done in South Africa exploring the career histories of 29 successful women in SET careers. On the one hand we indicate the difficulties they have encountered on their way to success in their careers, but on the other hand we also indicate what factors account for their success. It is suspected that self-efficacy mediates success in women in SET and that the mere breakdown of institutional or personal barriers is not sufficient for ensuring a successful SET career for women.

BACKGROUND

A metaphor frequently used to describe the under-representation of women in SET is the "leaky pipeline" (Pell, 1996). Women are lost for SET fields along the development of their careers from school and at various other crucial life milestone phases (cf. the funnel model of Cronin & Roger, 1999). In the remainder of this section we will give a brief overview of the usual reasons for leakage and also indicate what we think is responsible for success in a SET career.

Blickenstaff (2005) gave an overview of factors he found in literature for the under representation of women in SET (see also Cronin & Roger, 1999). He found that some factors provided good reasons why women are leaving SET fields whilst others did not stand up to scrutiny. He cited the following reasons from research why women leave SET: (a) biological differences between men and women, (b) girls are not well prepared for a science career, (c) girls have a negative attitude towards science and did not have positive experiences with science at school, (d) there are no role models in science for girls, (e) science curricula do not apply to girls, (f) pedagogical styles of science classes fit boys better than girls, (g) there is what Blickenstaff (2005, p. 372) calls a "chilly climate" in science classes towards girls, (h) it is expected that girls conform to traditional gender roles and (i) the worldview imbedded in scie4nce is masculine. These nine categories can be divided into individual factors and institutional or structural factors which are in principle external to the individual and are briefly discussed below (Fox, 1998; Sonnert, Fox, & Adkins, 2007).

Structural Barriers

a. Although SET fields are deplete of female role models, having more role models does not necessarily mean that women will be attracted or retained in SET careers (Blickenstaff, 2005). Empirically the effect of role-models still has to be unravelled (Buck, Clark, Leslie-Pelecky, Lu, & Cerda-Lizarraga, 2008; Downing, Crosby, & Blake-Beard, 2005; Murrell & Zagenczyk, 2006), and if women role models operate in a discriminatory environment it sends the message that women still have it tough and thus it is an environment to be avoided or to be accessed only by conforming to the male model of operating (Blickenstaff, 2005).

b. Blickenstaff (2005) is of the opinion that much has been done to eliminate gender bias from science curricula materials and design. The argument is that females experience the lack of female (or black or other minorities) examples in text books negatively.

c. Teachers can create an environment that fosters interest in science and mathematics (Roger & Duffield, 2000) but any negativity from their part against females can detrimentally influence girls interests and choices (Stewart, 1998) The way a teacher deals with children, for instance, by favouring boys in terms of reacting more positively and encouragingly to their questions and reactions can obviously send a clear message to girls (Alper, 1993; Blickenstaff, 2005). Even allowing boys more access to technological environments and discouraging girls can further strengthen the message that girls are supposed to conform to gender stereotyped roles (Mayer-Smith, Pedretti, & Woodrow, 2000; Plumm, 2008).

d. These issues above contribute to what Hall and Sandler (1982) called the "chilly climate" in the classroom, not only on a secondary but also on a tertiary educational level (Blickenstaff, 2005). To be constantly reminded in more ways than one that girls are expected to conform to certain roles, that science and in particular certain directions such as engineering is masculine (Barber, 1995; Evetts, 1998; Good, Aronson, & Harder, 2008; Henwood, 1996; Papastergiou, 2008; Siann & Callaghan, 2001; Stewart, 1998), cannot fail but to leave an indelible impression on a girl that SET is not for her. Morris and Daniel (2008) in a study on college undergraduates in the USA found that females experience the climate at college as more chilly than males. This applied to both traditional female majors (nursing and education) and non-traditional majors (engineering and IT). However, females with non-traditional majors experienced the environment as less chilly than their traditional majors counterparts. Morris and Daniel (2008) suggested that the reasons nontraditional career women stay despite experiencing the environment as discriminatory (albeit lower than traditional major women) are personality differences and the role of self-efficacy (see discussion below). Women in nontraditional female fields differ in personality characteristics from women in traditional females careers, such as lower scores on a femininity scale and less sensitivity to barriers created by the chilly climate.

e. It is argued that traditional gender roles and cultural factors and not biological factors influence men and women to make different career choices (Alper, 1993). What are the cultural factors that influence girls not to develop a positive attitude towards SET fields and consequently cause them not to be interested in these fields? Some find the difference in the perception that science or mathematics is for boys and not for girls. Blickenstaff (2005) mentioned issues such as dividing classes with male/female seating arrangements thus creating a climate of opposition between boys and girls. These ideas are probably facilitated very early in a child's development. Thus early socialisation has an important influence on a child's gender identity (Roger & Duffield, 2000; Stewart, 1998). The role of parents and teachers in forming the child's gender identity cannot be underestimated.

f. Underlying science, especially natural science, is a masculine world view, a view pointed out by feminist and related literature. Blickenstaff (2005) said that research on women in science rarely refers to feminist and related critiques but it is necessary to take cognisance of the political and social power underlying the types of questions scientists ask and that drive research agendas (Cronin & Roger, 1999; Davis, 2002; Kinzie, 2007; Phipps, 2006).

Blickenstaff's (2005) argument is that "external" or institutional aspects such as pedagogical and related factors are more responsible for filtering out women in SET fields than individual or "internal" issues such as capabilities and interest (Acker & Oatley, 1993; Fox, Sonnert, & Nikiforova, 2009; Rosser, 2003; Settles, Cortina, Malley, & Stewart, 2006). However, this conclusion should not be made too hastily because underlying Blickenstaff's own discourse could be the theme of what can and cannot be easily changed to attract and keep women in science. It is easier to change the classroom than personality, and the conundrum is if it is indeed true that a particular "type" of woman enters and stays in a SET career, the possibility that "other" women enter this career path becomes slim.

Individual Barriers

a. Although Blickenstaff (2005) did cite some studies that point to differences in psychological abilities, other studies found no significant difference between the mathematical and scientific ability of males and females (Alper, 1993; Good, et al., 2008; Roger & Duffield, 2000). However, this debate is still ongoing. Lynn (2008), for instance, contradicts views expressed in Ceci and Williams (2007) that boys do not differ in mathematical ability from girls. The point, though for Blickenstaff (2005) is that even if differences exist they are so insignificant that they are not able to explain why women do not move into SET fields.

b. Furthermore, Blickenstaff (2005) argued that in preparation for a career in SET women are sometimes better prepared than men but they still exit their SET studies. This point has also been made elsewhere that women have to work so much harder than men to prove themselves capable especially in their studies. Thus, biological differences and preparation for a career in SET cannot account for why women leave or leak from the pipeline.

c. Blickenstaff (2005) did refer to the research of Weinburgh (1995) who found differences in attitudes between girls and boys towards science and mathematics subjects. Girls were less likely than boys to be interested in science subjects but as Blickenstaff (2005) pointed out, the effect sizes were rather small. Weinburgh (1995) did find a correlation between liking science and achievement for both boys and girls. However, high levels of liking science were more important for girls' higher achievement in science (Weinburgh, 1995, p. 395).

d. Other barriers to allowing women in SET fields exist, such as the influence of one's personal life on work life and how to balance the two (Armstrong, Riemenschneider, Allen, & Reid, 2007; Hartman & Hartman, 2008; Morgan, 1992; Özbilgin & Healy, 2004). Women often make conscious choices to rather have time for their families and raising children than spending all their time at work. Although as we have found and discussed below that some women do manage to balance work and family life (Maree, Maree, Botha, & Gcabo, 2008), it remains a concern to women entering a SET field (Etzkowitz, Kemelgor, Neuschatz, Uzzi, & Alonzo, 1994). The issue of raising a family has an impact on a women's career continuity because a pregnancy must be scheduled carefully (Etzkowitz, et al., 1994) and even the availability of women at work and its impact on promotions become an issue (Armstrong, et al., 2007).

e. Another reason frequently cited for loosing women is lack of self-confidence: "In fact, a loss of self-confidence – rather than any differences in abilities- may be what produces the first leak in the female science pipeline" (Alper, 1993, p. 410). Although Penner and Paret (2008) found the contrary with regard to mathematics performance, Pell (1996) said that girls and boys mathematics performance are on par when they start school but as they progress the performance of girls become poorer. The poorer performance becomes more pronounced in adolescence were self-confidence reaches a low level (Pell, 1996). In a study amongst women enrolling in engineering and science, Brainard and Carlin (1998) found that women started the first year with high levels of self-confidence in their abilities in mathematics and science (Hollinger, 1983). However, during the first year there is a significant drop in confidence levels and the students struggle to recover their initial levels during their four years of study. Good, et al. (2008) found that stereotype threat, i.e. the idea that mathematics is not for women, can influence mathematics performance of women negatively, whereas those not exposed to stereotype threat did even better than their male counterparts. Thus, stereotype threat, when encountered, and a number of factors we have mentioned above probably combine to influence the self-belief and thus self-confidence of women which we would like to subsume under self-efficacy beliefs.

Recently the social cognitive perspective of Bandura (1977, 1986, 1997) proved to be profitable as a framework within which to analyse the problem of the under representation of women in science (Betz & Hackett, 1981). Bandura made a number of observations about self-efficacy and the role it plays in women's choice of non-traditional careers. Self-efficacy is defined as the belief that one has the capability of doing certain things (Bandura, 1986). It can also be combined with specific domains such as "careers" forming career, occupational or behavioural agency self-efficacy, thus denoting the congruency between capability and career choice (Ancis & Phillips, 1996; Hackett, 1997; Rivera, Chen, Flores, Blumberg, & Ponterotto, 2007; Sullivan & Mahalik, 2000). Thus, one is more likely to choose a career one believes one is capable of doing or mastering. This means that a woman will choose a career in line with her career self-efficacy beliefs and avoid those that she has low self-efficacy beliefs for (Ancis & Phillips, 1996).

According to Bandura (1986, 1997) self-efficacy beliefs are fed from four sources, namely (a) mastery experiences, (b) vicarious experiences, (c) supportive experiences and (d) emotional experiences. Mastery experiences refer to mastered activities and successes. Experiences of failure lower self-efficacy whilst successive mastery experiences raise self-efficacy. Bandura hypothesised that high self-efficacy in the light of a number of mastery experiences makes one resilient to subsequent failures. Failures would then not lower self-efficacy but the sources of failure would be sought in circumstances or in the lack of effort.

Vicarious experiences refer to seeing something done by someone else. If one sees somebody doing or achieving something that one would like to do it has an enabling effect on self-efficacy. Bandura (1997) included concrete and imaginal experiences under vicarious experiences. Maddux (2002) distinguished imaginal from vicarious experiences (thus forming a fifth source of self-efficacy knowledge) where the first refers to one's imagining of a successful act (or role model) and the second to observation of role models and successful behavioural acts. The beneficial effect of role models is built upon the idea of vicarious experience. The effectiveness of role modelling depends on prior experience. If one does not have previous experience of an activity or tasks then the achievement of a role model can increase self-efficacy. However, according to Bandura (1986) even low self-efficacy based on previous failures can be increased by role modelling. Despite that vicarious experiences are not as strong a source of self-efficacy information as mastery experiences, it can soften the impact of negative mastery experiences.

Supportive experiences (or verbal/social persuasion) entail verbal and related support for someone embarking on an activity. Verbal encouragement within particular significant relationships is important for enhancing one's self-concept and self-efficacy. Two aspects are important here. The first is that verbal encouragement will have no effect if it is not congruent to some extent with what one already believes about one self (Bandura, 1986), and secondly, it has a greater effect if it is received within a significant relationship such as from a peer, a parent, or a teacher (Zeldin & Pajares, 2000). It is also easier to discourage someone than encourage and thus increase self-efficacy. In other words, if a person is verbally discouraged he or she would rather avoid those activities that he/she doubts they can do (Bandura, 1986).

Emotional experiences refer to those physiological and emotional states that accompany actions. Thus, when doing something is accompanied by feelings of anxiousness it is taken to mean probable failure. Lack of negative feelings is associated with success (Bandura, 1986).

Social cognitive theory explicitly addresses the problem of women in non-traditional female occupations:

Societal practices require of women a robust sense of self-efficacy to pursue nontraditional vocations. In preparing for and entering careers dominated by men, they must believe strongly in themselves. Self-doubts are often difficult to override even in socially endorsed endeavors, but doubly so when non-traditional pursuits receive minimal support or are regarded with disfavor by many people. Stereotyping and discriminatory practices create additional obstacles. Progress in a career requires considerable sustained effort to produce the types of results that contribute to advancements and personal fulfillment. This is difficult to achieve if one has to fulfill the heavy demands arising from the dual workloads of career and household (Bandura, 1986, pp. 432-433).

Bandura (1986) continues with the observation that women require a high level of self-efficacy in social realities of work, namely the ability to establish good relations, leadership, organisational skills and so on. These remarks are amongst others built upon earlier research on self-efficacy and women's careers. More than 25 years ago Betz and Hackett (1981) investigated the relationship between self-efficacy and career choice and development focussing especially on women. Betz and Hackett (1997) found that men and women do not differ in levels of self-efficacy when entering in corresponding gender careers. In fact, males had high self-efficacy for both traditional and non-traditional careers. However, women had markedly lower self-efficacy for non-traditional female careers. Gender role socialisation and practices usually militate against women entering so called male vocations.

If women's self-efficacy is low for non-traditional careers then it could explain why so few women enter SET careers. However, it is necessary to investigate what characteristics women who are successful in SET careers have. In two qualitative studies the relationship between success in a SET career and self-efficacy were examined using Bandura's (1986) four self-efficacy sources as framework. In the first of these Zeldin and Pajares (2000) found that women in SET careers depend much more on vicarious and support experiences than performance experiences for high levels of self-efficacy. This was surprising given the predictions of social cognitive theory that mastery experiences have more weight in forming positive self-efficacy beliefs. Women thus depend much more on positive role models and verbal support from others in significant relationships. Both the role model and social support sources seem to involve *significant* others such as parents, peers, teachers and supervisors. This extends Bandura's theory slightly because it seems that if the role-model or supporter is not in some relationship with the woman then effects on self-efficacy will not be as marked. Zeldin and Pajares stress the context of relationships within which women find support and where their self-efficacy is sustained.

In the second study Zeldin, Britner, and Pajares (2008) compared females of Zeldin and Pajares (2000) to a group of males in SET careers. It was found that males depend much more on performance experiences than the females and less on vicarious and support experiences. Mastery of activities was important in forming men's self-efficacy beliefs and other sources of information largely corroborated their already formed beliefs about their capabilities. Thus it seems as if men and women differ on the way they utilize sources of information for selecting careers and for staying in those careers. Supportive social environments are much more important to women and if they can see that other significant women achieved their goals so much better.

Zeldin and Pajares (2000) found that the high levels of self-efficacy of the women in their sample had the result of building resilience to obstacles in their careers. According to Bandura's (1986) expectations women with established self-efficacy regarded obstacles not as failures but as challenges to overcome. Self-efficacy thus allowed them to develop resiliency and perseverance in the face of obstacles. Two types of resiliency were identified by Zeldin and Pajares (2000), namely, academic and social resiliency. The first refers to the women's perseverance in completing their studies whilst the second refers to the ability to overcome social discrimination.

PROBLEM

Although it is clear that women experience a multitude of problems within the SET environment and due to the leaky pipeline phenomenon leave SET careers, an important question to ask is what makes women stay in SET careers. Obviously there are a number of women making a success of their careers within SET and the question to be answered is what characteristics or factors make these women stay in their respective careers? The remainder of the chapter reports on a study done in South Africa amongst women that are generally regarded as successful in their careers (Maree, et al., 2008).

METHOD

The research was done in 2008 for the National Council for Innovation (NACI) who requested information on a number of women successful in their careers. The main research question was to find out what factors played a role in these women's career histories that made them successful. A qualitative approach was used by utilizing semi-structured interviews (Özbilgin & Healy, 2004). Although a narrative approach was used and women told their stories it did not amount to biographical research (see Birbaumer, Lebano, Ponzellini, Tolar, & Wagner, 2007). An interview schedule was drawn up and used as a guideline for the interviews. The interview schedule was based on the following questions:

- Tell me about your what your current career entails
- Why would you regard yourself as successful in your SET career?
- Did you find it difficult as a woman to achieve success in your career?
- What and who motivated you to choose a career path in Science, Engineering & Technology (SET)?
- Which factors contribute towards you remaining a professional in this field?
- Did you experience difficulties as a woman on your way to success and how did you overcome these difficulties?
- How did your personal life influence your professional life and vice-versa?
- In what way would you regard yourself as a role model for women and girls wanting to embark on the same road? In other words, what would you like to emphasise as a role model?

A time schedule was drawn up because these women worked across South Africa. Most were situated in the Gauteng area within which two of South Africa's largest cities, Pretoria and Johannesburg, are situated. Interviewers were trained in the interview procedure. Initially interviewers were paired until they have gained experience. Interviews were digitally recorded. One interviewer did the interview and the other handled the recording process. Interviewers quickly gained experience and then did interviews on their own. Each interviewee signed a consent form and completed a basic background and biographical questionnaire. Most of the interviewees were sent the background schedule beforehand to facilitate the process of interviewing so that time was not lost on completing the questionnaire during the interview. The interviews were mostly restricted to one hour. All interviews were conducted by women except for two interviews where a male stood in when schedules clashed.

The recorded interviews were transcribed fully by a team of persons not involved in the interviews. All the transcribed interviews were then checked by persons involved in the interviewing. The interviews were then perused by a person trained to identify the themes according to the interview schedule. These were analyzed by the researchers and summaries of the themes for each person in the sample were made. A final summary of themes were created from the profiles of women.

SAMPLE

The sample can be divided into two groups of women, namely those that were in their careers for a number of years and those that were in their careers approximately less than ten years. The more experienced group consisted of 19 women whilst the younger group consisted of 10 women. Thus a total sample of 29 women was utilized in the study.

The mean age (n = 19) for the older group was 42.5 with a minimum of 26 and a maximum of 68 (SD = 11.4). The mean age for the younger group (n = 10) was 32.2 with a minimum of 26 and a maximum of 39 (SD = 4.5). The mean age for the total group (N = 29) was 38.97 (SD = 10.7).

The mean number of years in their current posts for the older group (n = 19) were 6.1 with a minimum of 0.08 year (or one month) and a maximum of 35 years (SD = 8.3). There were a number of women that changed posts close to the time the interviews were held. The average number of years in current posts for the younger group (n = 10) was 3.16 with a minimum of 0.17 years and a maximum of 10 years (SD = 3.92). The average number of years in their current posts for the total group (N = 29) was 5.06 years (minimum 0.08 years and maximum 35 years with SD = 7.17).

RESULTS

What the Careers Entailed

The 29 women were involved in diverse careers (see Table 1 for a summary). The careers can be divided into fields of application and areas of expertise. Fields of application can be broadly divided into management, research, lecturing, and applied work. Some of the women were involved in more than one application especially when involved in academia. Some of these women lecture, do research and lead projects where they need to manage teams. The areas of specialization included telecommunications, chemical, electrical, metallurgical and mechanical engineering, nuclear physics, computer science, plant cell en seed biology, microbiology, astronomy, conservation biology, paleobiology, and geochemistry. One woman specialized in the human sciences, specifically psychology, and another in economic and management sciences.

Achieving Success in a SET Career

The question was put to the sample why they would regard themselves as successful in their SET career. Diverse answers were given to this question. Some women felt that they worked in a unique field and that they have the opportunity to make unique contributions. Other women said that they were recognized as experts, they publish and they are rated scientists. The reaching of concrete objectives such as the number of articles published, invitations to do keynote addresses at conferences, international recognition and collaboration and the supervision of post-graduate students were aspects indicated by others as well. Recognition as an expert in one's field seemed to be an important issue but one woman explicitly mentioned that she would regard herself as successful when people would look beyond her gender and race as a black woman and recognize her for the expert she is.

In a slightly different way some objectives were focused towards making an impact in one's surroundings. Aspects such as supervising post-graduate students so that they achieve success, and seeing that one's work is making an impact were mentioned. As an example of the last one woman said that a very concrete impact of her work was the reduction of CO_2 emissions by industry and seeing this achieved indicates success for her. Note how she summarises her success: *"I feel proud about that and that's what I feel is ... success."* Another woman regarded success as the ability to manage her team well and facilitate change.

One of the women said that she sees success not as concrete outputs such as publications but rather

Table 1. Fields of practice for the sample (N = 29)

Field	Number	Percentage
Plant cell and seed Biology/ Conservation Biology/ Microbiology/ Paleobiology	4	13.8
Chemistry/Chemical Engineering	8	27.6
Computer Science	1	3.4
Environmental Psychology	1	3.4
Management in engineering	3	10.3
Electrical Engineering	3	10.3
Mathematics Education	1	3.4
Metallurgical engineering	3	10.3
Business	1	3.4
Mechanical Engineering	1	3.4
Nuclear Physics/ Nuclear engineering	2	6.9
Tele-communications	1	3.4
Total	29	

regards growth as an indication of success. She specifically referred to intellectual and personal growth. Along this line one woman contrasted concrete indicators such as financial success with developing one's expertise and skills. Another woman confirmed the importance of relationships at work:

And also you have to be able to work with people – other people – you work as an individual, but part... mostly as a team, which is why all these international collaborations, because we cannot carry an idea by yourself, you need support. So which means you have to be able to have interrelations with other people, communication skills...

In a similar vein another mentioned that fulfilling one's potential was an indication of success. A number of women said that they enjoyed their work. Some women said that they regarded success as achieving the milestones in one's life such as completing studies successfully and obtaining a job, getting promoted and so on.

Another theme that came forward is that some women in the older group and most in the younger group did not regard themselves as successful yet but saw themselves as still developing and on their way to success. Goals such as obtaining a doctorate and becoming a manager were mentioned as some of the indicators.

The indicators of success can thus be divided into a number of categories. (a) Concrete and tangible indicators such as the number of publications, key-note addresses and general recognition as an expert in one's field. One probably could group an issue such as financial rewards and success in this category as well. (b) A second group of indicators are certainly also concrete but emanates from the desire to see a beneficial impact of one's work. Success is then measured against the extent that one's work benefits others which in an academic context means delivering students that are successful. (c) The third group of indicators is personal and involves personal development and growth. One can call these intrinsic indicators of success. (d) The fourth group refers to attaining the usual goals or milestones of one's life such as successful studies, attaining a job and so on. (e) The last group of indicators also deals with attaining goals but it is accompanied by the subjective feeling of not being there yet: one is still on one's way to success.

Motivation to Pursue a Science, Engineering and Technology Career

On the question on what motivated them to choose a career path in SET, a substantial number of women traced their original motivation to become interested in science to their parents or particularly their fathers. One woman said her father motivated her to do her best. Some women emphasized the support of their parents to study science as shown by the following comment: *"... my family always supported me in anything I chose to take on ... "* One Indian scientist said her father played an enormous role in motivating his daughters to be independent in their behaviour and thinking. Although she wanted to follow in his footsteps as a science teacher, he motivated her to do a degree in science and always go one step further than she thought she could. In this way she progressed to doing a PhD. One woman working as a chemical engineer attributed her interest in engineering to her mother's motivation and the stories she told about her own work. Her mother worked in sales in a technological environment and her encounters with engineers and related companies instilled an interest in science in her daughter.

Some women said they had an interest in science from school. Some of these women decided to study science but it only became clear what they wanted to do when they embarked on their studies. Others were lead by mentors who were teachers at school motivating them to study science and mathematics while some had mentors at university that encouraged them to follow a particular route. For instance, a nuclear physicist first studied medicine but then due to the motivation of one of her lecturers moved into physics. One earth scientist working in chemistry said that her university bursary required her to work at the institution that funded her in her vacation. She found this stimulating and was inspired to further her career in her particular field. Others mentioned significant events when they were at school such as attending a scientific expo, going for excursions to scientific institutions or meeting people. One woman said a female pilot made an impression her.

Some women mentioned that they had teachers that were negative towards females taking science and mathematics subjects. Interestingly this negativity motivated them to take science. One woman moved into a science study direction despite the fact that she observed that few women take science degrees. She regarded this as a challenge. It seems as if teachers at school played a significant role in a number of women's lives either negatively or positively. Whereas for some the resistance from teachers to girls taking science and mathematics just spurred them on to embrace this as a challenge, others found support from, for instance, their parents. One woman said that despite the situation at school her father took it upon himself to make science and mathematics interesting. Capabilities and interest also played a role: some took science and mathematics because they were good at it and enjoyed it. Other women said that they had good teachers that made science interesting. However, one electrical engineer mentioned that she did not really have and interest in mathematics until she realized how important it will be for her career.

Some women landed in science or the particular specialties they work in serendipitously. One person first wanted to move into the human sciences but quickly realised that she had to consider a viable alternative. Although she had an interest in science her choice was based on the chemical engineering department that agreed to take here on as a student. Despite some landing in science by chance, others were dependent on issues such as finances. One black engineer had an interest in geology but due to the availability of a bursary studied engineering. Another black specialist in nuclear physics related an inspiring story how her brother worked to provide for her school and university studies under dire financial circumstances.

Gender-Based Difficulties in Achieving Success

Although difficulties such as discrimination based on gender were experienced in various contexts in and in varying intensity, the life milestones marked the events were discrimination and difficulty was experienced. Significant stages such as studying, obtaining a career, job promotion and so on presented the difficult situations. It also seems that it was mostly at the start of each significant event that difficulties were encountered. For instance, one woman that studied engineering said that she was only one of two women in the field and had to put up with derogatory remarks from males and lecturers. This is best expressed in the comments of another respondent: *"... you know you would always find that one professor that thinks, 'You are a girl, what do you know?'"* However, this had the effect of her working harder and making sure she avoided mistakes lest eliciting further criticism from the class.

Even at school women experienced negativity against girls. The perception was that girls are not that smart and cannot do mathematics and sciences. As one black nuclear physicist put it: *"You have to believe 'No, I can do it. I don't need someone else to tell me that I'm not capable of doing it.' Because when I was at school, there was this idea that girls are not as smart... are not smart enough...".*

Women referred to difficulties encountered when they started their careers. Mostly they had to put up with comments from males and experienced that males felt threatened with well-qualified women in traditional male careers. In a related matter one woman mentioned the difficulties she encountered at university especially where the top structures were mainly male-dominated and where there were few female lecturers. However in her current organization she did not experience similar problems as there were many female managers. Another woman disagreed with the viewpoint about universities that are male dominated: she felt that universities accommodate female scientists well. A woman engineer in a mining environment, on the other hand, currently still finds male-dominated boardroom and management surprising.

Not only career entrance presented problems. Even when competing for promotions with males women struggle to make their voices heard. A physicist said that they as women over time learn to be assertive, and speak out.

Some mentioned that there are still negative perceptions about females' capabilities. As soon as it becomes known that a woman is studying science related subjects such as chemistry or mathematics, doubt is expressed about the woman's abilities. Although a number of women mentioned the perception at school as was said above, this perception is still alive at tertiary institutions.

It seems as if the intensity and frequency of discrimination from males depends on the particular organization and the time when a woman entered her career. One woman said that she did not experience overt discrimination but it was subtle and not as severe as she had expected or as she has experienced in the past. Some actually said that they did not experience problems of discrimination against them as women.

Some difficulties were restricted to the physical environment of traditional male careers especially in the field of engineering and mining. Basic amenities for females were lacking and some of the females had problems with issues such as wearing boots and uniforms.

One woman mentioned having difficulties in her studies because of her race and not so much because of her gender. In the time of apartheid in South Africa students of colour at the particular university had difficulty obtaining information, bursaries and in general had difficult transport and residential problems. However, she did acknowledge that the times have changed and that students experience less logistical problems. In this regard another woman who moved into the field of business said that she experienced race and gender discrimination from early on in her career. A senior female academic in a managerial position mentioned that racial discrimination was more of an issue than gender discrimination. She experienced discrimination not only from males but white females as well. Sometimes the difficulties faced by the women combine race and gender. More than one woman said that they find it difficult to be accepted as black females in a technological or science environment.

Another theme related to the one above was the distinction between gender and age. A number of women referred to the fact that in the beginning of their careers age was more of a factor than gender. Experiencing difficulties as a young scientist or researcher could be compounded by gender as well. One woman said that as a young academic researcher she got assigned the menial jobs in the department whilst she sometimes experienced that this happened more often to the females than males. Another woman in a technical environment said that older personnel find it difficult to acknowledge the expertise and knowledge of younger workers just entering their careers. Although she understands that younger persons lack experience they have better knowledge about technical issues that older personnel find hard to accept. A mechanical engineer also said that although one enters one's career with new ideas and enthusiasm, one quickly gets opposed by older and entrenched workers.

Quite a number of women said that negativity towards them as women only spurred them on to work harder. Some said that if someone had a problem with their gender they regarded it as a problem the other party had to deal with. A number of women said that they had to prove themselves and that they had to work had to gain the respect of others as evidenced by the following comment *"You always have to work two hundred percent to actually make your mark... to see that actually you can do that. You always, always have to prove yourself"*. Some black scientists found their career paths as black women lonely.

Another prominent theme that was raised was the discrimination women experience because of issues surrounding the starting of a family. One particular woman explicitly gave the example of her managers that were worried that she might be pregnant while she was merely sick for a few days. It seems as if the phase of starting a family and maintaining a family presented problems. Having women on the workforce that are highly trained and busy with projects while there is the possibility that they might take maternity leave is problematic for companies. Maternity leave and raising a family with young children was an issue for quite a number of the younger women. It seems as if there is not much support from some companies such as having facilities for small children.

Encountering Difficulties in the Career as a Woman and Overcoming Them

One of the more senior women in a university structure mentioned that she experienced a time when women did not voice their opinions openly. She had to find a way to get her ideas heard because she felt she had good ideas and solutions to problems. She eventually did manage this strategy but did not get acknowledgement for her ideas. However, she said this did not bother her. Another woman said that young female scientists with potential get promoted too early and managerial responsibilities can be overwhelming. In the end the perception is created that women are not good scientists but for the wrong reasons. Women scientists and researchers tend to disappear in the crowd when it comes to promotions. One woman said in this regard that one has to become visible in the company and more than one woman said that self-marketing is important.

The younger women mentioned that their age rather than their gender was a problem. People associate a young age with inexperience and one woman said that one cannot rely on one's title to gain respect. She mentioned that the only way to overcome prejudice is hard work and professionalism, such as being on time for meetings and preparing well.

One white chemical engineer said that she relates difficulties to the differences in men and women's personalities. She experienced herself as overtly carrying her emotions into the boardroom while men manage not to do this.

One woman said that she was very self-conscious as a woman in a male-dominated tertiary study environment. She overcame her self-consciousness with determination and stubbornness. Perseverance, hard work, focus, motivation and determination were qualities often mentioned by women to overcome discrimination as reflected in the following quote:

You have to be very focused, focus is very important because physics is about doing something... you're trying to create something sound that doesn't exist, but you have a belief in this idea that you have. So, to carry the idea out you have to be extremely focused, be very determined and more than that, you have to work very, very hard. You have to be a person who when you start something you have to see it through, because otherwise you won't succeed.

Experiencing resistance from others was seen by some women as mere motivation to work harder and grow. Another said that she had to overcome prejudice by developing social skills and taking on discrimination as a challenge. One woman mentioned that self-assertion is important. Self-belief is also crucial to overcome difficulties such as believing that one is able to do the work. In this regard one woman said that she believes people discriminate out of ignorance and that believing this is her manner of coping with difficulty.

Remaining in SET

Quite a number of women said that they are comfortable with their work and that they enjoy it. Their work affords them the opportunity to do what they love such as do research and publish. Two women used a metaphor of their projects as their children which they nurture. Some mentioned that their fields are interesting and one mentioned that the developments in her field, which is chemical engineering, keep her motivated. One said she gets satisfaction from her work whilst a chemical engineer said she finds her work a challenge. The opportunity to use and exercise one's skills was also mentioned. One woman said that she remains in the field of SET to show that she can do it although she also said she was passionate about science. One woman voiced this explicitly and said one should be passionate about one's goals. Two women mentioned, amongst others, the financial rewards their current jobs provide them which is in contrast to one woman in academia who mentioned the lack of financial reward. She, however, said the university allows other benefits such as the freedom to do what one loves. Another woman said that remuneration is certainly a problem in SET careers. Other women in private sector also mentioned that their job allows them freedom to pursue interesting projects. One woman mentioned that that she is loyal to the group that she works with.

Some women mentioned that making an impact with their careers is important. One particular black metallurgical engineer said she stays in her SET career in order to show her community what is possible. One woman who worked in a laboratory for many years now works at a scientific research institute motivating and advancing other women's careers in science.

Influence of Personal Life

A number of married women said that they had the full support of their partners. Their partners take care of children and household duties when the women have to work or travel. As one woman said: *"Yes, my husband is very supportive and he encourages me too, in fact, he is like a bedrock to whatever I'm doing ... "* Those with children have mostly some means of support such as a nanny or a husband that shares the childcare responsibilities. The single women said they focus more on their work and some said they struggle to find a balance between work and personal life. One woman said she doubts whether she would have been able to achieve what she did if she had a family. Even some of the married women acknowledged that they struggle to find a work-personal life balance. Other women were very adamant about the fact that they manage their time and work in such a way as to spend time with their families. One woman acknowledged that she is very career-focused which took its toll on her marriage. Another woman said that her family is taking some strain because of her work. One couple went to the extreme of each following their careers in separate cities with their child living with the father. One woman had to endure separation from her small baby and family when she was offered a position to study overseas. Luckily the family reunited later in her programme.

It seemed as if some have strong family support systems. Some mentioned friends that provide support whilst others struggle to find a work-personal life balance. Some also mentioned that they get involved in other activities such as hiking and church as to experience a balanced life. Others however are satisfied with work taking up most of their time. Another mentioned the importance of support at work:

So, I think that type of support is really important, rather than maybe just being alone and then you don't know what to do, and then it can get to you a lot and it can affect your performance. So, it does help to have those types of relationships at work – somebody that you can talk to about home or about work.

Acting as a Role Model

Only particular aspects the women regarded as important for following a career in SET are indicated. A number of women mentioned that girls and women should be showed that that there are a number of characteristics important for moving into a career in science. The first characteristic is to have the belief that one can do it. A number of women emphasized self-belief and self-confidence: "*You know, set, just set your goals and go for it. There should be nothing holding you back. Don't let someone else tell you that you can't do it. Because anything is possible.*" One should not heed negative beliefs that one cannot study nor do a particular job. One woman said that you *"cannot define your life by what people think of you."* Even though perceptions still exist and are even encouraged at schools such as science and mathematics are for males, these should be systematically broken down. Women should know that they can move into any career. Commitment and passion were frequently mentioned. One should realize that certain careers involve hard work but having passion and perseverance one would be able to do it. One woman said that one should not depend on others to further one's career and take an active role in promoting themselves. Another woman said that background and personal circumstances need not prevent one from achieving success.

DISCUSSION

The women's stories about their careers in SET and what aspects make them successful are as diverse as their backgrounds and their careers. Many different aspects came to the fore but even so there are some common threads running through their stories. Without claiming that the sample was representative in any way, the women's experiences reflect some of the barriers discussed above. The sources of self-efficacy provide a profitable framework to interpret the sample's experiences in terms of explaining why women are successful in SET.

Structural and Individual Barriers

Despite their different careers it seems as if they faced similar obstacles on their way to success. Although it was mentioned by Maree et al. (2008) that some of the women did not remember having difficulties at certain crucial stages of their development, the bulk of the sample eventually mentioned experiencing some difficulty or discrimination at some stage of their careers. It seems as if life milestones or developmental phases provided ample opportunity for difficulties and according to Pell (1996) the leakages in the SET pipeline occur in early child hood, adolescence, when women enter into undergraduate studies, in graduate studies, at the time when they enter careers and when they enter into relationships. For example, school was often mentioned in our sample and the perception that girls are inferior in terms of ability in mathematics and science subjects. When the women elaborated on what one can do to encourage girls and women to move into a SET career most of them emphasized eradicating negative perceptions about girls' abilities. One can only assume that apart from the few that had non-discriminatory circumstances at school, most of them encountered the negative perceptions surrounding girls' ability and interest in science. These negative perceptions seem to have a real effect on eventual performance and research found that the gap between boys and girls performance in mathematics tend to widen as they grow older (Pell, 1996; Stake, 2003).

Entering a tertiary educational institution presents another difficulty for women as they again have to battle negative perceptions from males or they experience a major struggle with self-confidence to maintain their positions within classes. Most of the women in the sample mentioned that females were in the minority in classes. The situation was even more difficult for black women (Tang, 1997). Although some said they did not have problems in class those that did mention obstacles said that they had to put up with comments from males and in some instances negative attitudes from lecturers.

Entering a new job also presented difficulties until the women had time to prove themselves and gain the confidence of their colleagues and superiors. Most of the young women mentioned that their age as qualified scientists, some already having PhDs, did not contribute to gain the approval of colleagues. Some mentioned the initial resistance they have encountered from older and more experienced males.

Within their career development some obstacles were also encountered. Some mentioned that promotion is not a given even if one performs well as a female (Gunter & Stambach, 2003). On the other hand it seems that having well qualified women in a SET environment afforded some of them the opportunity to be promoted early and some of them moved into management paths very early in their careers. One woman mentioned this as a particular problem for female scientists to prove themselves in their fields of specialty.

Family life also presented its own set of problems. Some were reluctant to start families, some did and found it difficult whilst others managed to find a balance. Balancing family and work seems to be a perennial issue for SET women (Morgan, 1992). It also seemed that some companies where these women work also made it difficult for them to start families. It seemed as if the mere fact of falling pregnant and leaving work and projects for some time presented severe problems especially for the younger women in the sample. They realize that they are "abandoning" projects and it takes constant hard work to stay at the top of their game.

In addition maintaining a family and balancing work and family life provided additional challenges. Some struggled with finding a balance, and others acknowledged that they spend so much time on their work at the expense of their private lives. Those that did find a balance between their family and work lives frequently mentioned the enormous contribution their partners made.

Each of these life phases or milestones discussed above provides ample opportunity for women to abandon their fields of study and careers. The leaky pipeline metaphor illustrates the situation well. It is with crucial phases in a woman's life and career development where the leaks occur and where women leave their careers (Maree, et al., 2008). The question is then how did this sample of 29 women manage to stay in their careers and why did they not exit at the crucial phases in the leaky pipeline?

Mastery Experiences

From the summary of the various responses one can see that there are a number of issues involved. For instance, they all get rewarded in some way: the indicators for success they have mentioned are what they get out of their careers and what they get for staying in those careers. These indicators of success can function as mastery experiences that inform self-efficacy beliefs (Bandura, 1986, 1997) A distinction can be made between extrinsic and intrinsic indicators of success (Maree, et al., 2008). First of all, the extrinsic indicators are the concrete fruits of their labour such as the number of publications, invitations to speak and hold keynote addresses, the number of students supervised, completed projects, financial rewards and so on. It must be noted that only about 25% of the women mentioned these concrete indicators which means that its importance is in line with the findings discussed above: women are less inclined to quote mastery experiences than men (Zeldin, et al., 2008). Another concrete indicator mentioned by a minor portion of women was the impact of one's work with the emphasis on its beneficial nature. Having students that complete their studies successfully is one aspect but seeing that one's work has an impact on the environment or on the particular research area one is working in provides the reward. This signifies an important aspect of how mastery experiences for women can be viewed. For instance, one can examine whether even some mastery experiences for women are relationally embedded as opposed to the clear-cut and concrete experiences of men. Women might find the *beneficial* impact of projects or students completing, both which signify relationships with others, as enabling or informing self-efficacy. The importance of relationships was evidenced by a physicist's following comments:

And physics itself, it gives an opportunity to go away to these places to interact with these people, not only professionally, also culturally and just to see how other people live. So, you establish relationships with people you never thought you'd ever have relationships with. So, in that sense it's also very rewarding, those relations.

The one source almost never mentioned in research utilising Bandura's (1986, 1997) social cognitive theory is that of emotional and physiological experiences. It is however, reasonable to think that beneficial acts towards others or perceiving the beneficial impact of a project will elicit feelings of well-being and satisfaction, thus providing another source of self-efficacy. As one woman said:

... even though you sacrifice so much... but I suppose in a personal sense you get satisfaction, because that's what essentially you want. And also that you become part of something bigger than yourself, which is what we're seeing already now about all these big experiments...

Another emphasised emotions in the experience of success:

For me, in my career, I'll say I'm successful if maybe one day when I get into a room and handle a problem, I wouldn't feel like I'm a woman, I wouldn't feel the barrier of being a black woman – I would feel accepted as a scientist. You know, that assurance, I would feel like... okay, then maybe I have come a long way here...

The weight of this source must obviously be investigated and it will be interesting to see whether men also experience these feelings in their work.

Intrinsic indicators of success that were referred to as those aspect that have a personal impact on persons, can also function as mastery experiences but experiences embedded within intra- and interpersonal relationships. First, there are indicators referring to personal development and growth as we have seen above. Some of the women obtain personal satisfaction by growing in knowledge and work-related skills whilst others grow psychologically by learning how to manage emotions, be assertive, and so on. Secondly, there are goals to be reached which can refer to the usual developmental goals such as getting a PhD or a job but it can also refer to completion of research and projects. Goals that have not been realised cannot function as mastery experiences except as imaginal experiences (Maddux, 2002). Thus, visualising the goals and achieving those goals can have an enabling effect on self-efficacy. The last group of indicators also refers to goal achievement but to those not reached yet. The women that mentioned this are those that feel that they are on their way to success and they are not "there" yet. The difference between the last two indicators is the satisfaction of having achieved a goal and the drive to achieve a goal.

The indicators for success the sample mentioned can be seen as a manifestation of the women's drive for achievement and also as a source of self-efficacy formation. Reaching certain goals and achieving particular outcomes play such a large role in their careers that they were mentioned as indicators of success. To be a worthy indicator of success it must appeal to a woman's motivation for achieving something and we would like to hypothesise that these women currently working and succeeding in SET careers have a high drive for achievement (cf. Etzkowitz, et al., 1994). One can argue that a high drive for achievement, clear goals and self-efficacy feed each other. Believing that one can do something or achieve a goal can be sourced from imaginal experiences but also requires goals or clear indicators of success, and is further strengthened by success.

Vicarious and Support Experiences

When looking at the question of what motivated these women to enter SET careers diverse answers were provided. However, a theme that constantly came to the fore was that of interest fostered within a particular set of relationships (Acker & Oatley, 1993). Parents, teachers, and mentors cultivated an interest in science (cf. Roger & Duffield, 2000). It is probably obvious that one would not move in a particular study field or career if one is not interested in it (cf. Siann & Callaghan, 2001, p. 93) but the question of course is whether interest is enough to guarantee success. As was stated above Weinburgh (1995) found that positive attitude is correlated with high achievement in science subjects for both boys and girls. However, interest and positive attitude is more of a requirement for higher achievement for girls (Weinburgh, 1995). Furthermore, the belief that one can do something is formed within these significant relationships, and research showed that self-efficacy more than interest can account for career success (Hackett, 1997).

The role of parents and especially fathers were emphasized in some of the women's narratives in developing a real interest in science confirming the importance of vicarious and support experiences within relationships (cf. Madill, Ciccocioppo, Stewin, Armour, & Montgomerie, 2004; Quimby & O'Brien, 2004; Scott & Mallinckrodt, 2005). This encouragement from parents and others served to overcome the perception that women are not capable of doing science and mathematics. It seems as if an active role of a significant person is important in fostering an interest in science and mathematics but also in conveying the message that girls and women are able to excel in it (Ferry, Fouad, & Smith, 2000).

A few women mentioned other reasons why the have developed an interest in science especially at school. It almost seems as if a streak of stubbornness and rebellion against the idea that girls cannot do science played a role in their interest development. It could be that those women who as young learners reacted against negative stereotypes had the drive to prove others wrong and themselves as capable of doing what people think they cannot do. Although a certain dynamic may be at play here, their reactions developed into real interests and successful careers. One might argue that a drive for achievement and a drive to prove one self capable of doing things are related.

The Role of Resiliency

How did the women overcome gender-related difficulties? From the results above one can certainly surmise that a number of women experienced gender-related difficulties and as our discussion on the developmental phases suggested, there were particular themes recurring such as doubt about women's abilities in particular SET fields (see Cronin & Roger, 1999, p. 650; Ellemers, van den Heuvel, de Gilder, Maass, & Bonvini, 2004), remarks from males, lecturers, and co-workers. What is also true is that in some instances age and race along with gender played a role in experiencing difficulties, emphasising the intersection of these issues in the problem of loosing women in SET (Carlone & Johnson, 2007).

Some women reacted by displacing the problem to the other party. This could show an attitude of high self-efficacy where a woman believes in her own capabilities and do not have the need to prove herself in anyway. Other women said that negativity just spurred them on to work harder and prove themselves. For some women proving oneself to gain respect from others was important, a theme constantly found in research (Etzkowitz, et al., 1994; Evetts, 1998; Henwood, 1996). Some said they did not experience discrimination and some of those that did mentioned it in passing. There are a number of ways to react to discrimination or negativity the main point is the following: whether these women rationalized about negativity towards them, said they ignored it or whether they exhibited confidence in their own capabilities, they somehow overcame the negativity and in such a way that it did not deter them from their goals. In a sense they showed resiliency, determination and tenacity in the light of discrimination. Of course, there were women that still expressed feelings of loneliness and discrimination to some extent but even they managed to carry on in their careers (Etzkowitz, et al., 1994).

The women in the sample mentioned a number of characteristics in terms of overcoming gender-based difficulties but they also complemented these with characteristics a girl or woman should have when considering entering a SET career. Personal characteristics such as perseverance, focus, motivation and determination were mentioned. Although this is an example in a medical context, 60% of a small survey Kass, Souba and Thorndyke (2006) amongst leading female surgeons (N = 10) also said perseverance and resiliency contributed to their success reflecting the findings of Zeldin and Pajares (2000), and Zeldin et al. (2008).

More than Self-Efficacy?

Passion for their work was mentioned in various contexts by the women but especially as an answer to why they remain in their SET careers. Being passionate about one's work expresses a sentiment that is much more forceful than being interested in a particular field. One can argue that passion and a characteristic such as the drive to achieve are related because both refer to inner energy focused on some goal or object. From the transcripts of the interviews it became apparent that these successful women were also very focused on their work. Their work played an important part in their lives in terms of the goals they wanted to achieve or have achieved. A number of women mentioned hard work as key to success. Focus as concentration on one's work in terms of cognitive resources and energy is closely associated with also putting in an effort into their jobs. Hard work means concentrating one's energy on the work, sometimes allowing it to interfere with one's personal life. The ability to channel and focus energy into one's work can also be associated with motivation. High levels of motivation to achieve goals and the ability to muster the energy to work hard are important. Determination and perseverance are also qualities that speak of the ability not to be put off by difficulties but maintain focus on goals.

The question then is whether we have here a sample of women that are successful due to what they are or due to what they have become (see Tang, 2003, p. 334)? In other words, what role does personality characteristics play in achieving success in science? The ability to stay focused, be passionate and to work hard seem like personality characteristics as are resiliency and high motivation. If this is the case then focusing on structural and some individual barriers will not help much in getting women into science. One must then only make it easier for those "naturals" to stay in the pipeline. However, we believe that although the influence of personality is a point that must be investigated, self-efficacy is the saving grace of getting women into SET. It seems as if high-levels of self-efficacy play a fundamental and mediating role in developing women's resilience and goal achieving ability. Because self-efficacy is not a personality characteristic (Maddux, 2002), several studies showed that interventions can enhance self-efficacy (Betz & Schifano, 2000; Sullivan & Mahalik, 2000). According to the sample one should believe that anything is possible and that, despite what others say or how difficult one's circumstances is, one should have a steadfast faith in one's ability to do the things that one wants to do. This self-efficacy belief probably is associated with the drive to prove one as capable of achieving certain goals and in a sense protects one from giving up when encountering difficulties. Self-efficacy belief is a mental attitude probably crucial to feeding the qualities we discussed above but also as functioning as protection against negative feedback (cf. Kass, et al., 2006). One can easily imagine someone not believing in her own abilities quickly giving up when encountering negativity from the environment. Negative feedback, such as women are not capable of doing, for instance, engineering, can easily increase the self-doubt of those with low self-esteem. Beasley, Thompson, and Davidson (2003) found that cognitive hardiness and aspects of coping style enable individuals to be resilient towards life stresses. They viewed cognitive hardiness as Kobasa (1979) defined it, namely, as (a) the belief that cognitive hardy individuals can influence and control events, (b) that they have a clear commitment to their own values and goals in life and (c) change is not viewed as a threat but as a challenge. These qualities of cognitive hardiness probably encompass some of the elements we have described about positive-self belief.

CONCLUSION

It is clear form our discussion that the women in our sample were exposed to similar difficulties as all women entering SET careers. They have difficulties with balancing work-life and personal-life and they have problems raising families. Some managed to overcome these difficulties with a strong support system while others still struggle. Most of the women in the sample were exposed to some form of discrimination during their lives. It seems as if the leaky pipeline phenomenon applied to the sample of successful women just as much as it applies to women leaving their SET careers. As compared to the brief literature overview of problems women face in SET careers our sample of successful women faced the similar problems. The question is then what made the women in our sample stay in their careers? Did they face less structural problems or individual barriers? Did they have a particular set of characteristics? From the sample's comments a number of characteristics were inferred and identified. We fully realize that one important limit of this study, namely, comparing our results with a sample of women that did exit their SET careers, precludes final answers on what makes women successful in SET careers. However, one can certainly take the clues from our study as supporting some existing trends and providing stimulus for further investigation. It seems as if personal characteristics played a role in making the women in our sample successful. Over and above the characteristics of focus, determination, and perseverance, a high drive for achievement was present. They were passionate, very goal orientated and achieving these goals was very important. Most importantly of all a strong sense of self-belief or a high level of self-efficacy was present. They believed that they are able to do the work and that anything they set their minds to is possible. It seems as if this attitude, which feeds resilience and cognitive hardiness, enables them to achieve and maintain focus. It provides a buffer against obstacles and difficulties.

This study showed that it is important to understand the personal qualities of successful women. Of course, one should beware of generalizing because as we have seen life and career stories differ. However, it is necessary to understand and investigate these qualities and investigate the best way to foster such characteristics in girls and women wanting to enter SET.

ACKNOWLEDGMENT

The National Council for Innovation (NACI) is gratefully acknowledged for their permission to use their report in this study.

REFERENCES

Acker, S., & Oatley, K. (1993). Gender issues in education for science and technology: current situation and prospects for change. *Canadian Journal of Education, 18*(3), 255–272. doi:10.2307/1495386

Alper, J. (1993). The pipeline is leaking women all the way along. *Science New Series, 260*(5106), 409–411.

Ancis, J. R., & Phillips, S. D. (1996). Academic gender bias and women's behavioral agency self-efficacy. *Journal of Counseling and Development, 75*(2), 131–137.

Armstrong, D. J., Riemenschneider, C. K., Allen, M. W., & Reid, M. R. (2007). Advancement, voluntary turnover and women in IT: A cognitive study of work-family conflict. *Information & Management, 44*, 142–153. doi:10.1016/j.im.2006.11.005

Bandura, A. (1977). Self-efficacy: toward a unifying theory of behavioral change. *Psychological Review, 84*(2), 191–215. doi:10.1037/0033-295X.84.2.191

Bandura, A. (1986). Social foundations of thought and action: a social cognitive theory. Englewood Cliffs, NJ: Prentice-Hall.

Bandura, A. (1997). Self-efficacy: the exercise of control. New York: W.H. Freeman.

Barber, L. A. (1995). U.S. women in science and engineering, 1960-1990: progress towards equity? *The Journal of Higher Education, 66*(2), 213–234. doi:10.2307/2943912

Beasley, M., Thompson, T., & Davidson, J. (2003). Resilience in response to life stress: the effects of coping style and cognitive hardness. *Personality and Individual Differences*, *34*, 77–95. doi:10.1016/S0191-8869(02)00027-2

Betz, N. E., & Hackett, G. (1981). The relationship of career-related self-efficacy expectations to perceived career options in college women and men. *Journal of Counseling Psychology*, *28*(5), 399–410. doi:10.1037/0022-0167.28.5.399

Betz, N. E., & Hackett, G. (1997). Applications of self-efficacy theory to the career assessment of women. *Journal of Career Assessment. Special Issue: Career assessment for women. Theory into Practice*, *5*(4), 383–402.

Betz, N. E., & Schifano, R. S. (2000). Evaluation of an intervention to increase realistic self-efficacy and itnerests in college women. *Journal of Vocational Behavior*, *56*, 35–52. doi:10.1006/jvbe.1999.1690

Birbaumer, A., Lebano, A., Ponzellini, A., Tolar, M., & Wagner, I. (2007). From the margins to a field of opportunities: Life story patterns of women in ICT. *Women's Studies International Forum*, *30*, 486–498. doi:10.1016/j.wsif.2007.09.001

Blickenstaff, J. C. (2005). Women and science careers: leaky pipeline or gender filter? *Gender and Education*, *17*(4), 369–386. doi:10.1080/09540250500145072

Brainard, S. G., & Carlin, L. (1998). A six-year longitudinal study of undergraduate women in engineering and science. *Journal of Engineering Education*, *87*(4), 17–27.

Buck, G. A., Clark, V. L. P., Leslie-Pelecky, D., Lu, Y., & Cerda-Lizarraga, P. (2008). Examining the cognitive processes used by adolescent girls and women scientists in identifying science role models: A feminist approach. *Science Education*, *92*(4), 688–707. doi:10.1002/sce.20257

Carlone, H. B., & Johnson, A. (2007). Understanding the science experiences of successful women of color: Science identity as an analytic lens. *Journal of Research in Science Teaching*, *44*(8), 1187–1218. doi:10.1002/tea.20237

Ceci, S. J., & Williams, W. M. (2007). Why aren't more women in science?: top researchers debate the evidence (1st ed.). Washington, DC: American Psychological Association.

Cronin, C., & Roger, A. (1999). Theorizing progress: women in science, engineering, and technology in higher education. *Journal of Research in Science Teaching*, *36*(6), 637–661. doi:10.1002/(SICI)1098-2736(199908)36:6<637::AID-TEA4>3.0.CO;2-9

Davis, K. S. (2002). Advocating for equitable science-learning opportunities for girls in an urban city youth club and the roadblocks faced by women science educators. *Journal of Research in Science Teaching*, *39*(2), 151–163. doi:10.1002/tea.10014

Downing, R. A., Crosby, F. J., & Blake-Beard, S. (2005). The perceived importance of developmental relationships on women undergraduates' pursuit of science. *Psychology of Women Quarterly*, *29*(4), 419–426. doi:10.1111/j.1471-6402.2005.00242.x

Ellemers, N., van den Heuvel, H., de Gilder, D., Maass, A., & Bonvini, A. (2004). The underrepresentation of women in science: differential commitment or the queen bee syndrome? *The British Journal of Social Psychology*, *43*(3), 315–338. doi:10.1348/0144666042037999

Etzkowitz, H., Kemelgor, C., Neuschatz, M., Uzzi, B., & Alonzo, J. (1994). The paradox of critical mass for women in science. *Science*, *266*, 51–54. doi:10.1126/science.7939644

Evetts, J. (1998). Managing the technology but not the organization: women and career in engineering. *Women in Management Review*, *13*(8), 283–290. doi:10.1108/09649429810243144

Ferry, T. R., Fouad, N. A., & Smith, P. L. (2000). The role of family context in a social cognitive model for career-related choice behavior: a math and science perspective. *Journal of Vocational Behavior, 57*, 348–364. doi:10.1006/jvbe.1999.1743

Fox, M. F. (1998). Women in science and engineering: theory, practice, and policy in programs. *Signs, 24*(1), 201–223. doi:10.1086/495325

Fox, M. F., Sonnert, G., & Nikiforova, I. (2009). Successful programs for undergraduate women in science and engineering: "adapting" versus "adopting" the institutional environment. *Research in Higher Education, 50*(4), 333–353. doi:10.1007/s11162-009-9120-4

Good, C., Aronson, J., & Harder, J. A. (2008). Problems in the pipeline: Stereotype threat and women's achievement in high-level math courses. *Journal of Applied Developmental Psychology, 29*, 17–28. doi:10.1016/j.appdev.2007.10.004

Gunter, R., & Stambach, A. (2003). As balancing act and as game: how women and men science faculty experience the promotion process. *Gender Issues, 21*(1), 24–42. doi:10.1007/s12147-003-0020-1

Hackett, G. (1997). Self-efficacy in career choice and development. In A. Bandura (Ed.), Self-efficacy in changing societies (pp. 232-258). Cambridge, UK: Cambridge University Press.

Hall, R. M., & Sandler, B. R. (1982). The Classroom Climate: A Chilly One for Women? (Report). Washington, DC: Project on the Status and Education of Women, Association of American Colleges.

Hartman, H., & Hartman, M. (2008). How undergraduate engineering students perceive women's (and men's) problems in science, math and engineering. *Sex Roles, 58*, 251–265. doi:10.1007/s11199-007-9327-9

Henwood, F. (1996). WISE choices? Understanding occupational decision-making in a climate of equal opportunities for women in science and technology. *Gender and Education, 8*(2), 199–214. doi:10.1080/09540259650038860

Hodgson, B. (2000). Women in science--or are they? *Physics Education, 35*(6), 451–453. doi:10.1088/0031-9120/35/6/313

Hollinger, C. L. (1983). Self-Perception and the career aspirations of mathematically talented female adolescents. *Journal of Vocational Behavior, 22*, 49–62. doi:10.1016/0001-8791(83)90005-2

Kass, R. B., Souba, W. W., & Thorndyke, L. E. (2006). Challenges confronting female surgical leaders: overcoming the barriers. *The Journal of Surgical Research, 132*, 179–187. doi:10.1016/j.jss.2006.02.009

Kinzie, J. (2007). Women's paths in science: a critical feminist analysis. *New Directions for Institutional Research, 2007*(133), 81–93. doi:10.1002/ir.206

Kobasa, S. C. (1979). Stressful life events, personality and health: an inquiry into hardiness. *Journal of Personality and Social Psychology, 37*, 1–11. doi:10.1037/0022-3514.37.1.1

Lynn, R. (2008). Review of Why aren't more women in science? *Intelligence, 36*(4), 380–382. doi:10.1016/j.intell.2007.06.005

Maddux, J. E. (2002). Self-efficacy: the power of believing you can. In C. R. Snyder & S. J. Lopez (Eds.), Handbook of Positive Psychology (pp. 277-287). Oxford, UK: Oxford University Press.

Madill, H. M., Ciccocioppo, A.-L., Stewin, L. L., Armour, M.-A., & Montgomerie, T. C. (2004). The potential to develop a career in science: young women's issues and their implications for careers guidance initiatives. *International Journal for the Advancement of Counseling, 26*(1), 1–19. doi:10.1023/B:ADCO.0000021546.17402.5b

Maree, M., Maree, D. J. F., Botha, C., & Gcabo, R. (2008). Changing perceptions of women in the science, engineering and technology industry: evaluating the career histories of role models in South Africa. Pretoria, South Africa: Institute for Women and Gender Studies, University of Pretoria.

Mayer-Smith, J., Pedretti, E., & Woodrow, J. (2000). Closing of the gender gap in technology enriched science education: a case study. *Computers & Education, 35*, 51–63. doi:10.1016/S0360-1315(00)00018-X

Morgan, C. S. (1992). College students' perceptions of barriers to women in science and engineering. *Youth & Society, 24*(2), 228–236. doi:10.1177/0044118X92024002006

Morris, L. K., & Daniel, L. G. (2008). Perceptions of a Chilly Climate: Differences in Traditional and Non-Traditional Majors for Women. *Research in Higher Education, 49*(3), 256–273. doi:10.1007/s11162-007-9078-z

Murrell, A. J., & Zagenczyk, T. J. (2006). The gendered nature of role model status: an empirical study. *Career Development International, 11*(6), 560–578. doi:10.1108/13620430610692953

National Advisory Council on Innovation (NACI). (2004). Facing the facts: Women's participation in science, engineering and technology. Pretoria, South Africa: National Advisory Council on Innovation (NACI) and the Department of Science and Technology.

Özbilgin, M., & Healy, G. (2004). The gendered nature of career development of university professors: the case of Turkey. *Journal of Vocational Behavior, 64*, 358–371. doi:10.1016/j.jvb.2002.09.001

Papastergiou, M. (2008). Are computer science and information technology stil masculine fields? High school students' perceptions and career choices. *Computers & Education, 51*(2), 594–608. doi:10.1016/j.compedu.2007.06.009

Pell, A. N. (1996). Fixing the leaky pipeline: women scientists in academia. *Journal of Animal Science, 74*, 2843–2848.

Penner, A. M., & Paret, M. (2008). Gender differences in mathematics achievement: Exploring the early grades and the extremes. *Social Science Research, 37*, 239–253. doi:10.1016/j.ssresearch.2007.06.012

Phipps, A. (2006). 'I can't do with whinging women!' Feminism and the habitus of 'women in science' activists. *Women's Studies International Forum, 29*, 125–135. doi:10.1016/j.wsif.2006.03.010

Plumm, K. M. (2008). Technology in the classroom: burning the bridges to the gaps in gender-biased education? *Computers & Education, 50*, 1052–1068. doi:10.1016/j.compedu.2006.10.005

Quimby, J. L., & O'Brien, K. M. (2004). Predictors of student and career decision-making self-efficacy among nontraditional college women. *The Career Development Quarterly, 52*(4), 323–339.

Rivera, L. M., Chen, E. C., Flores, L. Y., Blumberg, F., & Ponterotto, J. G. (2007). The effects of perceived barriers, role models, and acculturation on the career self-efficacy and career consideration of Hispanic women. *The Career Development Quarterly, 56*(1), 47–61.

Roger, A., & Duffield, J. (2000). Factors underlying persistent gendered option choices in school science and technology in Scotland. *Gender and Education, 12*(3), 367–383. doi:10.1080/713668300

Rosser, S. V. (2003). Attracting and retaining women in science and engineering. *Academe, 89*(4), 24–28.

Scott, A. B., & Mallinckrodt, B. (2005). Parental emotional support, science self-efficacy, and choice of science major in undergraduate women. *The Career Development Quarterly, 53*(3), 263–273.

Settles, I. H., Cortina, L. M., Malley, J., & Stewart, A. J. (2006). The climate for women in academic science: the good, the bad, and the changeable. *Psychology of Women Quarterly, 30*(1), 47–58. doi:10.1111/j.1471-6402.2006.00261.x

Siann, G., & Callaghan, M. (2001). Choices and barriers: factors influencing women's choice of higher education in science, engineering and technology. *Journal of Further and Higher Education, 25*(1), 85–95. doi:10.1080/03098770020030524

Sonnert, G., Fox, M. F., & Adkins, K. (2007). Undergraduate women in science and engineering: effects of faculty, fields, and institutions over time. *Social Science Quarterly, 88*(5), 1333–1356. doi:10.1111/j.1540-6237.2007.00505.x

Stake, J. E. (2003). Understanding male bias against girls and women in science. *Journal of Applied Social Psychology, 33*(4), 667–682. doi:10.1111/j.1559-1816.2003.tb01918.x

Stewart, M. (1998). Gender Issues in physics education. *Educational Research, 40*(3), 283–293.

Sullivan, K. R., & Mahalik, J. R. (2000). Increasing career self-efficacy for women: Evaluating a group intervention. *Journal of Counseling and Development, 78*(1), 54–62.

Tang, J. (1997). The glass ceiling in science and engineering. *Journal of Socio-Economics, 26*(4), 383–406. doi:10.1016/S1053-5357(97)90003-2

Tang, J. (2003). Women Succeeding in Science in the Twentieth Century. *Sociological Forum, 18*(2), 325–342. doi:10.1023/A:1024099831993

Weinburgh, M. (1995). Gender differences in student attitudes toward science: a meta-analysis of the literature from 1970 to 1991. *Journal of Research in Science Teaching, 32*(4), 387–398. doi:10.1002/tea.3660320407

Zeldin, A. L., Britner, S. L., & Pajares, F. (2008). A Comparative Study of the Self-Efficacy Beliefs of Successful Men and Women in Mathematics, Science, and Technology Careers. *Journal of Research in Science Teaching, 45*(9), 1036–1058. doi:10.1002/tea.20195

Zeldin, A. L., & Pajares, F. (2000). Against the odds: Self-efficacy beliefs of women in mathematical, scientific, and technological careers. *American Educational Research Journal, 37*(1), 215–246.

Profile 9
Engineering Career Spotlight:
Project Manager for a Major Power Strengthening Project in Regional Queensland

© 2008 Queensland Government Office of Women. Used with permission

Julie Martin, Electrical Engineer, Power Systems Team, TrackStar Alliance

Julie Martin has fifteen years postgraduate experience in electrical systems engineering, graduating with honours from the University of Queensland in 1994. She was born in Brisbane in 1973 to mother Bronwyn (doctor) and father Jim (teacher). Dux of Loreto College, Coorparoo in 1990, Julie was encouraged to pursue her own academic interests in Maths and Physics despite having grades to enter Medicine or Law. Believing that an Engineering degree best combined the theoretical study of these sciences and their practical application, Julie embarked on a Bachelor of Electrical Engineering course. After successfully completing her degree, Julie spent the next three years working in NSW coal mines and was the first female in history to sit and pass the written examination for the NSW statutory coal mining electrical engineer's certificate of competency. Leaving the mining industry after her Professional Graduate program ended, Julie worked in a variety of both operational and design engineering roles in London, Sydney and India. Her extensive knowledge and experience

Copyright © 2010, IGI Global. Copying or distributing in print or electronic forms without written permission of IGI Global is prohibited.

includes design, installation, operation, maintenance and renewal of electrical distribution systems, both high voltage and low voltage, in various industries, including rail infrastructure, mining and construction. Prior to accepting her current role, Julie cites her most rewarding work experience as the commissioning of all five high voltage substations designed and built to supply the new 13km twin bore Epping to Chatswood Rail Link tunnel in Sydney.

Introduction

Julie Martin is the Project Manager for TrackStar Alliance's coal rail power systems projects. An electrical engineer, she is responsible for the delivery of two power traction strengthening projects for QR in regional Queensland with a total project value of $50 million. These projects are part of a major overhaul of Queensland's coal network, and part of a program of infrastructure projects to be delivered as part of the COAL*RAIL* Master Plan.

In 2006, through the SEQIP*RAIL* program, QR first partnered with the TrackStar Alliance, who would design and construct an initial four rail projects in South East Queensland. It was through TrackStar's proven success to mobilise and get work underway on other projects including the Caboolture to Beerburrum Track Duplication Project and the Robina to Varsity Lakes Rail Extension Project that QR awarded the Alliance additional power strengthening projects. For QR, TrackStar Alliance has become a powerful means to attract and retain quality professionals, with the flexibility to link back into their partner companies and draw out expertise when the demand was there. It was through alliance partner company United Group that Julie was put forward to head up the coal power strengthening projects, a result of her extensive experience and proven results.

Julie Martin is TrackStar Alliance's first female Project Manager. She previously ran the commissioning of five high-voltage substations required to power the Epping – Chatswood rail tunnel in Sydney. At TrackStar, Julie is continuing to achieve, with the Alliance's first remote project outside of South East Queensland already underway, in record time.

Julie's role at TrackStar is multifaceted. She is an important interface with the separate teams responsible for delivering the project.

Julie is in regular contact with QR's Power Systems division to ensure TrackStar maintains consistency and compliance with the client expectations. She also is the conduit between the TrackStar construction team located on site at Bolingbroke (near Sarina) and the Alliance Program office in Milton, tracking construction progress, project costs and managing interfaces with other project stakeholders Powerlink and Ergon Energy. Julie will also project manage a similar power strengthening upgrade planned for Dalrymple Bay Coal Terminal (DBCT).

Julie was put forward to head up the coal power strengthening projects as a result of her extensive experience and proven results.

A Speedy Solution

TrackStar needed to price the Bolingbroke project for QR and get construction underway as soon as possible. The drivers were the coal producers calling for more efficiency in transporting coal to the ports.

The current electrical capacity on the Goonyella system limits the number of trains on a track section at any time, impacting train performance and rail network capacity. The Goonyella system services 17 coal mines in the northern and central areas of the Bowen Basin over a track length of 620 km. The significance of the Goonyella system to the Queensland economy is that it carries the highest tonnages in the QR network representing 55% of the total network coal tonnages. The power strengthening projects will help support expansion of coal mining activities in the Bowen Basin which underpins economic growth in the region.

Through TrackStar and Julie's smart guidance, QR is responding to the challenges of meeting this prospective demand for additional rail traffic by delivering new power systems at both Bolingbroke and Dalrymple Bay Coal Terminal. Bolingbroke is the first of these to start construction.

The Project

Julie's team is replacing the existing Bolingbroke Track Sectioning Cabin (TSC) with a new feeder station and two new TSCs which will be installed at approximately the mid-points between the three feeder stations (Oonooie, Bolingbroke and Wandoo).

Julie is responsible for:

- Overseeing detailed design of the Bolingbroke Feeder Station and associated sites comprising power and auxiliary systems, structural foundations and civil works
- Managing the procurement of all required materials and subcontract packages
- Managing and executing the site construction works
- Managing the installation of client issued power related equipment.

A New Approach

In the TrackStar environment, designers and constructors work together from the start and peer reviews and challenge teams provide a rigorous review process. Flourishing in this environment, Julie oversaw the development and sign-off of a project Target Cost Estimate (TCE) in just three months. Without the Alliance model, the project would have taken up to six months to get up to a costing stage. But with systems already in place, the designers and constructors already based at the Milton Program office with relationships and trust established, the Bolingbroke TCE was approved in early March 2008. In June 2008, eight months after first coming to TrackStar, considerable early works have already taken place at the three project sites established: Black Mountain, Bolingbroke and Balook.

Innovation

Most of QR's existing substations were built during the 1970s and 1980s when the metro and coalfield networks were first electrified. Through her position, Julie is project managing one of the first applications to use new Gas Insulated Switchgear (GIS) based technology.

The new equipment will be installed at the Bolingbroke substation and Julie's second project at the Dalrymple Bay Coal Terminal, both on the Goonyella network. GIS will modernise the existing system, providing increased ratings, high reliability, low maintenance and a higher degree of operator safety all at a lower price than QR's existing trackside switchgear.

New Ground for QR

Julie is treading new ground in her role. Traditionally QR was responsible for upgrading its power supply, but the infrastructure boom and growing Queensland population meant there simply weren't the resources to carry out the work. Julie is the first to sing the praises of the Alliance program approach to delivering projects, meaning information and resources are shared, lessons learnt are carried across projects and resources are on hand to allow faster delivery times. Under Julie's guidance, the Bolingbroke construction crew, with the support of the team at Milton have battled the logistics of remote working and established a presence at the Black Mountain site in regional Queensland (the first TSC location), mobilised the civil contractor and begun works.

The second TSC will be built at Balook and already the site's top soil has been stripped back and earthworks are underway. The two TSCs will connect with a new feeder station to be constructed at Bolingbroke to enhance power and boost capacity to support the projected coal industry traffic.

Creating a Legacy and Providing Community Benefit

Because the project is implementing the new Gas Insulated Switchgear, Julie's team also has a responsibility to upskill a QR team who will be maintaining the equipment, once handover is complete.

She must ensure that the handover considers project deliverables for the Operational Rail System including life cycle issues, training, spares, operational cost issues, post project use deliverables and includes an implementation strategy covering operator liaison, communication strategies and post implementation audits.

By implementing the new GIS technology, TrackStar is delivering value to the client:

- TrackStar Alliance power systems projects will provide assistance to QR with managing the contract and work with QR to integrate the new switchgear and associated operating systems
- TrackStar's power systems projects will provide support and the necessary knowledge transfer to build resource capability for QR teams using the new equipment
- Implementing the new GIS technology is an opportunity to further develop skills and maintain resourcing capability for QR teams using the new equipment.

The new GIS technology has been selected by QR for all future traction substation works.

Its application on the Bolingbroke and DBCT projects will set the benchmark for future upgrades, and help QR to establish the standard to which all subsequent traction substation work in QR's network will be designed and implemented. This standard will also be applied on future Metro Rail system upgrades in Brisbane, including the construction of a new feeder station at Roma Street station which is also being delivered by TrackStar Alliance.

Part of the Bigger Picture

This information sharing for the metro upgrades will improve passenger reliability for city travel. When faced with rising petrol prices, reliable public transport systems provide the community with added assurance and better options.

The two coal power strengthening projects Julie is overseeing are critical. If no action was taken to rectify these power system capacity problems, future growth would be hindered by traffic restrictions that would need to be imposed by QR. This would impact the coal export supply chain for the Goonyella system and in turn result in losses in revenue to QR, ports, coal companies and Queensland government royalties. Most importantly it would do irreparable damage to Queensland's position as a global leader in the export of coal. These projects are therefore making a significant contribution to the Government's plan for managing its growth challenges.

Chapter 10
Women Access to Computers and Internet:
A Malaysian Perspective

Maslin Masrom
Universiti Teknologi Malaysia, Malaysia

Zuraini Ismail
Universiti Teknologi Malaysia, Malaysia

ABSTRACT

Malaysia is a multi-ethnic, multi-lingual, multi-cultural society characterized by distinct values and norms about many issues including the role of women in society. Since Malaysia gained independence in 1957, Malaysian women have participated and contributed towards the social and economic development of the country. This was made possible by the policy standpoint of the Malaysian government which considers women are an important pool of resource that can be mobilized towards achieving the national development agenda. After fifty-two years of independence, Malaysia is one of the most developed countries in Southeast Asia. She has seriously geared its effort toward information and communication technology (ICT) by establishing the Multimedia Super Corridor (MSC) in 1996 with the widespread use of computer in Malaysia women have been found to have benefited from the modern technologies. They enjoy greater access to computers and Internet. Therefore, this chapter emphasizes accessibility, delivering access to the computers and internet info-structure to women in Malaysia. A framework for analyzing Malaysian women accessing computers and the internet is presented and discussed.

INTRODUCTION

Information and communication technology (ICT) has entered the organizations, office and home, and indirectly influenced employees, individual and family life. *Information technology (IT)* comprises of hardware (e.g. personal computer, printer, and scanner) and software (e.g. application and system program) that allow us to access, retrieve, store, organize, manipulate and present information by electronic means. Meanwhile, *communication technology* consists of telecommunication tools that allow information to be accessed and sought, for example telephones, facsimiles, and modems.

The Malaysian government has heavily invested in ICT centers all over the country. IT and in par-

DOI: 10.4018/978-1-61520-657-5.ch010

Copyright © 2010, IGI Global. Copying or distributing in print or electronic forms without written permission of IGI Global is prohibited.

ticular the internet have become an important part of the Malaysian life. Malaysia has established its Multimedia Super Corridor (MSC) in 1996 (NITC, 2004). The provision of ICT platforms resulted in mega implementation such as e-government and e-governance, besides the surge of multimedia industry and the existence of biotechnology valley. The very presence of IT has lead to ICT literacy and further accelerate human resource development.

Malaysia is a developing country that has integrated ICT into its grand narrative of modernity. The national development plan Vision 2020 is aimed at giving all Malaysians a fully developed and modern society by 2020. The Malaysian government has been an enthusiastic supporter of internet technology since the early 1990s, and has employed a range of policies to encourage Malaysian businesses to venture online. ICT is highlighted as a media to eradicate poverty and facilitate progress.

The story of the commercial internet in Malaysia began in 1990, when the Malaysian Institute of Microelectronic Systems (MIMOS, now MIMOS Berhad) launched JARING (Joint Advanced Integrated Networking), the first Malaysian ISP. It was not until 1992, however, with the installation of a satellite link between Malaysia and the United States of America, that Malaysian users gained easy access to the internet. Subsequently, Malaysia's second internet service provider (ISP), TMNet, launched in 1995. Since then, the market for both commercial and residential internet access has grown steadily. There are now eight ISPs within Malaysia offering both dial-up and broadband connectivity, namely TIMENet Central, Jaring, TMNet, Streamyx, Nasionet, Maxis Net, Silicon Central and Malaysia Online.

Digital Divide

Digital divide researchers have pointed out the critical gaps in society among different groups in the context of their access to technology (Dutta-Bergman, 2005). Most definitions of the term 'digital divide' refers to access to ICTs (Harris *et al.*, 2007). Harris *et al.* (2007) reports that Malaysia has made considerable advances to bridge the digital divide. According to Malaysian Communications and Multimedia Commission (MCMC) reported that as of June, 2007, 59% of the Malaysian population are internet users. The majority are in urban locations, with 26% and 24% of internet subscribers from Selangor and Kuala Lumpur respectively, and less than one percent in less developed states of Perlis and Kelantan (Shah, 2004). Due to the disparity, the government has initiated many promising e-pilot projects to ensure people in remote areas have access to computers. The digital divide becomes an issue of access to ICTs in order to improve the social and economic well being of underserved sections of the Malaysian community. They include women, the poor, the disabled, the elderly, youth, those in the rural areas, and indigenous group.

Recognizing that ICT can contribute to the empowerment of women and to reduce gender inequalities, the Malaysian government has taken initiative to improve women's access to technology. Joshi (1997) relates that men are considered the first beneficiaries of innovations, while women follow and take over the positions vacated by them. The progress of women is hampered by widespread prejudices regarding their ability to learn and to apply new technology. Technological work including ICT are often regarded as masculine and mastery, and control are often in the hands of men (Joshi, 1997). According to Wajcman and Pham Lobb (2007), such opinion obstructs women from gaining access to information about how to use technology. Nevertheless, women too are considered to play an important in the ICT revolution empowerment that transcends stereotype beliefs.

The International Women's Day of 1997, establishes all Malaysians regardless of gender, recognizes full gender equality, hence instate the role of women in ICT development (Srinivasan,

2004). Prior to this, the 1995 MIMOS internet user survey released in August 1996 showed that the internet in Malaysia is predominantly male with women representing only eleven per cent of all users. Similarly, women in many Asian countries do not have as much access to computers and the internet as men (Srinivasan, 2004). In this context, gender is found to be an issue in internet usage. In line with the demands and needs of current industries, more educational and training opportunities are being provided to enable women to meet the demands of the knowledge-based economy and improve their upward mobility in the labour market (Ali, 2005).

Additionally, the Malaysian government has taken efforts to improve technology accumulation through education, experience and skills development by optimizing the benefits and incentives (Hashim, 2008). For example, every working Malaysian can withdraw their savings from the Employee Provident Fund in order to ensure the success of the government's "One-House One-Computer" campaign in 2003. Computer courses or subjects also were introduced in 1980s, as early as in pre-school in order to ensure that every Malaysian is able to acquire the necessary ICT skills.

Surprisingly, the gender disparity ratios observed in IT-related higher education and in the IT-related sector of Malaysian industry is obvious. For example, for the 2001/2002 academic year at the School of Computer Science at Universiti Sains Malaysia, women makes up 65% of the student enrollment (Goh Beng, 2002). Similarly high percentage of women in the professional IT related-sector. Despite these two observations, in general, working Malaysian women still possess average ICT skills and seldom use ICT at their workplace (Hashim, 2008).

Therefore, Malaysia's approach to facilitate women participation towards modernization is characterized into three-phases:

1. *Access*, delivering access to the ICT information structure to everyone.
2. *Adoption*, promoting regular and widespread use of ICTs in everybody's daily life.
3. *Inclusion*, ensuring the socio-economic value of ICTs is achieved by all underserved Malaysians.

Given the Malaysian scenario, the aim of this chapter is to present and discuss the computers and internet accessibility concept, namely delivering access to the computers and Internet information-structure to women in Malaysia.

BACKGROUND

Women and Computer Access

In Malaysia, a number of programs and initiatives have been introduced to increase computer ownership and ensure universal access among the population. A survey among households was conducted by Jaafar and Sulaiman (2005) to study the impact of ICT and computers on Malaysian society. The survey was carried out to examine the level of domestic computer usage and identify the common activities associated with computers among the households. The study found differences in terms of age and income groups towards their level of computer usage. Malaysian teenagers are heavy domestic computer users in comparison to the other age categories. As expected, the high income people are also heavy domestic computer users. Interestingly, the study has found that computers are being used domestically for personal and household purposes such as entertainment rather than enhancing the flow of information among households and business organizations. Besides this computers can be accessed through cyber or internet cafes. Compare to other public internet access place, cyber cafes are found to be an important role towards computer and internet access in both urban and rural areas. They come to enjoy the unique, upscale, educational, entertaining and innovative environment provided at the cyber café (Shah Alam, *et al.,* 2009). Based on their survey study conducted in two cities (i.e. Malacca and Sarawak) in Malaysia, showed that male are more frequent patronize cyber café in comparison to women.

Do Malaysian Women have Access to Computer and ICT?

Access is defined as the ability to use a computer for prolonged period of time (Spennemann, 1996). According to Spennemann, the use of the World Wide Web (WWW), however, does not require owning a computer, but merely access to use. If access to and use of IT is directly linked to social and economic development, then it is very necessary that women in developing countries understand the importance of these technologies and learn to use them (Balasingam, *in press*).

Women access to computer can help them acquire the skills for their jobs and market know-how to conduct successful at-home businesses. The recent developments in communications technology have progressively reduced the barriers to communicating over distances. At the same time, computers have enormously increased the capacity to accumulate and access information sources and data. Combined with communications technology, the possibilities for information access are almost infinite. In today's "information society", the rapid access to relevant information is indispensable for timely intervention in most realms of human activity; access to this technology is therefore a key to successful action.

Information and communication technologies include a variety of technologies. The knowledge and skills of women should be enhanced through the use of information and communication technologies which could enable a better preservation and transference of traditional knowledge and skills. With the advent of information and communication technology, women are ensured not left behind, this time in the information revolution.

Recognizing that information and communication technology can contribute to the empowerment of women and to reducing gender inequalities, efforts have been taken by the Malaysian government to improve women's access to such technology, in order to achieve an informed and computer-literate society. Women's access to information and communications technology has improved through training programmes.

The study done by Hashim (2008) revealed that the working women in Malaysia possess only average level of ICT skills. They seldom use the internet and e-mail at their workplace or at home, they do not face serious learning barriers with respect to ICT, and they have complexity as their innovation characteristics. However, they are innovators in the adopters' categories. The combined variance in the characteristics of learning barriers, ICT skills, and ICT usage amounted to 70 per cent in innovation characteristics. The Malaysian government has invested highly in ICT centers all over the country. Yet there is a sharp and growing "digital divide" between male and female computer literacy, especially for poor, less educated women.

Women and Internet Access

The internet has proved be the most fastest growing technology and has become essential in the development of information society. In most countries, developing in particular (Teltscher, 2002), the typical internet surfer used to be male, young, well-educated and well off. However, there was a significant gap between male and female internet users, where growth rates of female internet users are steeper than those of male users. They used internet for reading online news, chatting, shopping, entertainment, visiting educational sites and others

The Malaysian National IT strategy includes a strategy to ensure that there is no high-tech gender gap in Malaysia and that women in Malaysia are

Table 1. Internet usage and population growth in Malaysia (Source: Internet World Stats – www.internetworldstats.com/stats3.htm)

YEAR	Users	Population
2000	3,700,000	24,645,600
2005	10,040,000	26,500,699
2006	11,016,000	28,294,120
2007	13,528,200	28,294,120
2008	14,904,000	25,274,133

fully represented in IT development, the fastest growing and most important sector of the economy in the new millennium. Table 1 illustrates the Malaysian internet usage and population growth. It is noted from the year 2005 to 2008 there is approximately one million additional internet users yearly. This indicates a steady growth of internet usage.

A study by Economic and Social Commission for Asia and the Pacific (ESCAP) shows that Malaysia is one of the top 5 most connected countries in Asia-Pacific, alongside with New Zealand, Japan, South Korea and Singapore (ESCAP, 2008). These countries all have between 55% to 80% of their populations using the internet by 2007, while the average in this region is at 20%. A quick check at Internet World Stats does confirm that 59% (14.9 million) of our population are Internet users. By this ratio, Malaysia is number 5 in Asia behind Japan (73.8%), South Korea (70.7%), Hong Kong (69.5%), Taiwan (67.2%) in 2008. In terms of total internet users, Malaysia is number 9 in Asia as illustrated in Figure 1. This, of course is weighted by the total population and does not represent how *connected* the country is.

According to the Malaysian household use of the Internet Survey 2006 conducted by the Malaysian Communications and Multimedia Commission (MCMC), women comprise 47% of home Internet users. The survey found that as of September 2006, there were 13.5 million Internet users in the country. The top five activities on the internet were getting information (84.5%), communication by text (for example, e-mail, chat room, instant messaging (80.7%)), leisure in the form of music, video or games (52.6%), formal or informal education (45.9%), and financial activities such as e-banking (23.6%). Till 2008, as reported by Za (2008), in Malaysia women still make up nearly half of Internet users (47%).

Malaysian women are important contributors to the country's economic and social development. In the year 2000, 46.7% of Malaysian women contribute to the labour force (Shah, 2004). There is a "digital gap" between urban and rural women, and between women of different economic status that can be addressed by policy, particularly policy aimed at providing entrepreneurial skills. In line with this, Malaysia specifically committed itself to strengthen the implementation of the National Policy on Women in order to among others, ensure access to and benefits of development particularly in science and technology in the enhancement of economic growth. Poor information flows, combined with low levels of literacy and limited skills development and training act as barriers to women's economic advancement and continuing participation in this area.

In most countries, especially developing countries such as Indonesia and Thailand (Teltscher, 2002), the typical internet surfer used to be male, young, well-educated and well off. However, there was a significant gap between male and female internet users, where growth rates of female inter-

Figure 1. Asia top ten internet countries (Source: Internet World Stats – www.internetworldstats.com/stats3.htm)

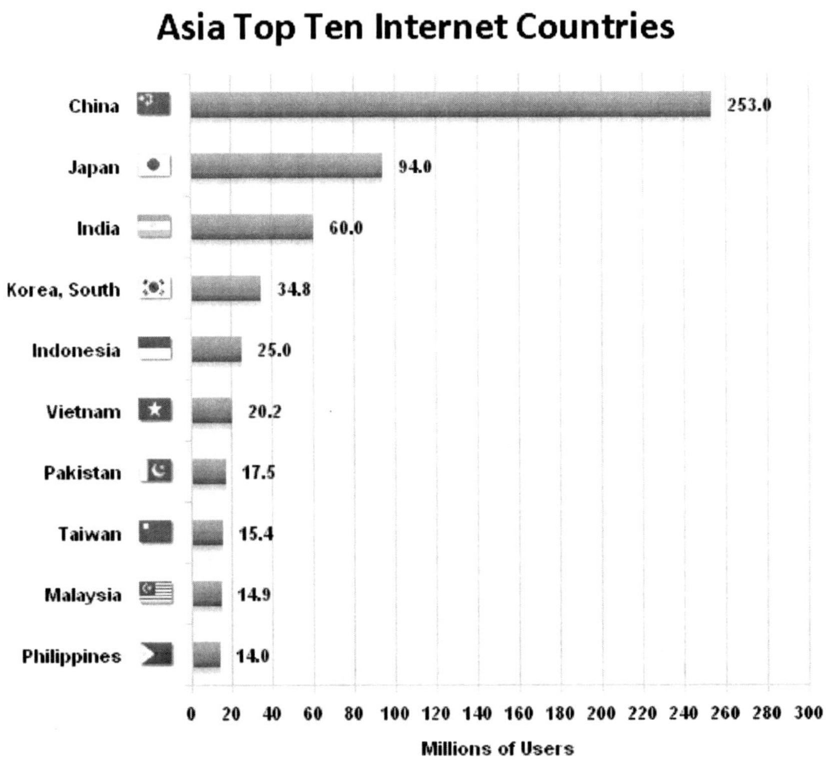

net users are steeper than those of male users. The Malaysian National IT strategy should include a strategy to ensure that there is no high-tech gender gap in Malaysia and that women in Malaysia are fully represented in IT development which is the fastest growing and most important sector of the economy in the new millennium.

The authors realized that learning about computers and the internet requires an enormous amount of incremental knowledge. For example, women needed to know the basics of modems and computer connections, high-speed connections, servers and networks. In other words, women need to be a computer-literate person (i.e. have the ability to understand and to use computers). Despite their interest in computers, an appropriate training in computer use and internet is needed. Further, one of the most significant obstacles faced by women in learning about computers and the Internet was use of English language.

According to Teoh and Chong (2008), in its effort to turn Malaysia into "Information Society" as part of the efforts to become a knowledge-based economy, the government has pledged to see that all segments of the society, including women entrepreneurs in urban and rural areas have access to new IT, particularly computers and internet facilities. Formal and informal training programmes and seminars were jointly conducted by the Government and the non-governmental organizations (NGOs) to ensure that women entrepreneurs are not marginalised in this era of ICT.

Education is considered to be by far the most important policy intervention to improve the ability of women in developing countries to participate in the information society (Teltscher,

2002). Malaysia in particular, the government have stated incorporating ICTs in the educational agenda, specifically equipping schools with computers and other basic infrastructure needed to access the internet. This provides an essential pre-requisite for effectively using many new ICT and the internet.

Realizing the importance of ICT, in 2003, the Ministry of Women, Family and Community Development set up an urban centre, the e-Women Centre targeting single mothers and homemakers, where they can receive training and have access to the computer and internet. Women in Malaysia have been involved in small businesses as petty traders, family businesses or as operators both in rural and urban areas. A research component in "Networking Women" project examined accessibility and usage of computers and internet by urban and rural women in Peninsular Malaysia and by women on the East Coast of Malaysia which has a more rural population. The women in Peninsula Malaysia were quite diverse, including women of different ethnic composition.

Survey data on access to computer and internet and the use of these technologies from 3000 respondents confirmed the presence of digital divide between groups of women (Gibb, 2001). About 80% of urban women from in and around Klang Valley had computers at home, compared with 60% of women in Peninsula Malaysia and 37% of rural women in the East Coast. This is largely directly related to income, educational status and profession (Gibb, 2001). In addition, the use of internet and e-commerce seems to be particularly attractive to women owning small enterprises provides flexibility. Women are able to tap into new markets while saving time and money but allowing them more time with raising a family. The technology to communicate inexpensively with customers, handling accounts and online processing are attractive options for home-based businesses owners.

A FRAMEWORK FOR ANALYZING WOMEN'S ACCESS TO COMPUTERS AND INTERNET

Based on the previous studies (Mustafa, 2003; Balka, 2005; Ndubisi, 2005; Cross and Linehan, 2006; Wilson-Kovacs et al., 2006; Harris et al., 2007; Loh-Ludher, 2007; Hashim, 2008), we develop a framework that relates to the access of women to computers and internet. The aim of the framework as shown in Figure 2 is to help analyze Malaysian women access to computer and internet.

As shown in Figure 2, the factors that influence Malaysian women access to computer and internet consist of government intervention, attitudes, training, ICT infrastructure, language, educational status and income. The explanation of the factors included in a framework is given below.

Access to Computer

The presence of computers in home and offices is now commonplace. Women are at all levels of IT in Malaysia have opportunities to access to computers. However, women who are not using a computer in their works, do not have an opportunity to make experiences with and are also not interested in the internet. If they could get access to computer, women could communicate with each other easily through e-mail.

Access to Internet

Access to internet was widely available in Malaysia, except in East Malaysia, where the internet was often not available beyond urban centers. The issue faced by the women's with respect to the adoption of computer networking technology is access. Access becomes an issue at several levels and a key element that needs examination. The first relates to communication constraints imposed by the infrastructure of data lines and value-added carriers (Balka, 2005). Access to

Figure 2. A framework for women's access to computers and Internet

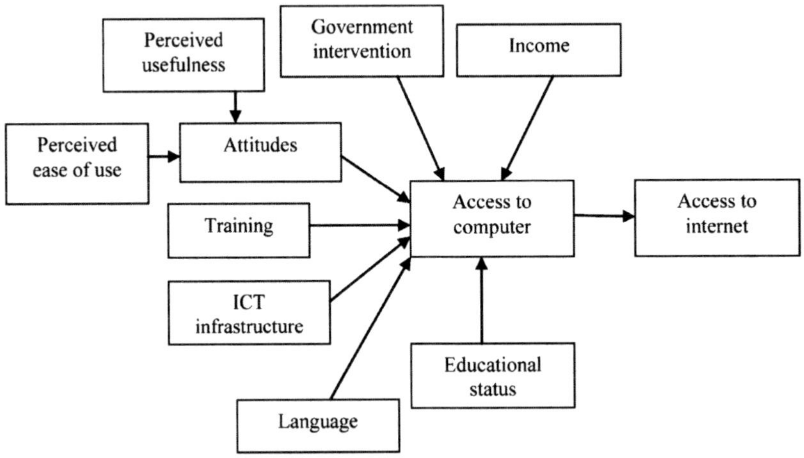

computer networks is also determined by the location of networks and terminals: whether they are located in a public place and available for use free of charge, or whether they are located in a private home or office.

Government Intervention

Harris *et al.* (2007) state that in the Ninth Malaysia Plan (2006-2010), increasing emphasis is placed on equitable access to affordable PCs and online services. The government will continue to play a key role in bridging the digital divide by initiating and providing online services throughout the country to enable Malaysians to access and utilize the internet as a part of everyday life. To keep abreast with the anticipated increasing demand, the number of PCs installed is expected to increase from 21.8 to 40.0 per 100 population by the end of the plan period. With a population of 27 million, Malaysia's increasingly affluent and skilled workforce have readily adopted the use of ICTs in business and personal activities. Mobile phone use is substantial with penetration rates of around 75 per cent of the population and demand for internet, especially broadband, is growing. Despite this, only about 14 per cent of national households have broadband. Exacerbating this is limited market demand in the less lucrative rural and peri-urban markets. Examples of targets for network and broadband connectivity for institutions include:

1. Broadband service subscriptions for organizations paid by government.
2. Connection of approximately 900 Federal, state and local government departments.
3. Networking around 10 000 primary schools through the Schoolnet programme.
4. 74000 connections for public universities and research institutions.
5. 4000 hospitals and clinics networked and connected to internet resources.
6. 1700 libraries and a multitude of community centres with a focus on educational, entertainment and e-commerce applications. (ICT Policy & Regulations Researches)

Attitudes

An individual's attitude toward using a technology reflects instrumentality and intrinsic motivation to

Table 2. Women's view about technology

No.	Women's View
1	Fantasize about it as medium
2	See it as a tool
3	Want to use it for communication
4	Are impressed with its potential for creation
5	See it as expressive
6	Ask it for flexibility
7	Are concerned with its effectiveness
8	Like its ability to facilitate sharing
9	Are concerned with integrating into their personal lives
10	Talk about wanting to explore worlds
11	Are empowered by it

use technology. In the case of accessing computers and internet, the internet provides a limitless source of information, therefore the skilful use of the internet and the right and positive attitude ensure that women will make the correct selection of relevant and appropriate information for their needs and practical realities. Brunner et al. (1998) summarized how women viewed technology as shown in Table 2.

One of the earliest research that examined attitudes toward computers was conducted by Lee (1970). Two dimensions of attitude were identified, namely (i) beliefs in the computer as a beneficial tool and (ii) beliefs that the computers are autonomous entities. Lee's conclusion has been supported by Cancro and Slotnik (1970) and Gardner, Young and Ruth (1989). Lee (1970) hypothesized that lower autonomous entity attitude scores would result in an increase in access to computers which would in turn result in higher beneficial tool attitudes. He then concluded that the driving factor in determining computer access was beneficial tool attitude.

Perceived Usefulness

Perceived usefulness is defined as "the degree to which a person believes that using a particular system would enhance his or her productivity" (Davis, 1989). Studies comparing the salience of perceived usefulness and ease of use between men and women users of technology have shown that perceived usefulness is more important for men users while women users emphasize ease of use in technology usage decisions (Ndubisi, 2005; Bagozzi, 2007).

Perceived Ease of Use

Perceived ease of use is defined as "the degree to which a person believes that using a particular system would be free of effort" (Davis, 1989). Men display higher instrumentality (i.e. outcome) compared to women who display higher process orientation (ease of use or difficulty) determinants of technology adoption (Venkatesh *et al.*, 2000; Venkatesh & Davis, 2000; Venkatesh & Brown, 2001).

Training

Women require a variety of training programs to satisfy their diversified needs. For example, basic computer training for daily use and professional IT training for employment, self-employment and certification. Basic internet training course for full-time housewife is also needed to enable them to access the internet in their daily lives such as searching useful information for children's education. Therefore, ICT training is required to optimize use of digital resources. Small scale ICT training can be offered on community level.

ICT Infrastructure

Although it may be assumed that the internet is a global phenomenon, the reality of the situation for some users is that access may either be unavailable or not considered a priority. For those that have access, the availability may not be as balanced or equal as assumed. Users are sometimes denied access because of high costs of providing computer facilities and bandwidth. In Malaysia, the heavy investment into ICT infrastructure and its related programmes to promote the widespread availability of ICTs, has yet to narrow the digital divide. The Communications and Multimedia Commission recently initiated the Collective Internet Access program to promote installation of adequate Internet access technologies in all communities (Mustafa, 2003).

Mustafa (2003) states that internet access depends on four critical players. First, providers of telecommunications infrastructure (bandwidth capacity) for international access to the global Internet backbone. Second, providers of national long-distance telecommunications transmission capacity (such as leased lines) to connect Internet service providers (ISPs) with one another and with international connectivity nodes. Third, providers of local loop access (narrowband and analogue, such as traditional copper wire connections, or broadband and digital, such as digital subscriber lines (DSL), cable television modems, and fixed wireless service). Fourth, ISPs, which provide Internet services to customers using these layers of networks.

Increasing women to IT and communication in developing countries like Malaysia needs increasing availability of communication in the areas where women live. Extension of facilities like wireless and satellite communications to rural and semi urban areas is essential to increase women's access to IT. Common use of facilities such as tele-centers, phone jobs and other forms of communications in convenient and accessible places for women is essential.

Educational Status

As widespread use of, and dependency on, the internet increase especially in education sector. It was found that not everyone has equal opportunities to its access and usage (Papastergiou & Solomonidou, 2005). It is certainly a cause for concern as it is a part of the digital divide problem. The educational level or status has been attributed to this disparity (Norris, 2001).

Education is considered to be by far the most important policy intervention to improve the ability of women in developing countries to participate in the information society (Teltscher, 2002). The key strategy would be to focus attention in generating demand for IT education from the women themselves through awareness programmes. Exposing women to computers and the internet should be initiated by governments, particularly education ministries. In Malaysia, as mentioned in previous section, the government has started incorporating ICTs in the educational agenda, specifically in equipping primary and secondary schools with computers and other basic infrastructure needed to access the internet. These could be useful measures to enhance digital literacy and stimulate the use of ICT among girls or women at early stages. Some research shows that women are influenced very early (Ahuja, 2002; Adya & Kaiser, 2005).

Language

The dominance of English in the internet is apparent. The ability to use English facilitates enormous accessibility for knowledge and entertainment. English is the second official language in Malaysia considering she was once under the British colony. English remains to be part of the education system and English is used by varying proportions of the population. As such being educated and knowing English go hand in hand (Gupta, 1997). In comparison, among the ASEAN countries, only Filipinos uses English when online. For example, countries like Malaysia and Indonesia, non-English language are more widely used (Rao, 2009). Contradict to a study done by Shah Alam et al. (2009) on cyber café found that English is the respondents first preferences to browse the internet. The UNDP report draws the conclusion that the complex imbalance of internet access is such that "the typical internet user worldwide is

Table 3. Correlation between computer ownership and household income (Source: Computed by Rao, 2009 based on data provided by UN)

Country	Significance Level
Austria	/
Belgium	X
Brazil	/
Chile	/
Denmark	X
India	/
Indonesia	/
Malaysia	/
Philippines	/
Singapore	X
Sweden	X

Note: / = significant, X = not significant

English speaking (Ngini et al., 2002). Women prefer technology that is audio in nature not purely because they are illiterate or semi-literate, but largely because they do not speak English.

Income

There is a correlation between income and technology (Frank, 1995). Technology utilization usage especially internal and broadband facilities require the individual to bear the expenses incurred (Rao, 2009). Generally, women from upper-income family have no problem in accessing computers and Internet compared to women from lower-income family. Most women use Internet at work. Despite lack of reliable figures, it can be estimated that in general women use computer at work as a tool of routine office work, data entry, manufacturing, and programming However, of late, computer may become a tool of communication with the introduction of social networking sites like Facebook and better accessibility due to lower cost of broadbands. Hence, income influences in the digital divide could be higher if the cost of accessing technology is high proportionately to income.

Table 3 depicts the correlation between computers per 1000 with the household income. The participating countries comprises of developed and developing countries. Countries like Belgium, Denmark, Sweden and Singapore illustrate computer ownership does not really depend on household income. In countries where income is low like Malaysia and Indonesia, computer ownership is highly significant (Rao, 2009).

Scrutinisizing, Malaysian women who live in rural areas are lacking access to internet (Loh-Ludher, 2007). This is primarily due to the social development differences coupled with geographical makeup. One way of addressing this problem is to provide such women access to ICTs that help them acquire the skills and market know-how to conduct successful at-home businesses. Many Malaysian women work from the home to increase family income. Several governmental and private agencies have sponsored computer centers in rural area, which offer Internet access and other computer services.

Women regard the value of computer technology as a way to simplify tasks and to increase productivity, and also they have become more comfortable with technology, removing a stumbling block to opportunities related to technology (Ray, Sormunen & Harris, 1999).

CONCLUSION AND RECOMMENDATIONS

The internet is an indispensable resource for economic growth, prompting the attention of individuals, groups and governments all over the world. This phenomenon changes the role of women in Malaysia. The women in Malaysia have to accept ICT, and be willing and able to respond to change in order to ensure that they are not left out of the ICT era. They can acquire ICT skills and become competent in accessing computers and internet through appropriate and systematic training approaches. With the government aggressive intervention in providing opportunities, financial aids, infrastructure developments and continual awareness programs it is expected the usage of ICT in all aspects of their daily life will contribute to minimize the gender inequalities.

With an increasingly significant value of ICT as agents for development, it is necessary to include women in the area of ICT policy and regulation (Phipphitkul, *in press*). There are certainly some arguments of cyber-feminism that cyberspace is an area where women are taking control and subverting the technology towards their own ends (Adam & Green, 2001) Therefore, women need to enhance their potential, ability and confidence in order to take advantage of opportunities offered by information technology. Women in Malaysia need to fully realized the importance of computers and the internet in their work and family lives. ICT is an imperative tool to promote women's empowerment, rights, dignity and full participation in the information society. They need to gain access to computers and the internet so that could use e-mail and understand their children's computer use.

For Malaysian women who work from the home to increase the family income, need extension agents who will teach them in their homes how to use the internet and other computer applications to improve their job prospects (Gaither, 2008). Those with home businesses need to be technology literate. Providing at-home education for these business women would help them network with friends, suppliers, and customers to improve their home businesses.

Since IT and particularly the internet have become an important part of the women's life, research to elucidate the extent to which women use the internet and the way they react to it is appropriate and necessary. In work environments in which women work with computers it is important to consider the ways of computer use, risks connected to computerized work settings and impacts of computers on work practices.

FUTURE RESEARCH DIRECTIONS

For future research direction, the framework proposed in this chapter could be used to analyze Malaysian women's access to computer and internet. Also, it would be interesting to discover the reasons for low ICT access among Malaysian women (working and non-working) and how this affected their job performance. Furthermore, future research could examine what actions non-government organizations (NGOs) have taken to improve women's access to computers and internets. Also, Malaysian women's roles in computer and Internet should be examined and discussed.

REFERENCES

Adam, A., & Green, E. (2001). Gender, agency, location and the new information society in Alison Adam & Eileen Green. In Virtual Gender. London: Routledge.

Adya, M., & Kaiser, K. M. (2005). Early determinants of women in the IT workforce: a model of girls career choices. *Information Technology & People*, *18*(3), 230–259. doi:10.1108/09593840510615860

Ahuja, M. K. (2002). Women in the information technology profession: a literature review, synthesis, and research agenda. *European Journal of Information Systems*, *11*(1), 20–34. doi:10.1057/palgrave/ejis/3000417

Ali, H. (2005). *Permanent mission of Malaysia to the United Nations*. Revised Version. Available at http://www.un.int/malaysia/GA/60thGA/6030.PDF

Bagozzi, R. P. (2007). The legacy of the technology acceptance model and a proposal for a paradigm shift. *Journal of the Association for Information Systems*, *8*(4), 244–254.

Balasingam, M. (n.a.). *Female involvement in information technology in developing countries*. Retrieved from http://trincoshanmuga.sch.lk/article/Female_Involvement.doc/

Balka, E. (2005). *Women's access to on-line discussion about feminisme*. Retrieved from http://feminisme.eserver.org/gender/cyberspace/feminist-use-of-cyberspace.txt

Brunner, C., Bennett, D., & Honey, M. (1998). Girl Games and Technological Desire. In J. Cassell & H. Jenkins (Eds.), From Barbie to Mortal Kombat: Gender and Computer Games, (pp. 72-88). Cambridge, MA: MIT Press.

Cancro, R., & Slotnik, D. (1970). Computer graphics and resistance to technology. *American Journal of Psychotherapy*, *24*, 461–469.

Davis, F. D. (1989). Perceived usefulness, perceived ease of use, and user acceptance of information technology. *MIS Quarterly*, *13*(3), 319–339. doi:10.2307/249008

Dutta-Bergman, M. J. (2005). Access to the internet in the context of community participation and community satisfaction. *New Media & Society*, *7*(1), 89–109. doi:10.1177/1461444805049146

ESCAP. (2008). *Internet penetration in Malaysia*. Available at http://malaysiacrunch.blogspot.com/2008/11/internet-penetration-in-malaysia.html

Gaither, M. (2008). *Loh-Ludher on Homeschooled Malaysian Women*. Available at http://gaither.wordpress.com/2008/

Gardner, E., Young, P., & Ruth, S. (1989). Evolution of attitudes toward computers: A retrospective review. *Behaviour & Information Technology*, *8*, 89–98. doi:10.1080/01449298908914542

Gibb, H. (2001). Gender mainstreaming: good practices from the Asia Pacific Region. The North-South Institute, Canada, and APEC.

Goh Beng, L. (2002). Modern dreams: An enquiry into power, cultural production and the cityscape in contemporary urban Penang, Malaysia. New York: Cornell University Press.

Gupta, A. F. (1997). The Internet & the English Language. First Conference on Postcolonial Theory.

Harris, R. W., Yogeesvaran, K., & Lee, L. W. Y. (2007). *Telecentres for national e-inclusion in Malaysia.* Available at http://14donline.net/ATF/2007/fullpapers/Roger_ATF07ABS113.pdf

Hashim, J. (2008). Learning barriers in adopting ICT among selected working women in Malaysia. *Gender in Management: An International Journal*, *23*(5), 317–336. doi:10.1108/17542410810887356

ICT Policy & Regulations Researches, February 3, 2009, Malaysia's broadband plan – stimulating the private sector. Available at: http://ictpolicyn-regulation.blogspot.com/2009/02/malaysias-broadband-plan-stimulating.html

Jaafar, N. I., & Sulaiman, A. (2005). Domestic computer usage and activities in West Coast Malaysia: age and income differences. *Information Development*, *21*(2), 128–137. doi:10.1177/0266666905054507

Joshi, I. (Ed.). (1997). Asian women in the information age: new communication technology, democracy, and women. Singapore: Asian Media Information and Communication Centre.

Lee, R. (1970). Social attitudes and the computer revolution. *Public Opinion Quarterly*, *34*(1), 53–59. doi:10.1086/267772

Loh-Ludher, L. L. (2007). The Socioeconomic Context of Home-Based Learning by Women in Malaysia. *Distance Education*, *28*(2), 179–193. doi:10.1080/01587910701439233

Mustafa, M.A. (2003, March). Internet access. *The World Bank Group Private Sector and Infrastructure Network,* (pp. 256).

National, I. T. Council (NITC). (2004). *Multimedia Super Corridor*. Available at http://www.nitc.org.my/initiatives/msc.html

Ndubisi, N. O. (2005). Effect of perception and personal traits on computer technology adoption by women entrepreneurs in Malaysia. *Journal of Asia Entrepreneurship and Sustainability.* Retrieved from www.asiaentrepreneuralshipjournal.com

Ngini, C. U., Furnell, S. M., & Ghita, B. V. (2002). Assessing the global accessibility of the Internet. *Internet Research: Electronic Networking Applications and Policy, 12*(4), 329–338. doi:10.1108/10662240210438399

Norris, P. (2001). Digital divide: civic engagement, information poverty, and the Internet worldwide. New York: Cambridge University Press.

Papastergiou, M., & Solomonidou, C. (2005). Gender issues in Internet access and favourite Internet activities among Greek high school pupils inside and outside school. *Computers & Education, 44*(4), 377–393. doi:10.1016/j.compedu.2004.04.002

Phipphitkul, W. (n.a.). *Gender justice: digitally empowered women through information technology.* Available at www.wsisasia.org/materials/wil.doc

Rao, R. (2009). Digital Divide: Issues Facing Adult Learners. *Computer and Information Science.*, *2*(1), 132–136.

Ray, R. M., Sormunen, C., & Harris, T. M. (1999). Men's and women's attitudes toward computer technology: a comparison. *Office Systems Research Journal, 17*(1), 1–8.

Shah, F. H. (2004). Mainstreaming potential women exporters in international markets through ICT: Malaysia. Melaka, Malaysia: APEC Committee on Trade and Investment.

Shah alam, Syed., Abdullah, Zaini, & Ahsan Nilufar. (2009). Cyber Café Usage in Malaysia: An Exploratory *Study. Journal of Internet Banking and Commerce,* 14(1).

Spennemann, D. H. R. (1996). Gender imbalances in computer access among environmental science students. *Electronic Journal of Instructional Science and Technology, 1*(2). Retrieved from http://www.usq.edu.au/electpub/e-jist/docs/old/vol1no2/contents.htm

Srinivasan, S. (2004, January 12). *Computer access unequal in most Asian nations*. Available at http://www.asiaitsummit.com

Telterscher, S. (2002). Gender, ICT and Development. *UNCTAD Electronic Commerce and Development Report.* The presentation given at the World Civil Society Forum Geneva, 14-19 July.

Teoh, M-Y, W. & Chong, S-C. (2008). Improving women entrepreneurs in small and medium enterprises in Malaysia: policy recommendations. *Communications of the IBIMA, 2,* 31-38.

Venkatesh, V., & Brown, S. (2001). A longitudinal investigation of personal computers in homes: adoption determinants and emerging challenges. *MIS Quarterly, 25*(1), 71–102. doi:10.2307/3250959

Venkatesh, V., & Davis, F. D. (2000). A theoretical extension of the technology acceptance model: four longitudinal field studies. *Management Science, 46*(2), 186–204. doi:10.1287/mnsc.46.2.186.11926

Venkatesh, V., Morris, M. G., & Ackerman, P. L. (2000). A longitudinal field investigation of gender differences in individual technology adoption decision making processes. *Organizational Behavior and Human Decision Processes, 83*, 33–60. doi:10.1006/obhd.2000.2896

Wajcman, J., & Pham Lobb, L. A. (2007). The gender relations of software work in Vietnam. *Gender, Technology and Development*, *11*(1), 1–26. doi:10.1177/097185240601100101

ADDITIONAL READING

Abdul Jalil, S. (2007). Women empowerment and gender mainstreaming, In Syed Arabi, I. (Eds), Malaysia at 50: Achievements & Aspiration (pp. 69-74). Kuala Lumpur: Thomson Learning and International Islamic University Malaysia.

Ariffin, J. (1995). Reviewing Malaysian women status: Country report in preparation for the fourth UN world conference on women. In Koshal, M. Gupta, A. and Koshal, R. (1998). Women in management: A Malaysian perspective. *Women in Management Review*, *13*(1), 11–18.

Brayton, J. (1999). Women's love /hate relationship with the internet. In I. Markovic (Ed.) Cyberfeminism [ver 1.0]. Zagreb: Centre for Cultural Studies. 193-202.

Lagesen, V. A. (2008). A cyberfeminist utopia? Perceptions of gender and computer science among Malaysian computer science students. *Science, Technology & Human Values*, *33*(1), 5–27. doi:10.1177/0162243907306192

Norazah, Mohd. Saki. (2001). Malaysian Internet users motivation and concerns for online shopping. *Malaysian Journal of Library & Information Science*, *6*(2), 21–33.

KEY TERMS AND DEFINITIONS

Access: The ability and ways necessary to store data in, to retrieve data from, to communicate with, or to make use of any resources of a system (computer or internet).

Information Technology (IT): Information technology refers to both the hardware and software that are used to convert, store, protect, process, transmit, retrieve, and manipulate information. In other words, information technology is the use of computer and hardware to manage information. It involves the development, installation, and implementation of computer systems and applications.

Information and Communication Technology (ICT): Information and communication technology encompasses communication devices or applications (for example, computer, network, hardware, software, television, radio, cellular phones, satellite systems and etc.), the services and applications associates with them (for example, videoconferencing). The importance of ICTs depend less on the technology itself than its ability to create greater access to information and communication in underserved populations.

Internet: A worldwide system of computer networks. A network of networks in which users at anyone computer can, if they have permission, get information from any other computer, and sometimes talk directly to users at other computers.

Digital Divide: The differences in opportunities or chances available to people who have access to information technology (IT) or information and communication technology (ICT) and those who do not. Also, can be defined as a gap between those with regular, effective access to IT or ICT, and those without.

Information Society: Also called knowledge society, is a term for a society in which the creation, distribution, and manipulation of information has become the most significant economic and cultural activity. A society characterized by a high level of information intensity in daily life of most citizens, in most organizations and workplaces; by the use of computers and telecommunications, and by the ability to transmit, receive and exchange digital data rapidly between places irrespective of distance.

Cyber-Feminism: Cyber-feminism is a feminist community that uses the internet to organize and communicate, including philosophy and set of practices concerned with feminism interactions with and acts in cyberspace. It is broadly concerned with expressing and developing feminism in the context of online interactions and online art.

Profile 10
Technology Career Spotlight:
Cooperative Education in USQ Networking Programs with Leading Industry Organisation and Service to South-East Queensland Regional IT Professionals

© 2008 Queensland Government Office of Women. Used with permission

Dr. Yan Li, USQ, Toowoomba

Yan Li is an Associate Professor in the Department of Mathematics and Computing at the University of Southern Queensland (USQ). Yan is the founder of USQ Cisco Networking Academy and the winner of 2008 Queensland Government Smart Women-Smart State Award in the category of Community/Public Sector – ICT, for her innovation and leadership in incorporating Cisco Certified Networking Associate (CCNA) courses into USQ Networking Program. Having recognised the need to reconstruct the major after teaching the networking courses for several years, Yan was determined to make every effort to bring the project to fruition.

Yan was born in China. At the age of 15, she was admitted to one of the top universities, HuaZhong

Copyright © 2010, IGI Global. Copying or distributing in print or electronic forms without written permission of IGI Global is prohibited.

University of Science and Technology (HUST), in China. She later became a lecturer in HUST, and studied in Hong Kong Polytechnic University. Yan obtained her PhD degree from The Flinders University of South Australia in 2003.

Introduction

Since 2001, I have taught various Computer Science courses in the Department of Mathematics and Computing at the University of Southern Queensland (USQ). I am particularly passionate about the Computer Networking and Communication courses. Over the past two and half years, in order to make the Networking Programs at USQ more professionally relevant and consequently more attractive to students, I have been leading a major team project to redevelop the courses in these major programs by incorporating an international leading industry based training program from Cisco Networking Academies. This was a significantly ambitious task given the size of the project. It involves teaching staff being training and assessed to become certified instructors for four Cisco Certified Networking Associate (CCNA) courses, the establishment of a fully equipped Cisco Network Laboratory for practicals, and, of course, the project leadership and management tasks required to make the project happen.

In early 2006, when the project idea was initiated, the head of our department was very supportive to the project but the department did not have the budget for it. Furthermore the department could not afford to allocate the extra workload to the project due to the prevailing heavy teaching load in the department. Having recognised the need to reconstruct the major after teaching the networking courses for many years, I was determined to make every effort to bring the project to fruition. This was while I still performed a full workload of the normal teaching and other academic commitments, including an active research program. I am a Chief Investigator for an ARC Discovery Project and Chief Supervisor for three PhD and two Masters students, and an Associate Supervisor for another two PhD students.

I quickly started to work on identifying the facilities required and to lead efforts to obtain the required support, including the necessary funding, for the project. At the same time, the proposed teaching team did the four Cisco training courses during the USQ teaching breaks over the year of 2006. At that time, I was the only woman who successfully completed all four training courses. I engaged in countless meetings with various levels of USQ staff. I had discussions with the national, regional and local leaders of Cisco Networking Academies Australia. Because of my persistent and persuasive efforts, I eventually obtained their strong support. I successfully applied for $80,000 from the Vice-Chancellor's Discretionary Fund through 2006 USQ Strategic Development Fund Program. I eventually obtained about $266,500 worth of actual and in-kind (equipment) funding from either USQ or Cisco for the project. Furthermore I also obtained an ongoing technical support kindly provided by the Division of Information, Communication and Technology (ICT) at USQ.

Once the budget was secured, in order to set up a Cisco Network laboratory I took initiatives to arrange meetings with related USQ ICT leaders and network technicians, and staff in the Faculty of Sciences involved in the project. These meetings were to analyse the Network Lab requirements and identify possible locations. I arranged for our teaching team and ICT network support staff to visit other Cisco Networking Academies in Brisbane, such as QUT, UQ and several TAFE campuses where we discussed our plans for advice.

After about one year of hard work, we established an advanced Cisco Network laboratory at USQ. In Semester one (S1) 2007, USQ network students were able for the first time to do real practicals. By the end of 2007, 13 students had successfully completed all Cisco CCNA assessments and obtained their industry accreditation. In 2008, IT professionals in South-East Queensland region have the opportunity to undergo their Cisco CCNA training locally (in Toowoomba).

In addition, I actively promote the Networking Major Programs and ICT Careers to potential students. Each year starting from 2003, I have delivered a presentation to promote ICT Careers in Toowoomba and at USQ *Tertiary Expos*. In order to help our students better comprehend the links between their current study and future ICT careers, I regularly invite network engineers from industry, such as Cisco, to give seminars and to provide opportunities for the students to communicate with those professionals. I was a mentor for the girls in *The Girls in Mathematics and Sciences Summer School* from 2003-2005.

The Project

I took the initiative to explore the industry-base Cisco Network training system and successfully embedded an international leading industry training based courses into USQ networking major. The significant project leadership and management tasks that I carried out are outlined above. The following discussions refer to the educational initiatives, innovation, outcomes and benefits involved in the project.

1. Incorporate Cisco Network Training Courses into USQ Networking Major

In order to incorporate Cisco Network training courses into USQ networking major, I led our teaching team to develop two network core courses, CSC3407 Network Fundamentals and Routing and CSC3427 Routing, Switching and WAN Technologies. We developed the course materials (Introduction Books and Study Books), set assessments for the courses, and conducted the course delivery. I have been the examiner for both CSC3407 and CSC3427.

In S1 2007, we started to offer the first two Cisco Certified Network Associate (CCNA, version 3.1) courses which were mainly covered by CSC3407. We provided all on-campus students with the opportunity to have the Cisco CCNA course training. Subsequently 13 students successfully completed all the assessments required. For external students, we developed the extra course materials and provided them with solid network knowledge, so that they can take the Cisco CCNA assessments in their local training centres. In S2 2007, we delivered the new course CSC3427 which covers the second two CCNA courses. The course attracted 33 students in its first offer. Finally, the13 students completed the full training and obtained their leading industry certificates. Of these, six received a letter from the Present of Cisco Systems Inc. for their excellent work (grade over 90%).

The teaching staff for CSC3407 and CSC3427 have to run two separate assessment and grading systems, USQ course teaching system (Assignments and exams) and Cisco CCNA training system (Module exams, final on-line exams, case studies, practicals and their tests etc). Only the teaching content is in common in the two teaching systems. In order to offer the extra Cisco CCNA course training, we were required to learn Cisco Networking Academy teaching system on how to create the courses, activate the teaching materials, take assessments and exams, grade the final results and conduct the weekly hands-on lab practical work.

2. In line with Cisco New Curriculum, Cisco CCNA Exploration, and Update the Cisco Lab

We, at USQ, established our Cisco Networking Academy at the time of the industry turning point. In August, 2007, Cisco launched two new CCNA curricula in response to the rapid development of networking technologies. In order to provide the best teaching materials and latest technologies to our students, I again took the initiative and led our network teaching team to redevelop CSC3407 and CSC3427 courses for 2008. With the support from ICT, we updated the USQ Cisco Network lab with more equipment to be ready for one of the new Curricula for universities, *Cisco CCNA Exploration version 4.0.* In S1 2008, as the course examiner I developed and delivered the first two courses in CSC3407 which is now renamed as Network Fundamentals and Routing. In S2, I redeveloped CSC3427 and will deliver the course which includes the second two Cisco CCNA Exploration courses. Once again, I learnt the new teaching system and conducted the Cisco CCNA Exploration training for on-campus through Cisco Networking Academy systems and provided the training. 28 students completed all the assessments for the first two Cisco courses in S1 2008.

The Innovation of the Project

To my knowledge, this project is the first USQ degree program to be integrated with a leading industry organisation. While there are other institutions offering the Cisco CCNA training and incorporating the industry courses into their university degree programs, the development of this project is significant both because very few of these are regional and also because of how successfully and seamlessly this was achieved at USQ. My tireless leadership was critical to making this project a success. We made this significant and complex project a reality on the top of our usual work when, at other institutions, it normally requires one full-time job for a year in other organisations to establish a Cisco Network Academy. For example, this was the case at UQ's ICT Cisco training centre.

As USQ is a regional university, we have a significant number of Distance and e-learning students (external students) who do not come to USQ campuses but are spread out all over the world. That means that the students enrolled in one course, such as CSC3407, are both on-campus students and external students. We need to develop flexible systems to accommodate two groups of students without any assessment differences between on-campus and external students. We successfully embedded the Cisco training courses into our existing teaching systems. We developed a Cisco CCNA training structure in parallel with USQ degree courses (CSC3407 and CSC3427). The structure enables our students to obtain an extra benefit of industry accreditation while completing their USQ degrees.

The Outcomes of the Project

1. The Cisco collaborative initiative has attracted more student enrolments into courses CSC3407 and CSC3427, and other networking major courses, such as CSC3412, CSC3413, CSC8407 etc. Based on USQ official enrolment records, CSC3407 enrolments have been increased by 12% in 2007 and 30% in 2008 in comparing with 2006. The new course, CSC3427, had 33 students in its first offer in S2 2007. The project attracted more students to enrol into CSC3407 and CSC3427. Starting from S1 2007, the two Cisco courses attracted more students from other faculties. Those students took CSC3407 and CSC3427 as their electives over other courses because of the additional benefit of extra industry accreditation. The enrolments in the other network related courses have also increased because of the industry collaboration and the two new Cisco courses.

2. Cisco training program will boost USQ's reputation in the Southeast Queensland region and service to the communities. The Cisco Networking Academy at USQ will be the incubation centre for Networking professionals in Toowoomba. This project is an important regional initiative. We are planning to offer full fee-paying Cisco CCNA Exploration training courses to local IT professionals at the end of S2 2008, during the summer semester. Traditionally, people in Toowoomba had to go Brisbane to take the Cisco training courses, which involves extra time, inconvenience and additional costs.

In S1 2008, without even advertising our training program, we have three IT professionals who are not USQ students enrolled into the Cisco CCNA Exploration training courses in parallel with USQ courses, CSC3407 in S1 and later CSC3427 in S2. They had IT degrees and were pleased for having the opportunities to be trained at USQ, in Toowoomba.

3. USQ will have more opportunities to get in touch with IT industry for potential collaborations. We are in the process of working out several Internships for networking students in Cisco Systems Inc.. At the same time, our network staff are establishing a closer relationship with the industry partner which will enhance and facilitate research collaborations with interested parties.

4. Benefit to the students. With IT Network Administrator and IT Network Engineer two of the job categories with good and growing employment demand, we have successfully targeted an identified skills need in the booming economy by establishing this integrated academic and process accreditation offering. The offering makes it more convenient and quicker to get the two types of qualifications that are needed for such employment.

The academic and professional industry training mix improves both the understanding of theory and practical skill of students as well as their employability. By seamlessly integrating higher education with professional training provided Cisco, the courses

- enhance our students' work readiness for a leading industry immediately after graduation;
- are more industry oriented and therefore more attractive to employees;
- expose our students to the cutting edge of technology;
- ensure students are well informed about possible changes in Standards and Regulations, and,
- familiarize our students with industrial practices.

Summary

I have lead and managed a complex and challenging project to seamlessly integrate professional industry training into academic courses for the mutual benefit of students and potential employees. Through the provision of industry training options, the project outcomes are also beneficial regionally. For my tireless effort and outstanding achievement, I was honoured with the Dean's Award in December 2007 from Faculty of Sciences at USQ.

Chapter 11

SET Women and Careers:
A Case Study of Senior Female Scientists in the UK

Susan Durbin
University of the West of England, UK

ABSTRACT

Very few studies of senior female scientists have been conducted in the UK. This chapter explores the careers of 13 senior female scientists in a male-dominated, UK public sector organisation. These women operate within a context which reinforces, 'think management, think male' (Schein 2007). Despite this, they have followed traditional career paths of science qualifications and employment. Whilst many parallels can be drawn with other women in science, engineering and technology (SET) these women buck the trend in that they have achieved senior positions and unusually long lengths of service, post-childbirth, despite little or no help from mentors and support networks. Although these women's experiences of gender relations were variable and they have limited strategic leadership career options, they have sustained a positive work orientation and a strong desire to reputation-build. Notwithstanding their failure so far to break through the glass ceiling, they hold the potential to be role models for others in SET.

INTRODUCTION

Women comprise almost half the paid UK labour force but this has not resulted in gender equality in terms of the jobs that men and women perform (horizontal segregation) and the levels at which they operate (vertical segregation). In the case of the latter, women remain under-represented in senior management positions in the UK (Sealy *et al.* 2008) where vertical gender segregation is most pronounced and where women often exist as 'tokens' (e.g. Kanter, 1977). Whilst a third of managers and senior officials are women, only a few have broken through the glass ceiling into senior management. In 2008, women held just 12% of Directorships in FTSE 100 companies, an increase from 7% in 1998, indicating that progress has been extremely slow (Sealy *et al.*, 2008).

This persists, despite the regulatory route for transition to a public gender regime in the UK,

DOI: 10.4018/978-1-61520-657-5.ch011

Copyright © 2010, IGI Global. Copying or distributing in print or electronic forms without written permission of IGI Global is prohibited.

one in which women's access to employment is facilitated by the legal removal of discrimination, regulation of working time so that it is compatible with caring and policies to promote social inclusion (Walby, 2004). This poses a problem, as it is at these senior levels that crucial decisions are made and leadership exercised around critical strategic issues. If women are not promoted to senior posts this will continue to act as a barrier in defining career choices for women leaving higher education (European Commission, 2006) and will do nothing to improve the already relatively low numbers of female science graduates taking up SET employment.

This chapter analyses the career progression of 13 female scientists who have progressed to senior positions within a single case study, male-dominated organisation engaged in scientific research and development. The careers of these particular women are of interest because they represent a group who have accessed senior management positions within their scientific disciplines, despite their organisation being male-dominated, at every level and within almost every function. The women interviewed straddle three occupational groups: senior management, SET management and science professionals, reflecting their scientific status and progression into managerial roles within their own scientific disciplines. All have reached senior positions but have yet to reach the highest echelons of their organisation.

BACKGROUND: WOMEN IN SET

In their study of SET employees, Wynarczyk and Renner (2006) report that just two percent of women hold senior managerial level jobs in SET, which is far lower than for women in other sectors. In addition, while the gender division at lower levels of organisations may be improving, this, in common with other sectors, has not resulted in proportionate numbers of women progressing into strategic leadership levels. With an estimated shortfall of 300,000 SET recruits over the next few years, this will have huge ramifications (The Independent, 2008).

In 2008, SET occupations[1] represented 3,142,127 of the UK workforce, of whom 595,606 were female and 2,546,521 male. In the same period, there were a total of 1,112,483 SET managers[2], of whom 153,951 were women and 958,532 men and 133,665 'Science Professionals'[3] of whom 52,615 were women and 81,050 men (UKRC, 2009). Men are six times more likely than women to be employed as SET managers and one and a half times more likely to be employed as Science Professionals. Across 17 EU member states, just three can boast a rate of over 40% for female representation on scientific Boards (Norway, Finland and Sweden) while the UK figure is 31%. Cyprus and Poland have the lowest rates, at just 7% (European Commission, 2006).

The women interviewed for this research have been successful in securing senior positions and this chapter aims to establish how they have achieved this status and what they still need to do to progress further. The main objectives are therefore to explore the career histories, challenges and strategies of thirteen senior female scientists. To achieve these objectives, three related themes are examined. Firstly, career backgrounds, including career challenges and career progression strategies, are explored. Secondly, the extent to which mentors have played a role in the progression of senior female scientists is examined, in the context that mentoring is perceived by the government and SET networks, such as the UK Resource Centre for Women in SET, as an important activity for women's progression in SET (Helsinki Group on Women in Science, 2002). Thirdly, the potential for senior women to act as mentors and role models to other women in SET organisations is examined. The aim is to gain a sense of the challenges faced by senior female scientists and the social conditions that underpin these.

PROBLEMS FACING WOMEN IN SCIENCE, ENGINEERING AND TECHNOLOGY

A review of the current literature reveals an uneven coverage of published research on women in SET. While much has been written on the working lives and careers of women in SET, this tends to be biased towards women and engineering (Evetts, 1998; Miller, 2004; Powell *et al.*, 2004) and women and technology (e.g. Cockburn, 1983, 1991) with much less attention given to senior women in science, with some notable exceptions (Ellemers *et al.*, 2004; Harding, 1986). The following review of current research pulls together a range of research literature and government initiatives aimed at women in SET as well as a consideration of gender stereotyping and the nature of gendered organisations. There is a general consensus that occupational segregation within SET is often the result of a masculine SET culture and organisational cultures that clearly mark women as 'different' (Evetts 1998). The following review highlights some of the current thinking and some problems associated with this.

A starting point is the consideration of the various government initiatives aimed at tackling women's under-representation in SET. The reasons for government interest in this is that it recognises SET as important for economic growth in the UK, that women's skills are under-utilised in this area of the economy and because of an increasing feminisation of the UK labour market. Several reviews of the situation in the UK have been conducted (The Rising Tide Report 1993, the Helsinki Group on Women and Science in 2002, (after which the 'Promoting SET for Women Unit' was set up at the Department of Trade and Industry in 1994) and The Roberts Review, 2002). More specifically, the Greenfield Report (2002) included the setting- up of working groups to tackle different stages of SET careers for women, including senior management.

As a result of these reviews, the UK Resource Centre for Women in SET, was set-up in 2004. Recommendations included, encouraging more women into science teaching and addressing the under-representation of girls in science subjects at school. This could be an important initiative early in a woman's career especially as careers in science tend to begin at an early age, with an educational course that forms a progression of study (Fox and Stephan, 2001). Recognising the 'vital contribution' that women make to UK competitiveness, the government claims that it is committed to raising the profile of women in SET careers and to ensuring that women reach the highest levels (Greenfield Report 2002). Yet in 2009, little has changed at the strategic leadership level and women remain poorly represented.

Another important reason for women's under-representation is the low retention rate, post-childbirth, of women in SET. Figures cited by the Independent (2008) clearly demonstrate this retention problem: after giving birth to their first child, just 8,000 women returned to SET employment, compared with 24,000 women in the general workforce[4] which has been partly blamed on a lack of proactive HR policies. It is claimed that women are considerably more likely than men to exit professional scientific jobs in their first two years of employment, the median tenure for women in their early thirties in professional scientific occupations being just under four years, compared with ten years for men. Returning to work is more difficult in science-based employment than in other fields (Blackwell and Glover, 2008). One explanation for this is a lack of a family-friendly culture within SET which does little to encourage women back.

If, as Wynarczyk and Renner (2006) and Blackwell and Glover (2008) claim, family formation is likely to affect retention, then resources should be directed at changing the attitudes and perceptions of employers towards female employees with children, although evidence suggests that women engaged in scientific fields tend to be younger and relatively fewer of them have children compared with other female employees. An additional explanation for the lack of returners could be that the nature of the work and the pace at which science evolves, means that women may feel that they have lost touch with their scientific disciplines and that their skills have become outdated.

There are some initiatives to encourage women back into SET including, 'return to work' courses (Herman and Kirkup, 2008) one of the main aims being to help women to advance their careers and continue practising their professions. There are also a wide range of networks aimed at women in SET, with over seventy of these in the UK alone (Greenfield *et al.,* 2002) e.g. the Promoting SET for Women Unit (PWSET). These organisations offer career development advice, job search links and networking hints and tips. Mentoring is another important initiative as mentors are recognised as being able to identify opportunities for training and development, explain and raise awareness of the 'political landscape' of the organisation, provide access to privileged information, make introductions into important decision making networks and convey norms of behaviour and the values of the organisation. Mentoring relationships may open up networking and career opportunities for the person being mentored.

MENTORING: A CAREER PROGRESSION STRATEGY?

Mentoring is important to women's career progression (Luethge and Byosiere, 2007) and acts as a channel for the exchange of tacit knowledge and information that is often linked with promotion opportunities (Durbin, forthcoming; Swap *et al.*, 2001). A lack of access to mentors is a significant barrier to women's career progression because they are denied access to important information shared by the dominant group (Groysberg, 2008; Vinnicombe and Singh, 2004). At a more strategic level, mentors give 'credibility' to their mentees and introduce them to their important (and often powerful) networks. Access to these critical relationships is the most important service a mentor can provide. In her study of female analysts on Wall Street, Groysberg found that 'star' women were more likely to have had mentors than other women but less likely to have done so when compared with their male counterparts. Critically, many of the women who have made it to the top have had supportive bosses (Greenfield *et al.,* 2002). The gender of the mentor can also be important: Groysberg (2008) found that when men mentor women they are often unable to offer pscychosocial support, for example, when dealing with sexism. In contrast, when women act as mentors, they frequently may not be in a position to facilitate integration into the firm's culture, presumably because they are often not integrated into this culture themselves.

Most SET networks offer mentoring schemes. For example, the review of women and science (2002) by the 'Helsinki Group' identifies a three-year pilot mentoring scheme that was set up by the Women's Engineering Society and the Association of Women in Science, Engineering and Technology. The aim was to increase the numbers of women who maintain SET careers after career breaks and to help women to achieve their potential for promotion and engagement in policy making circles. Patricia Hewitt MP (2003) - UK Secretary of State for Trade and Industry - reiterates the importance of mentors (and networking) in her strategy for women in SET, reflecting that many of these women work in isolation in occupations that are male dominated within organisations that are male-lead. In reality, however, many organisations do not offer mentoring programmes and women are left to find their own mentors, often struggling to find a senior female mentor.

Gender Stereotyping and Women in SET

Despite the initiatives cited above, women remain under-represented at senior levels within SET employment and this indicates that the problem needs to be tackled at several levels. Reasons for this under-representation have been identified as, a lack of career guidance counselling prior to university education, early differential socialisation of males and females, lack of support from family, friends and professional engineers and cultural and occupational barriers (Dryburgh, quoted in Powell *et al.,* 2004). The negative perceptions by SET doctoral students about their prospects for careers in research universities compound the situation (Fox and Stephan, 2001). While government initiatives demonstrate that efforts are being made to promote women (e.g. Greenfield Report, 2002) it is not clear how these will address the underlying issues of the gendered division of labour, gender stereotyping and the persistence of the, 'think manager-think male' scenario' (Schein, 2007). Overall, 'the main thrust of government activity is at the level of giving advice from a supply-side perspective to employers and employees about recruitment and retention issues in SET' (Blackwell and Glover, 2008). The Greenfield Report (2002) is an exception to this, as it identifies the negative outcomes of gender stereotyping and cultural barriers for women in SET.

The environment surrounding SET employment renews and extends male hegemony over women and this negates any prevailing belief in 'neutrality' in terms of gender. If organisations are to cultivate a work environment supportive of employees' multiple roles (e.g. mothers with domestic responsibilities) it is necessary to go beyond policies and laws and create a more supportive culture within the organisation (Swanberg, 2004). In other words, government white papers, support networks and mentoring schemes may raise awareness and stimulate policy, but they fail to tackle the underlying issues of gendered organisations and gender stereotypes. As Phipps (2006, p. 126) points out, 'women in SET activity is largely focused on compensating for the shortcomings of schools, universities and SET workplaces and/or retrieving girls and women who have been lost to SET disciplines and professions'.

It is only when feminist thinking about science is adequately theorised that we will have a clearer grasp of how scientific activity is gendered. Harding (1986) argues that those who focus on 'individual gender' often fail to see the division of labour by gender in the larger society and gender symbolism that are equally responsible for the small numbers of women in science. Therefore, simply implementing 'equal opportunity' programmes does nothing to change the *status quo*. There is also reluctance by some women activists within SET to claim an allegiance with feminism, instead opting for an equal opportunities (liberal feminist) approach. For these women, identification with feminism would generate personal, professional and political problems. Phipps concludes, '...the relative lack of progress on the issue of women's participation in SET in the past thirty years seems to suggest that the 'equal opportunities' approaches which have dominated the field so far are not adequate' (2006, p. 133).

Gendered Organisations and Gendered Careers

Gender stereotyping affects the ways in which people estimate the potential of men and women and may result in a biased evaluation of actual achievements. Ellemers *et al.* (2004) found no evidence for differences in the level of commitment and research outputs between male and female research students yet perceptions that women were less committed than their male colleagues were found amongst senior faculty members, especially older women. This is the outcome of the stereotyping of women, which results in women having more difficulty than men in having a successful academic career and, because of their own precarious position, women are more likely than men to engage in gender stereotyping in this context.

Women's careers are frequently positioned as 'different' to those of men (Halford *et al.,* 1997). For example, men are often viewed as the 'ideal' worker due to their unencumbered and uninterrupted relationship with the paid labour market (Swanberg, 2004). This results in a division of labour maintained by organisational ideologies around recruitment and promotion practices (Ozbilgin and Woodward, 2004) in which women become the 'out' group. Employers will make assumptions about the commitment of female employees who have maternal responsibilities (Wilson, 1998). Research indicates that male senior managers often hold stereotypical perceptions about the skills, ambitions and commitment of their female workers and in turn, this contributes to a lack of career progression for women (Durbin, 2006). Formal workplace policies and informal practices can prevent employees from managing work, family and personal responsibilities and reproduce or sustain gender inequities (Swanberg, 2004).

Cultural, structural and action dimensions each form the context in which women's career choices are made and which differentially affect the occupational destinations and career trajectories of women and men (Evetts, 2000). While women make choices in the labour market and organisational contexts, they do so in constrained settings. The culture of management itself, which is usually perceived as strongly masculine, reflects the belief that what constitutes 'good' management is in conflict with women's roles and responsibilities around caring, relatedness or connectedness (e.g. Schein, 2007). The structural dimension is closely related to that of the cultural, as reflected in the division of labour and departmental systems, as well as promotion and career paths. Acker (1990) postulates that organisation structures consist of promotion ladders and hierarchies of work positions that determine women's paid work and career experience. Women's under-representation in management stems from inadequate career opportunities, gender-based stereotypes and 'old boy's networks' (Oakley, 2000).

The managerial role is sex-typed as a male occupation, the 'think manager-think male' attitude having held constant over time (Schein, 2007) which may be exacerbated by an even more pervasive masculine culture in SET organisations (Powell *et al.,* 2004). Those women who are promoted to senior levels may also end up 'managing like men' (e.g. Wajcman, 1998) and reinforcing the existing 'male' culture, thus deterring other women from entering (Powell *et al.,* 2006) as well as shaping the gendered nature of work at a foundation level. This 'masculine' culture also permeates the culture and curriculum of higher education and has a negative impact upon female students (Powell *et al.,* 2004). Women engineers in a high technology organisation in the UK deployed three strategies to cope with their token status: 'fronting it out', 'playing the little women' and 'building a reputation', the latter being the most preferred strategy, based upon being a 'good engineer', building a reputation and earning respect (Evetts, 1998).

Women in SET employment face a number of personal and professional barriers. In their research on women in SET in the UK private sector, Wynarczyk and Renner (2006) found that just over half of the women they interviewed had experienced barriers in their personal lives that were likely to hinder professional development, e.g. family commitments and dependant children. Sixty seven per cent had also experienced professional barriers, including institutional sexism, gender stereotyping, lack of support from senior managers and male biased career development opportunities. Such barriers lead to gendered attrition, which in turn has also been linked to a lack of contacts and networks for senior women, the 'mens' club' and lack of broad-based expertise in relation to resources and strategy issues (Greenfield *et al.*, 2002).

SET is an occupationally segregated area in which women are under-represented, despite attempts by the government and SET support networks to encourage more women into the profession. Science careers tend to start early but there exists a dual problem: there are relatively low numbers of women entering SET and many of those who do so, do not stay. Mentors are perceived as one of the answers to the problem but one that has not resolved it. Equal opportunities policies are mainly ineffective but some women in SET are reluctant to engage with more radical feminism, through fear of being 'labelled'. The outcome is a highly gendered profession, within a gendered organisation (Acker, 1990) where careers are different for men and women (Evetts, 2000). The preceding review of current research highlights a gap in the SET literature due to the paucity of published research on senior women in SET. There is even less on women who progress their careers within SET and how this has been achieved.

METHODS

This research is based upon 13 semi-structured interviews with female scientists who have reached senior management positions in a single case study, male-dominated public sector organisation (Scienceorg) engaged in scientific research in the South of England. Scienceorg is a leading global research institution that is government-owned and supplies services to both government as well as commercial organisations.

The organisation comprises a predominantly scientific workforce, employing a total headcount of 1900 of which 143 work part-time (mainly women). In four of the five largest and most important departments in the organisation, women represent just one fifth of employees. The majority of the workforce is aged between 30 and 49 years with very few under the age of 20 and over the age of 60 years.

The organisation operates on the basis of a hierarchical job structure, ranging from Level A (clerical grades) to Level E (senior grades) and the Board. The majority of employees are clustered in Level C (scientific officer grades) (213 women and 669 men) with 341 located in the senior management grades (Levels D and E) of whom just 67 are women. There is just one female member on a Board of six.

As a public sector organisation, Scienceorg is subject to the UK Equality Bill, which is currently being debated in UK Parliament. The Bill aims to simplify, harmonise and extend existing equality duties by introducing a new Equality Duty on the public sector. The aim is to end age discrimination, extend the scope of positive action, strengthen enforcement of current legislation and require transparency in the public sector through the publishing of clear information about progress on important equality issues (The Equality Bill, 2008). For example, it is proposed that employers take account of the under-representation of disadvantaged groups at the point of recruitment and selection, when selecting between two equally qualified candidates, such as women and ethnic minorities. It will be interesting to see the impact of this legislation on the position of senior women in SET, once it becomes law.

Table 1 sets out the profile of the thirteen women interviewed.

The women interviewed are all senior managers at grades D and E. Ten are employed on a full-time and three on a part-time basis. Three are Heads of Department, seven Managers and three Senior Project/Project Managers. People management responsibilities varied. Four had no people management responsibilities as they were employed as project managers. For those with people management responsibilities, the average number of employees for which they were

Table 1.

Respondent	Job title	FT/ PT	Age	Ethnic Origin	Marital Status	Children	Qualifications	Annual Salary
A	Head of Department	FT	46-50	*WB	Married	2, aged 15 & 13	MSc, PhD (Science)	£55,000
B	Head of Function	FT	41-45	WB	Married	1, aged 6	Maths, Physics, Chemistry 'A' Levels; HNC Computer Science; MA Leadership	£48,000
C	Senior Manager	FT	51-55	WB	Widowed	No	Physics Degree; MSc; PhD	£46,200
D	Operational Applications Manager	FT	56-60	WB	Married	No	Maths, Physics, Chemistry 'O' & 'A' Levels; Maths Degree	£46,700
G	Senior Manager	FT	36-40	WB	Divorced	1, aged 9	Maths Degree; PhD; PGCE; Post-doc	£38,000
I	Senior Manager	FT	36-40	WB	Married	2, aged 10 and 14	Science Degree	£37,000
J	Project Manager	FT	36-40	WB	Single	No	Physics Degree	£35,000
K	Head of Department	FT	41-45	WB	Married	2, aged 8 & 11	Maths, Physics, Chemistry 'A' Levels; Chemistry Degree	No data
M	Senior Project Manager	FT	56-60	WB	Married	2, grown up	Maths, Physics, Chemistry 'A' Levels; Physics Degree	£37,000
N	Senior Manager	FT	36-40	WO (Spanish)	Married	1, aged 23 months	Science Degree; MSc; PhD	£37,000
O	Senior Manager	**PT**	36-40	WB	Married	1, aged 4.5	Maths Degree; PhD	£39,000 (PT)
P	Senior Manager	**PT**	36-40	WB	Living with partner	1, aged 2	Maths & Physics Degree; MSc	£22,500 (PT)
Q	Senior Project Manager	**PT**	46-50	WB	Married	Twins, aged 8	Physics Degree	£22,000 (PT)

* White British

Note: Respondents E, F, H and L were excluded from the sample as they are not qualified scientists.

responsible was 15 (the highest 35, the lowest 5). The average length of service was 17 years, the longest being 37 years and the shortest one year.

The women were asked to give their approximate ages: six were aged 36-40, two 41-45, two 46-50, one 51-55 and two 56-60. This compares with the majority of the workforce being aged between 30-49 years. The average salary for full-time employees was £41,690.00 and for part-time, £27,800.00. This compares with the average mean gross annual salary of £30,205 for female Science Professionals generally (UKRC, 2009). Five of the women hold PhDs, two a Master's Degree and six a Degree. All PhDs are held in science-based subjects and Master's qualifications are held in science and management subjects. First degrees are held in maths, chemistry and physics. Ten women were married/living with partner, one widowed, one divorced and one single. Ten have one or two children, their ages ranging from 23 months to 15 years.

The interviews were conducted between March and September 2008 after the interviewees were approached by HR and asked whether they would be interested in taking part in the research project. Of 20 who expressed an interest, 17 were interviewed, 13 of whom were qualified scientists. Interviews were conducted either face-to-face or by telephone, and lasted approximately one hour. All interviews were tape recorded and fully transcribed. Interview questions explored the nature of the interviewees' work, length of service, people management responsibilities, marital status, domestic circumstances and formal qualifications. Questions on career included career background, career challenges and strategies, satisfaction with career progression and future career aspirations. Finally, the role of mentors was explored, as well as the advice that senior women would offer other women who aspire to progress into senior management positions. Although this is a small sample that is not representative of the wider female SET population, the findings offer valuable insights into the career paths and career aspirations of a group of senior female scientists working in gendered roles, in a gendered sector in a gendered organisation.

FINDINGS

Career Background

The career backgrounds of the women interviewed were explored to establish how they had progressed to their current positions and the length of time taken to achieve senior status. Eleven interviewees had worked for Scienceorg only, indicating that their careers had progressed within an internal labour market. Of the remaining two, one had spent 17 years working in another public sector organisation, the other a year in administration, before joining Scienceorg. Length of service averaged 17 years, the longest 37 years and the shortest one year. This could indicate that family formation has not affected retention, contrary to Blackwell and Glover's (2008) findings. Interviewees had joined Scienceorg as qualified scientists, although some were more qualified than others, indicating that science careers started at an early age through an educational course that forms s steady progression of study (Fox and Stephan, 2001).

Four interviewees had undertaken quite substantial further study since being recruited. For example, Respondent A joined with a Master's degree and has now acquired a PhD and Respondent H joined with A' Levels and now has a Degree. Of those who had undertaken further study since being recruited, two are Heads of Department, one a senior manager and one holds a strategic resourcing role. The remainder are managers and project managers and have slightly less seniority. Most had been in senior roles for an average of four and a half years and one had joined at a senior level.

Career structures are very similar for these women, usually beginning with a science or maths degree, some with a specialism in their scientific subject. This indicates a traditional route from education to science-based jobs. Just one interviewee had joined Scienceorg at a senior level, the remainder joining at Scientific Officer/Assistant Scientific Officer levels. Career progression has been based upon working on special projects, acquiring additional responsibilities and the development of expertise (and further academic qualifications) over a period of time. Amongst the 10 women who have children, career breaks for maternity leave averaged just over one year, with one interviewee taking a career break of several years.

Not surprisingly, the four women who had undertaken further academic study had taken longer to progress to their current positions (an average of 15 years). This compares with an average of six years for those who had joined Scienceorg with the requisite science qualifications. All 10 who had taken a career break for maternity leave had returned to Scienceorg, indicating a high retention rate amongst this sample of thirteen women. It would seem that the most rapid route to senior levels is through joining the organisation with the requisite science qualifications, indicating an educational course that starts early and forms a progression of study (Fox and Stephan, 2001) although childcare breaks will have extended the time taken to progress for these 10 women. The average length of service is long and in contrast to Blackwell and Glover's (2008) findings that women were exiting professional scientific jobs in their first two years of employment, with a median tenure of just under four years. This long length of service could be linked to the specialist scientific nature of the job, the nature of science and the lack of alternative organisations engaged in similar research in the UK.

Career Challenges

Interviewees were asked about the most significant challenges to their career progression so far. For some, the challenges were gender-related (Wynarczyk and Renner, 2006). Examples include having to juggle work and home responsibilities, a challenge that was recognised as not being a consideration for many male colleagues:

I suppose with my private life in mind, it's juggling and giving enough time and attention to my children and giving enough time and attention to my job, and I think that balance is very, very difficult; never successful with that, you can never please both sides, you know. So I think that's definitely the biggest challenge (Respondent K, Head of Department)

This inherent conflict may indicate why so few women return to SET careers but may not be a problem just for SET women. Another interviewee felt that being a female scientist in a male dominated working environment was a challenge in itself, although this view was not expressed directly by any other interviewee:

Probably getting to where I am today has been difficult because I think if you're female actually, it's not that easy to be honest...I've attended meetings with other people in which...I think it's all been done with the best of intentions really, but sort of commenting on such and such at very different levels and making comments like, 'she's only saying that because she is female' type of thing. Not talking about me, but talking about somebody else (Respondent N, Senior Manager)

Lack of confidence was another challenge cited by two interviewees:

All the way through my career, I've pushed myself to develop some sort of way of being more interactive with people, more confident with people to put forward my views. It's obviously worked because I have been put in this position, but it was something I struggled with. I still sometimes, I am having conversations with people and I still have to remind myself that actually now I am the boss. I have the authority and the responsibility and ...I'm not bad at making decisions, it's not that exactly, it's just that I have to pinch myself that this is down to me now. I can do this and I have the authority to do this (Respondent P, Senior Manager, PT)

For others, challenges were more directly related to progressing into new work areas, sometimes through project work, and taking on people management responsibilities. Successfully completing the assessment centre, which is essential for promotion to a senior level, was also a challenge for several women:

I think it is difficult to measure yourself, I think that to me, to really know whether or not I was worth getting through the system, for me it was very important that I did pass the [promotion] for me it was important that I passed it because it makes me feel that I'm ok (Respondent M, Senior Project Manager)

Responses to this question highlighted challenges that are similar for other SET women (Wynarczyk and Renner, 2006). There were some gender-related issues, such as managing work and home life, being a woman in a male dominated area and a general lack of confidence to progress, whilst the challenge for some women was getting through the assessment centre, a critical step in the promotion process.

Career Progression Strategies

Interviewees were asked about the extent to which they had been pro-active in their career development and the type of strategies they had adopted. Many interviewees stated that, 'being good at my job by building a reputation' was an important strategy (Evetts, 1998). Other strategies included developing new skills, taking on new challenges and 'visibility'. Researching a role before deciding to take it on was also a popular strategy:

Being good at my job I think and making sure that people, kind of, know that I am here, I'm willing to do extra things rather than just my job, I'm well into getting involved in corporate things, and put myself out to do that (Respondent B, Head of Function)

Other than making yourself visible. And I don't think the only strategy, I think that's one of the strategies, people won't recognise you otherwise (Respondent D, Operational Applications Manager)

Three interviewees said they did not have a strategy, although in hindsight, two indicated that they wished they had been more strategic in their approach:

I'm starting to think now more in terms of strategies in thinking where I want to go. But no certainly I started off, I was very much a come in, do my job, hopefully somebody will recognise that I'm wonderful but since then it's yeah, it's the way the organisation works, you have to be able to throw yourself into another strategy of where you want to try and get to (Respondent J, Project Manager).

Interestingly, the three most senior women in the sample all said that they had a career progression strategy which could partially explain why they have achieved their more senior status. This could be a key learning point for those women who aspire to senior positions.

THE ROLE OF MENTORS

At the time of the research at Scienceorg, no official mentoring scheme was in place, which is perhaps surprising given the government initiatives aimed at women in SET and Scienceorg being a public sector organisation. Official policy was that all interviewees should be allocated a mentor while going through the assessment centre for promotion. This relationship normally ended upon completion of the assessment centre and promotion to senior levels.

Just three interviewees said that they had been allocated a mentor while going through the assessment centre. Views on the effectiveness of this varied: Respondent L felt that being mentored by a scientist in another area of the organisation was advantageous and although the formal mentoring relationship had now come to an end, she had decided to continue with it on an informal basis. Respondent H had decided to change her mentor as she felt the relationship was not 'fruitful'. She describes her current mentor as a 'sounding board' who can help her to think through difficult decisions. Respondent P, although maintaining the relationship with her mentor post-assessment centre, had not felt the need to maintain regular contact:

I have a nominal mentor but we don't really meet. As part of this assessment and development centre where I got my promotion, we were assigned a mentor. I met with her two or three times and at crucial times I have gone to her for some advice, say when I was doing my application form for this job, I went to her and that was very helpful. We've had one or two general chats, but I haven't felt particularly that I needed to do that regularly and so I haven't. She is still there (Respondent P, Senior Manager, PT)

This indicates that the process of assigning mentors to those going through the assessment centre is perhaps not successful and that Scienceorg should consider an alternative mentoring programme both pre- and post-assessment centre. Those who expressed a desire to be mentored said that they would prefer to be allocated a mentor outside of Scienceorg. The importance of choosing the 'right' mentor was made apparent by one interviewee who said that she would prefer to 'wait and see', whilst another interviewee complained of the lack of female mentors available at a senior level:

I don't see the value from somebody internal because the way mentoring works, I don't see what anybody internal could tell me or give me, that I don't already do or know myself. You know because I mentor other people, so from that point of view, I don't know what value it would give me personally, otherwise I think mentoring is a good thing (Respondent B, Head of Function)

The desire for external mentors was evident amongst the women interviewed. Another interviewee, who was studying for a management qualification, had been advised to find a mentor to help her through this period of study. She also expressed a preference for a mentor external to the organisation, although she was also keen to take up the offer of mentoring from a more senior male colleague within the organisation, as she recognised that both could offer different benefits:

So I'm going to choose him as well. But that's a different kind of mentor to the one outside because you can't talk about quite the same things with somebody who's your line manager as well (Respondent A, Head of Department)

She had identified a suitable mentor outside of the organisation but saw this as much more of a short-term arrangement:

I've identified somebody...I mean I will probably only meet with her a couple of times, I won't make it a long term sort of thing. But she's quite, she's female for a start... ....you know but she's quite a forthright sort of personality and I'm interested to know where that works and where it doesn't because I'm quite similar.... it's like "this is what I think" and I just thought it's quite interesting to see how that kind of leadership style works (Respondent A, Head of Department)

One interviewee could see the value of mentoring in terms of career development but also recognised the potential impact of the lack of female mentors within the organisation:

...most meetings I go to I'm the only female and it would be very hard to say, "well, is there another female here other than [senior female colleague] that I could actually aspire to and say mentor me, you've got a lot to give?" (Respondent B, Head of Function)

Another was planning to be pro-active as part of her training and development programme and find herself a mentor:

...ironically in my training and development plan that I just sent off yesterday, one of the things I am going to do is (a) train to be a mentor and (b) find somebody who can be a mentor for me. So I've just started thinking about, "ooh, what does that actually mean? What do I...what would I want to get from it? What sort of person am I looking

for?" And I haven't answered those questions yet (Respondent G, Senior Manager)

Despite the importance placed on the role of mentors for career progression touched upon earlier, the women interviewed had mostly achieved senior status without the assistance of a mentor. For those who had been allocated a mentor, their experiences were varied and not altogether positive. Those who expressed a wish to be mentored said that they would prefer to find a mentor outside of the organisation, although one interviewee could see the value in having a mentor both internally and externally to the organisation. Interestingly, just one interviewee articulated the advantages of being mentored by a scientist which could indicate that the use of non-science, external mentors may be preferable. Most interviewees had received a good level of support from past and present line managers, who had encouraged them to progress their careers and to think about longer-term career strategies. The statement by Greenfield (2002) that many women who had made it to the top have had supportive bosses would appear to apply here.

SENIOR WOMEN AS POTENTIAL MENTORS AND ROLE MODELS

Interviewees were asked about the advice they would offer to women looking to progress within the organisation. Most felt that at the most basic level, it was important to be 'doing a good job'. However, it was also recognised that this in itself was not sufficient to move up to senior levels. Interviewees said that they would advise other women to attempt to gain more exposure to senior staff within the organisation in order to influence a wider group of people, perhaps through project work, to pick up and develop a specialism, to enhance their skills, to raise their profiles by becoming recognised as the best at what they do, to gain a broad understanding of how the organisation operates and to network with others within the organisation. Self-promotion was perceived to be important:

Get a wider knowledge....and if you want to be noticed, you've got to get into things like processes and politically bits and that's the way you get progress is that you've got to be..... it's not just programming because if you just sit down and program people won't take any notice of you. You've got to be good; you've got to put yourself about a bit....raise your profile. And get in with the right people and seeing what they're good at and working from there. You know, as to really show them the best way, what will be the best way at that time of getting them to get there. If there was a particular piece of work that might be useful for her, something like that (Respondent D, Operational Applications Manager)

You've got to get experience in different areas of the office, you really have got to get out there and learn about how the different parts of it...because it's a big complicated organisation, it's not just [science discipline] you've got to understand how it all fits together, so that would be my overarching advice, depending on where that person was working at the time. Don't get stuck in one area. Ok, you may....if it's a scientist who's in the [work area] you may want to be a [grade] two or a one in that area, but you've got to get out, gain other experiences in order to go back in (Respondent G, Senior Manager)

There was a clear recognition that the onus is on the jobholder to be pro-active, to put themselves forward and to get themselves noticed. Personal career strategies were much more about being good at the job, seizing opportunities and taking on new challenges, with some mention of the importance of visibility and self-promotion at the executive level. Other women aspiring to senior positions in Scienceorg could learn from these strategies.

SATISFACTION WITH CAREER PROGRESSION AND FUTURE CAREER ASPIRATIONS

Finally, interviewees were asked how satisfied they were with their careers so far and about their future career aspirations. The majority expressed a high level of career satisfaction, irrespective of their job level and employment status. Comments ranged from feeling 'reasonably ok, 'quite satisfied' to 'very satisfied'. The following quotes represent the general consensus on career satisfaction:

I've enjoyed very much the work that I've done, and I do feel satisfied generally speaking with what I've achieved and what I've got. I'm ambitious, I want to go further, so there's always that kind of, "well you're not there yet." But nonetheless, I wouldn't dismiss what I have achieved (Respondent G, Senior Manager)

Very, actually. Very. I think if I had given it any thought, that first day when I joined, I'm not sure that I would have expected to be here. So yes, I'm pleased that I've managed to do that (Respondent Q, Senior Manager, PT)

The interviewees recognised that future career progression was a limited option, given the relatively small numbers of women employed at these levels and the presence of a male-dominated Board with just one female role model. The structure was described by one interviewee as a 'pyramid shape that is very narrow at the top'. Three possible options for career progression were mentioned by interviewees: an expansion of the current senior role with increased responsibilities; making a success of the current role (although eventually, most would wish to move up to the next level) and leaving to join another organisation (although this option would appear limited, largely by choice.

I probably didn't think much about where I wanted to go beyond where I am now. I was probably at the point where I did talk about it with my managers, I think at some level to go to the next stage; you do have to decide whether you are going to focus more on the science or more on the management side of things. So how you spend your time in your current post can sort of dictate that a little bit. But then in another way, going to the next level, it is really about what opportunities come up. There are obviously less posts at the next level (Respondent O, Senior Manager, PT)

Some interviewees expressed concerns about limited career options because of their gender. One believed that if she were to have another child, it would be very difficult for her to go any further in the organisation. This view was based upon her experiences so far:

I guess that does depend whether I have more children or not. If I were to have another child I think it would be very difficult to get anywhere. And I'm fighting to keep it at this level to be honest (Respondent N, Senior Manager)

Another interviewee felt that being a woman working in a predominantly male scientific area was perhaps holding her back:

I mean, it's partly, there aren't that many [women] because it's much more mathematical and programming side as opposed to soft science on [research area]. And a lot of the [non-senior] women have sort of left to have families. But in general at the moment there aren't that many [senior women] in the area I'm working in (Respondent C, Senior Manager)

When asked whether she thought it was a lack of numbers of women at the present level or something preventing the women who are there from progressing to that next level, she replied:

It's probably more the lack of women. Some directors will support women more than others. For some reason he [director] doesn't give me the opportunities that he would say give some of the male managers, which in the past I found frustrating. And essentially the way I'm dealing with it now is to get involved with activities outside of the [organisation] (Respondent C, Senior Manager)

Some expressed a desire to stay where they were for the time being, for a variety of reasons. Respondent I had recently attained her senior status within the past two years, a post that had taken her ten years to acquire. Two other interviewees (Respondents P and Q) wished to make a success of their current roles before moving on. Both are employed on a part-time basis, Respondent P having been promoted to her senior position within the past year and Respondent Q thirteen year's ago. All have a long length of service with the organisation:

I think I'm quite happy with where I am at the moment. I certainly want to stay at the [company] I don't know if I want to do my current role until I retire. I don't know what else I can do actually at the moment because I've wanted to get to this position for about 10 years (Respondent I, Senior Manager)

There is a high level of career satisfaction amongst the majority of the interviewees. However, when career aspirations were discussed, it became evident that some were frustrated with the potential lack of career progression opportunities past their current levels and were uncertain about the future. Two interviewees raised the problem of their gender and the implications of this for their own progression. Three interviewees were happy to stay where they were for the time being, either because they had no ambition to go any further or because they wanted to make a success of their current role. Lack of any clear progression path for interviewees beyond a certain level was compounded by relatively few organisations to which to go, due to the nature of their jobs.

CONCLUSION AND FUTURE DIRECTIONS

This chapter has analysed the careers of 13 senior female scientists who are employed in a UK, public sector organisation engaged in global scientific research. The objectives of the chapter were to explore the career histories, challenges and strategies of these women and how they have achieved and continue to maintain senior roles within a male-dominated organisation and industry. These objectives were linked to three related themes: career background, the role of mentors and the potential for the women interviewed to act as role models and mentors to other women.

The chapter makes a contribution to the already-established SET literature but also extends and develops knowledge and understanding of a relatively under-researched group of women who are not only under-represented in their chosen professions but also within their own organisations. The reality for these women is that at each stage in their careers, they are operating within a context which continually reinforces the, 'think scientist, think manager, think male' scenario – from education through to employment and beyond. While government initiatives exist to encourage more women into SET, these fail to tackle the underlying problems of gender stereotyping and the ways in which careers are gendered. While it is acknowledged that these underlying problems are not easily surmountable, this conclusion discusses three main points that have emerged from the research and possible action that could be taken by SET organisations.

The first point to note is that although these women's experiences varied, most dealt with working in a male-dominated organisation by identifying with their work and attaching importance to reputation-building, a strategy that Evetts (1998) found amongst female engineers. This is evidenced by the ways in which these women were determined to perform well and remain motivated. These women also became managers of scientists, despite having work-family responsibilities, a finding contrary to Evetts (1998).

Secondly, what also separates this particular group of women from other women in SET is their unusually long length of service, coupled with an unusually high retention rate post-childbirth, which is contrary to the findings of Blackwell and Glover (2008). This may be a reflection of the supportive, public sector environment in which they operate, with decent terms of employment and generous flexible working practices. It may also be the case that given the geographical location of Scienceorg and the specialist nature of the research, there are limited career opportunities outside of the organisation, aside from a few small, private sector organisations.

This means that the women interviewed had developed their roles within an internal labour market over an extended period of time, within an organisation that they felt was supportive but lacking in career opportunities past a certain senior level, such as the Board. Some interviewees said that they felt 'stuck' when discussing career progression and that their options appeared limited. A long length of service amongst employees generally means that turnover is not sufficient to facilitate the movement of women into more senior roles once they become vacant. In light of the advice by senior women that women who aspire to senior positions should gain experience in different areas of the business and raise their profiles, Scienceorg should consider more use of secondments into projects across the business to extend the experience and motivation of those 'stuck' on particular grades. A second, longer-term option would be for Sciencorg to consider adopting a broadbanded job structure that would facilitate lateral career development with enhanced responsibilities and salaries.

Thirdly, mixed views were expressed by interviewees on the role of mentors. There was a desire to be mentored by other senior women outside of Scienceorg and not specifically by scientists. Most interviewees were positive about the support from line managers. This is perhaps an area where Scienceorg, as well as government and support networks, could focus their efforts in supporting women in SET. Where women had been specifically allocated mentors, when going through the assessment centre, this tended to be less successful and this would indicate that senior women should be given the opportunity to seek out their own mentors, both inside and outside of Scienceorg. The women interviewed have significant potential to act as mentors and role models for other women in science, both within their own and other SET organisations, which fits with their desire to reputation build and improve their personal visibility. However, that few of these women had active mentors seemed to influence their perspective on becoming a mentor themselves.

The efforts by various bodies to encourage women into SET employment, largely through the UK government, have been largely unsuccessful and appear to have little bearing on the Scienceorg women. During discussions about networking membership, none of the interviewees made reference to any of the UK-wide networks cited in this chapter. In general, networking tended to be carried out informally and within Scienceorg.

SET organisations need to become involved in not only encouraging more women into the profession but also nurturing them once they are there. This could extend to programmes to encourage mentoring relationships inside but especially outside of the organisation, which could enable women to see the bigger picture and to consider other career options within their own organisations. Scienceorg could consider implementing a mentoring programme, initially as a pilot project, for its senior female scientists. The aim should be to encourage senior women to not only identify their own mentors, especially outside of the organisation, but also to mentor other, non-senior women. Scienceorg should also be active in approaching other organisations to take part in the pilot mentoring scheme. If successful, this could be rolled out to other women in the organisation and become a permanent part of organisation life and career development for female scientists. Clearly, the buy-in of organisations is paramount as government initiatives are not effective on their own.

REFERENCES

Acker, J. (1990). Hierarchies, Jobs, Bodies: a theory of gendered organization. *Gender & Society*, *4*(2), 139–158. doi:10.1177/089124390004002002

Blackwell, L., & Glover, J. (2008). Women's Scientific Employment and Family Formation: a longitudinal perspective. *Gender, Work and Organization*, *15*(6), 579–599. doi:10.1111/j.1468-0432.2007.00385.x

Cockburn, C. (1983). *Brothers: Male Dominance and Technological Change*. London: Pluto Press.

Cockburn, C. (1991). *In The Way of Women: Men's Resistance to Sexual Equality in Organisations*. London: Macmillan.

Durbin, S. (2006). Gender, Skills and Careers in UK Call Centres. In Burgess, J., & Connell, J. (Eds.), *Developments in the Call Centre Industry: analysis, changes and challenges* (pp. 117–135). Oxon, UK: Routledge.

Durbin, S. (forthcoming). Creating Knowledge through Networks: a gender perspective. *Gender, Work and Organization*.

Ellemers, N., van den Heuvel, H., de Gilder, D., Maass, A., & Bonvini, A. (2004). The Underrepresentation of Women in Science: differential commitment or the queen bee syndrome? *The British Journal of Social Psychology*, *43*, 315–338. doi:10.1348/0144666042037999

European Commission. (2006). *Women and Science Statistics and Indicators: SHE Figures 2006*. Brussels: European Commission.

Evetts, J. (1998). Managing the Technology But Not the Organisation: women and career in engineering. *Women in Management Review*, *13*(8), 283–290. doi:10.1108/09649429810243144

Evetts, J. (2000). Analysing Change in Women's Careers: culture, structure and action dimensions. *Gender, Work and Organization*, *7*(1), 57–67. doi:10.1111/1468-0432.00093

Fox, M. F., & Stephan, P. E. (2001). Careers of Young Scientists: preferences, prospects and realities by gender and field. *Social Studies of Science*, *31*(1), 109–122. doi:10.1177/030631201031001006

Greenfield, S. (2002). *SET Fair: a report on women in science, engineering and technology*. London: Department for Trade and Industry.

Groysberg, B. (2008). How Star Women Build Portable Skills. *Harvard Business Review*, *86*(2), 74–81.

Halford, S., Savage, M., & Witz, A. (1997). *Gender, Careers and Organisations*. London: Macmillan.

Harding, S. (1986). *The Science Question in Feminism. Milton Keynes*. UK: Open University Press.

Helsinki Group on Women and Science. (2002). *Women in Science: review of the situation in the United Kingdom*. London: Department for Trade and Industry.

Herman, C., & Kirkup, G. (2008). Learners in Transition: the use of ePortfolios for women returners to science, engineering and technology. *Innovations in Education and Teaching International*, *45*(1), 67–76. doi:10.1080/14703290701757468

Hewitt, P. (2003). *A Strategy for Women in Science, Engineering and Technology*. London: Department for Trade and Industry.

Indepdendent. (2008). *Women and Science: how to achieve the right balance*, (September).

Kanter, R. M. (1977). *Men and Women of the Corporation*. New York: Basic Books.

Miller, G. (2004). Frontier Masculinity in the Oil Industry: the experience of women engineers. *Gender, Work and Organization*, *11*(1), 47–84. doi:10.1111/j.1468-0432.2004.00220.x

Oakley, J. G. (2000). Gender-based Barriers to Senior Management Positions: understanding the scarcity of female CEOs. *Journal of Business Ethics*, *27*, 321–334. doi:10.1023/A:1006226129868

Ozbilgin, M. F., & Woodward, D. (2004). Belonging and Otherness: sex equality in banking in Turkey and Britain. *Gender, Work and Organization*, *11*(6), 668–688. doi:10.1111/j.1468-0432.2004.00254.x

Phipps, A. (2006). I Can't do With Whinging Women! Feminism and the habitus of 'women in science' activists. *Women's Studies International Forum*, *29*, 125–135. doi:10.1016/j.wsif.2006.03.010

Powell, A., Baglihole, B., Dainty, A., & Neale, R. (2004). Does the Engineering Culture in UK Higher Education Advance Women's Careers. *Equal Opportunities International*, *23*(7/8), 21–38. doi:10.1108/02610150410787882

Schein, V. E. (2007). Women in Management: reflections and projections. *Women in Management Review*, *22*(1), 6–18. doi:10.1108/09649420710726193

Sealy, R., Vinnicombe, S., & Singh, V. (2008). *The Female FTSE Report 2008*. Cranfield University, School of Management.

Swanberg, J. E. (2004). Illuminating Gender Organization Assumptions: an important step in creating a family-friendly organisation: a case study approach. *Community Work & Family*, *7*(1), 3–28. doi:10.1080/1366880042000200271

UK Resource Centre for Women in Science. *Engineering and Technology.* (n.d.). Retrieved 30.3.2009 from http://www.ukrc4setwomen.org/html/research-and-statistics/statistics/

Vinnicombe, S., Singh, V., & Kumra, S. (2004). *Making Good Connections: Best Practice for Women's Corporate Networks*. Cranfield University.

Wajcman, J. (1998). *Managing Like a Man: Women and Men in Corporate Management*. Oxford, UK: Blackwell.

Walby, S. (2004). The European Union and Gender Equality: Emergent Varieties of Gender Regime. *Social Politics*, *11*(1), 4–29.

Wilson, E. M. (1998). Gendered Career Paths. *Personnel Review*, *27*(5), 396–411. doi:10.1108/00483489810230334

Wynarczyk, P., & Renner, C. (2006). The 'Gender Gap' in the Scientific Labour Market: the case of science, engineering and technology-based SMEs in the UK. *Equal Opportunities International*, *25*(8), 660–673. doi:10.1108/02610150610719128

ENDNOTES

1 SET occupations include: SET Managers; Science Professionals; Engineering Professionals; ICT Professionals; Teaching Professionals; Research Professionals; Building Professionals; Science and Engineering Technicians; Draughtspersons and Building Inspectors; IT Service Delivery Occupations (based on the 4 Digit SOC2000 Definition)

2 The 'SET Manager' definition includes research and development managers

3 The 'Science professionals' definition includes physicists, geologists and meteorologists

4 Although the time period in which this took place is not stated

Profile 11
Science Career Spotlight:
Pollinating Insects in Natural and Agricultural Ecosystems in Australia's Tropics

© 2008 Queensland Government Office of Women. Used with permission

Romina Rader, School of Marine and Tropical Biology, James Cook University, Cairns

Growing up in inner city Sydney, I looked forward to every school holidays when my parents took us hiking in the great outdoors. Nearly every summer my eyes widened and my interest deepened in response to the myriad of animals and plants we encountered. After completing a Bachelor of Science (zoology), my next career move was inspired by a holiday to Cairns with my sister. The minute my eyes met those rainforest clad hills from above, I wanted to find out more about what lay beneath that seemingly impenetrable rainforest canopy. Six months later I moved north and enrolled in a Master of Science at James Cook University to study small rainforest mammals. Deciding that I really liked research and the opportunities it presented to study new things in greater depth, I went on to do a PhD, this time on insects in both natural and modified ecosystems.

Copyright © 2010, IGI Global. Copying or distributing in print or electronic forms without written permission of IGI Global is prohibited.

Project Objectives

This research study examines insect pollinator assemblages in four different land uses (dairy, orchard, cropping and remnant forest) with the aim of determining the impact of land use on biodiversity and the fundamental ecosystem service of pollination. The project consists of monthly sample collections of ALL insect groups visiting crop flowers over a 12 month period on the Atherton Tablelands, north Queensland. The project uses Flight Intercept and Sticky Traps to capture pollinating insects and examines the pollen carried on their abdomen. This project is significant in that it will provide practical information to farmers regarding where and when pollen carrying insects will be and what factors pollinating insects respond to in the environment. It will ulitmately help to make farming more sustainable by improving existing pollination services provided by native insects.

The outcomes of the completed study will encourage and contribute to:

1. Knowledge and understanding of the biodiversity of insect pollinators in Australian natural ecosystems. The distribution of biodiversity in the landscape is little known in modified environments and yet may contribute to as well as benefit from, vital services such as pollination and pest control.
2. A greater understanding which insects are beneficials or pests to agriculture and when they are most abundant. If used effectively this information regarding the seasonality of insects may reduce the need for pesticides to be applied as frequently to orchards and crops in the longer term. This will ultimately result in benefits to the community from reduced chemical application.
3. A greater understanding of the contribution of flies and beetles to crop and orchard pollination.
4. A more detailed understanding of the contribution of remnant native vegetation in providing (currently unvalued) pollination services to a range of land uses.
5. Collaboration between CSIRO, James Cook University, Crop and Food Research New Zealand, the Australian Museum, the Queensland Museum, Department of Primary Industries Queensland, cropping farmers, orchardists, dairy farmers and owners of privately owned remnant forest. This will facilitate knowledge and information sharing between research scientists, government land managers and policy makers, industry and the general public.

Project Rationale

Pollination is an ecosystem service that maintains function in natural systems as well as enhances the production of many crops. Of the world's flowering plants, between 70-90% are pollinated by animals; bees in particular pollinate one or more cultivars of at least 66% of the world's 1500 crop species. The value of these services is estimated to be in the order of 3 billion dollars in Australia and New Zealand and 300 billion dollars globally, many of which are provided by native insect species.

Land use modification by humans is one of the primary drivers of biodiversity loss yet the distribution of biodiversity among human modified land uses is unclear. Insect pollinators have been identified across several insect orders as well as several families within those orders, hence it has been difficult to predict responses to land use change without a specific pollinator study. Pollinator decline is now apparent on a global scale, largely due to unprecedented changes in land use and declining honeybee populations.

When this project is complete, the information concerning pollen carrying insects will contribute to the study being the first of its kind, since studies to date predominantely resolve taxa at the level of order or are specific to one location. A knowledge of the composition of the insect communities distributed amongst these land uses is necessary in order to understand and make recommendations regarding specific groups that are potentially valuable in providing the ecosystem service of pollination.

Current (and Pending) Progress and Achievements

1. Preliminary research results suggest flies and beetles as well as native bees carry pollen in agricultural landscapes.
2. Insect trapping method (Flight Intercept and Sticky Traps) has been developed and carried out to ensure regular insect capture
3. Insects have been collected from traps monthly and stored for processing
4. Samples of insects have been processed and sorted into different insect Orders
5. Pollen collected from insect bodies
6. Pollen mounted on slides for microscopic identification
7. Insects carrying pollen identified and pinned to enable higher resolution identification.
8. Dissemination of project objectives and preliminary findings into wider community via media and workshop presentations
    - Presentations at the Organic Producers Association of FNQ and Mitchell River Watershed Catchment Management Field days
    - Presentation at Crop and Food Research Institute and Foundation For Arable Resources, Christchurch New Zealand.
    - ABC radio interview 21 June 2007 and local newspaper media releases 3 July 2007.

Progress and Achievements in the Next 3-6 Months

9. Insect families/genera/species identified with assistance from expert taxonomists
10. Workshops held to assist farmers in identification and diversity of pollen carrying insect species
11. Presentation of final data analysis to local community and Industry groups, literature produced and published and media notified of project results October/November 2009

Innovative Aspects of Project

This project is innovative because it has looked beyond the traditional services of the honeybee for pollination services. This research has demonstrated that other insects can carry pollen around north Queensland crops and orchards, namely native bees, flies and beetles. By examining all insects visiting agricultural areas, this project has found other potential native insect pollinators.

The project is also innovative in that it is a large landscape scale study that samples both human modified as well as natural systems to determine which land use hosts the most effective pollinators. Most studies of pollinator communities to date have been local, small scale field based studies, that fail to include human modified environments.

Aspects of Project Demonstrating Community Benefit

This project will provide primary industry (particulary orchard and cropping farmers) with a greater understanding which insects are beneficials to agriculture and when they are most abundant. This will enable them to identify and encourage insects that the project ascertains are pollinators and not pests. This will effect longer term beneficial change in modified agricultural environments and flow-on effects to nearby natural systems in the same landscapes by enabling targeted pest control (as opposed to generalist insect control) potentially reducing chemical use and associated effects of run off.

If used effectively this information regarding the seasonality of insects may reduce the need for pesticides to be applied as frequently to orchards and crops in the longer term. This will ultimately result in benefits to the community from reduced chemical application.

Environmental Benefits

As stated above, the project will benefit the environment by its capacity to identify beneficial insects in the environment. This will enable more targeted pest control in agricultural areas. A more detailed knowledge of the insects carrying pollen will also improve our understanding of the factors influencing biodiversity in Australian natural and modified ecosystems. The distribution of biodiversity in the landscape is little known in modified environments and yet may contribute to as well as benefit from, vital services such as pollination and pest control. In addition this project will provide a more detailed understanding of the contribution of remnant native vegetation in providing (currently unvalued) pollination services to a range of land uses.

Economic or Business Viability

This project has economic viability in that it reduces the reliance of farmers on the introduced honeybee. In some crops, honeybee hives have been introduced (as a service paid for by farmers) as the contribution of native insect pollinators was previously unknown. By understanding and identifying the extent of pollination services provided by native insects, farmers will be better informed as to which insects provide the service and should be encouraged in their crops/orchard. This may also reduce the need for frequent pesticide application if the beneficials and pest insects can be differentiated and managed separately, thereby reducing costs.

Chapter 12
The Career Challenge of the Gendered Academic Research Culture:
Can Internet Technologies Make a Difference?

Anne Manuel
University of Bristol, UK

ABSTRACT

It is clear from many of the contributions to this volume that there are career advancement challenges that are specific to women in the science, engineering and technology (SET) disciplines. In this chapter I will be looking at one of the career challenges that face not just women in those disciplines (although the problem here is particularly acute) but women seeking an academic career in any discipline, and that is the gendered nature of its predominant research culture. I go on to consider how the use of Internet technologies (ITs) by academics in their research might be enabling women to meet this particular challenge and whether women in a scientific academic setting have a different experience in this respect from colleagues in social sciences and humanities. Findings are presented from 25 in-depth interviews and content analysis of 750 academic web profiles. Results would suggest that although there are areas where the gendered research culture is being circumvented by the use of ITs for women in all disciplines, yet there are areas where women are not taking full advantage of the potential of the web to increase their visibility and research profile. Moreover, the opportunities presented by ITs are not necessarily tied to discipline and women in SET disciplines appear to be no more or less likely to take advantage of them. Finally in some important respects, the gendered cultures and structures that exist in 'real' society are continued into the 'virtual' one, rendering the gendered research culture little changed by increasing use of ITs.

DOI: 10.4018/978-1-61520-657-5.ch012

Copyright © 2010, IGI Global. Copying or distributing in print or electronic forms without written permission of IGI Global is prohibited.

INTRODUCTION

This chapter is set within the context of the academic environment of UK higher education and in particular explores the barriers experienced by many women in their careers as academics, owing to the operation of a gendered research culture. In this chapter, Acker's model of a gendered organisation (1990, 1998, 2006) is employed to show the extent to which the UK academic research culture can be seen as a gendered organisation and, drawing on empirical research, goes on to investigate whether and to what extent, this culture is being disrupted through the increasing and widespread use of Internet technologies for research activities. The study also considers the effects of discipline to determine whether women in SET have different experiences in this respect from colleagues in other disciplines.

Women in academia have long suffered from the 'glass ceiling' effect, finding it difficult to progress beyond a certain point in their career or gaining promotion more slowly than male counterparts (Brooks, 1997; Knights & Richards, 2003). This is particularly so in SET disciplines. In 2006/7, statistics show that 92% of professors and 81% of senior lecturers in the physical and biological sciences were men (Higher Education Statistics Agency, 2008) compared to figures for the sector overall at 83% and 64% respectively. Theories for the persistence of gender inequality in academia in recent years have tended to revolve around the gendered nature of the academy with its male dominated history and patriarchal structures and systems which put women at a disadvantage when it comes to career progression (Bagilhole & Goode, 2001; Benschop & Brouns, 2003; Deem, 2003). In SET disciplines, the persistent association of all things technical with men is of course a problem (Clegg, 2001; Faulkner, 2001; Wajcman, 2004) but the sub-text or organisational culture of male dominated workplaces has also been identified as being a major barrier for women looking for careers in SET disciplines (Sappleton & Takruri-Rizk, 2008). In particular, the academic research culture has been shown to be problematic for women in SET and indeed in all academic disciplines (Fletcher, Boden, Kent, & Tinson, 2007; Harley, 2003; Katila & Meriläinen, 1999; Knights & Richards, 2003; Smith, 2008).

The research culture can be defined as the circumstances and values surrounding the production of research within a department or institution (Deem & Lucas, 2007) but is also used by commentators to describe the wider context of research generally (Fletcher, Boden, Kent, & Tinson, 2007; Knights & Richards, 2003). The primary focus of the analysis in this study is organisational research culture as experienced by individual academics within their institutions as will be seen later. However, reference is also made by interviewees and in the commentary to the more general sense of how research is perceived by the academic community and thus in this chapter the term will be used in both senses, with its meaning clarified through the context in which it is used.

Research has traditionally been vital for a high-flying academic career, with 'publish or perish' being the mantra of academic staff for more than a century (Lucas, 2006). But over the past twenty years, research productivity has become increasingly important to both departments and individual academic's careers in the UK, with the introduction of the Research Assessment Exercise (RAE) (currently being replaced by the Research Excellence Framework). This determines to a large extent, the amount of government research funding that university departments attract, based on the quality of individuals' research output. Research is therefore vital to the continued financial health of a department, and individuals who are not seen to be pulling their weight in terms of research output, face career stagnation, early retirement, even redundancy (Court, 1999) - a situation which is exacerbated by the current recession.

Given the importance of research publication to the academy (Lucas, 2006), it is fundamental to the careers of academic women that they are seen to be as research active as their male colleagues. However, reports and research show that women may not be publishing as much as men (e.g. Brooks, 1997; Leahey, 2006; Xie & Shauman, 1998). Possible reasons for this have been suggested as being excessive teaching loads and pastoral care of students, (Acker & Feuerverger, 1996; Halsey, 1992), greater home-care responsibilities, (Deem & Hillyard, 2002; Probert, 2005) and general lack of departmental support (Bagilhole & Goode, 2001). Furthermore, the research culture of UK academe generally has been argued to privilege male academics over female through its reinforcement of a highly competitive, single-focused, achievement-oriented way of working (Fletcher et al., 2007; Priola, 2007) - a situation which the RAE has compounded. (Harley, 2003; Smith, 2008).

At the same time as the research culture has been gaining in pre-eminence in the UK, the use of Internet technologies amongst academics in their research has been steadily climbing to the point where it is now a normal part of the way we work in academia and indispensable in many disciplines.

However, despite much attention being focussed on new technologies in teaching and learning in HE, few studies to date have looked at how the Internet is impacting on the research process in the UK.

This study looks at the ways in which the Internet might be disrupting the structures and practices of the gendered research culture at the micro levels of the organisation, through interactions and identities and takes an overall view of how the disruption might work differently in different disciplines.

The research presented in this chapter is divided into three sections. Firstly I examine the extent of the gendering of the research culture in UK academe through use of aspects of Acker's model of the gendered organisation (1990, 1998). Then I move on to describe some of the findings from a content analysis of academic websites and in-depth interviews with 25 male and female academics in the UK that comprise the empirical study into academic Internet use for research in the UK. Finally I discuss the findings, showing where androcentric practices and behaviours of the research culture are being circumvented and whether discipline is a factor in this and, importantly where the potential exists for women and men to further disrupt the gendered assumptions, attitudes and practices that underpin the UK academic research culture.

BACKGROUND

The Gendered Research Culture

The gendered nature of the research culture, as mentioned above, has been noted and examined in a number of different studies in recent years (Fletcher, et al., 2007; Harley, 2003; Smith, 2008). Although they are useful additions to our understanding of gender inequality in academia, what has been lacking to date is a detailed understanding of how different aspects of the research culture are differently gendered and, more importantly, what can be done to make the research culture less 'oppressively gendered' (Britton, 2000).

In this study I use Acker's theory of the gendered organisation to examine closely where gender inequality is manifest in the academic research culture. Acker's article 'Hierarchies, Jobs, Bodies: A Theory of Gendered Organizations' (1990) is seen by many as ground breaking, (Barry, Berg, & Chandler, 2006; Britton, 2000; Martin & Collinson, 2002) setting out for the first time how organisations are predicated on gendered lines with the ideal worker clearly being male. Gender is 'done' throughout the processes of the organisation (West & Zimmerman, 1987) and Acker identifies five areas where this is to be seen which Britton (2000) summarises as structure, symbols and values, interaction, identity and policy.

Structure concerns the construction of divisions in the organisation along lines of gender and is perhaps the most obvious place to see gender being 'done'. Examples of this include use of physical space, remuneration, and the exercise of power. Symbols and Values are defined in the way that images and implicit organisational values (used by the press, the public and in language), explain or reinforce the structural divisions. Interaction refers to the ways that workers behave towards each other. Studies involving conversation analysis or ethnography for example, show gendered patterns that are reinforced through accepted behaviour in the organisation and in general it is a story of male dominance and female submission (Coates, 1995). The fourth aspect of the gendered organisation is seen in the internalisation of the gendered values of the organisation through presentation of identity. Thus what men and women choose to wear, what jobs they apply for and who they associate with are expressions of the gendered identity that prevails in the organisation. Finally the fifth process concerns policy or 'organisational logic' - how the organisation is run, how evaluations are carried out, contracts drawn up, communication effected, worth recognised. These are the gendered 'understructures' (Acker, 1992) which run deeply throughout the organisation, reinforcing gender inequality in everyday policy and practice.

Acker's account of gendered organisations is valuable because it reveals clearly how gender inequality in organisations cannot be eradicated merely by increasing the number of women in the organisation or by introducing new recruitment and pay policies. Gendered practices run so deep in an organisation and in such complex layers, that no one solution can liberate the organisation from its discourse of dominance and subordination (Sappleton & Takruri-Rizk, 2008). But Acker's approach is not without its critics. Britton (2000) for example notes that the theory lacks a detailed consideration of context and points to studies that show how different organisations in the same field are differently gendered, depending on the make-up of the senior management. She cautions against the blanket bracketing of organisations as uniformly gendered which leads to the rather depressing view that nothing can be done. By looking at the ways in which different organisations are differently gendered, and by focussing on local situations, one can start to identify the means by which organisations might be made less oppressively gendered. Her recommendation is that research should focus on gaining a better understanding of how organisations are gendered at each of Acker's five levels in specific contexts and through studies 'from below'. This should provide an insight not only into how different organisations are gendered differently but also highlight areas of resistance to the dominant culture. This study takes up this challenge and looks in detail at how the gendered organisation of the UK academic research culture operates at two of Acker's levels and goes on to consider the use of Internet technologies as the mechanism of resistance.

Elsewhere in my research I have considered in detail how Acker's model of the gendered organisation can be mapped onto the research culture of UK universities (Manuel, 2008). In this chapter I will be focussing on the two levels of the gendered organisation which can be said to operate at the 'micro' level – interaction and identity. It is at these levels that equal opportunities policies and legislation struggle to have most effect because individual beliefs and assumptions are hard to shift (Deem & Morley, 2006; Saunderson, 2002). It is at these levels then, that the potential exists for significant change.

Micro Levels of the Gendered Research Culture: Interactions and Identities

Interactions

Acker (1990) identifies that gender is 'done' in interactions (or absence thereof) between men and women, women and women and men and men, in both verbal and non-verbal communication. The gendered nature of university interactions has a long history and whilst overt sexual harassment occurs less frequently thanks to equality legislation and equal opportunities policies (Deem & Morley, 2006), nevertheless an examination of the 'micropolitics' of the academy reveals that gender inequalities are reproduced in the everyday encounters between men and women through behaviour, conversation, and calculated actions (Benschop, 2009; Morley, 1999; Saunderson, 2002).

In the context of the research culture we can think of interaction in two ways – interactions regarding collaboration over the conceptualisation, production and dissemination of research, and interactions in the research community to build and maintain reputation amongst one's peers (Becher and Trowler, 2001, Halsey, 1992). Both these strands require individuals to operate in academic networks. Invitations to collaborate and bid for research funding, to present work at conferences and to edit journals are offered to those who are 'known' and who operate in the correct networks. This becomes a gendered concept because for a number of reasons outlined below, women in academia have found it difficult to participate in these networks (Bagilhole & Goode, 2001; Mavin & Bryans, 2002; Travers & Pemberton, 2000) and have therefore not benefited from them in the same way that men do.

Networks in academia can be formal, as in disciplinary associations, or informal as in groups of colleagues working in the same building or institution but whatever their manifestation, they have been shown to operate best where there is a high degree of similarity amongst the participants, particularly important for communication, predictability of behaviour and building relationships based on trust and reciprocity (Ibarra, 1992). Gender is one of the main characteristics with which individuals identify and therefore also use to differentiate themselves from others (Lincoln & Miller, 1979). Thus women attempting to network with men may come across suspicion and hostility if men feel that the trust on which their network is based is being compromised (Alemany & Diaz, 2007).

Networks of all types are often developed and perpetuated through social and semi-social interactions and events but since these often occur at the end of the working day or in the evenings, women more than men may find this timing problematic because of the increased likelihood that they will have the bulk of domestic responsibilities at home (Singh, Vinnicombe, & Kumra, 2006), whether that be related to the primary care of children (Probert, 2005) or other caring/housekeeping activities (Ramsay & Letherby, 2006).

Tonge (2008) has recently identified the barriers that men and women face in their networking efforts in the service industry and found that women perceive more than twice the number of barriers than do their male colleagues, including their age, appearance, gender, a discouraging culture, family pressures and tiredness. Overall there was more reluctance amongst women to carry out networking activities than amongst men and this was connected by the author with the larger number of barriers experienced by women.

Lack of participation in networking activities is a vicious cycle. Women who do not engage in networking activities or who are excluded from them, may not be aware of the importance of such networks for career progression. Thus making the considerable effort required to break into a long established male dominated network may not be seen as worth the bother (Tonge, 2008). Sometimes they are simply not aware that an informal network exists in their organisation (Gray & Kurihara, 2006). Moreover, some research has revealed that where women do network informally, their goals are different from men. Male networking tends to be more 'utilitarian' or work-focussed while for women networking provides more of a social and supportive role (Ibarra, 1992). Thus even where mixed networks do operate, the different goals operating for men and women mean they achieve different ends out of their networking efforts with women gaining friends and men gaining promotion.

The contribution of the Internet to networking issues for women has been recently been researched by Donelan et al (2009). They found that although women were seeking the support, advice and collaboration of other women through online networks, very few credited their career progression and job opportunities to their networking activities. This was despite the advantages of online networking for women over face-to-face networking identified by the authors (timing, lack of traveling required, fewer confidence-related issues). The study only looked at women operating in women-only networks in SET and thus it is difficult to draw conclusions about their activities compared to male colleagues.

Mavin and Bryans (2002) also report on a women-only network which was set up specifically to support and encourage women in academia to publish research, seek promotion and tackle gender-related workplace issues. Deliberately set up as a formal networking opportunity, some women involved voiced concern about being categorised as a 'whingeing feminist' and wanted to be clear about aims and objectives to prevent the group becoming a 'knitting club'. Such negative terms highlight an extra problem for women: if they do create networks to create the supportive environment that most men are able to tap into, they risk being pigeonholed as men-haters or gossips and losing academic credibility as a result (Bagilhole & Goode, 2001). This problem of how academics are perceived and perceive themselves is further explored in the other 'micro' level of the gendered organization, concerning identity.

Identity

Acker describes this level as the way we internalise the other levels of the organisation and the norms of the world around us. The research culture has been described in terms of a 'discourse that features competition, individual achievement, striving for continuous improvement and placing of responsibility for success in ones own hands' (Acker and Armenti, 2004). It requires self-promotion, self-confidence, and self-assertion and yet is situated within a culture that prides itself on its egalitarian meritocratic systems (Deem, 2007; Knights & Richards, 2003). Thus the culture is seen to be gender-neutral but the qualities required to succeed within it are largely associated with socially constructed 'masculine' characteristics making it another problem area for women.

Bagilhole and Goode (2001) carried out 37 interviews in one university, trying to establish what cultural and institutional barriers exist that might explain why women do less well in academic careers. One of their key findings was that women have difficulty with the concept of promoting their own research work, something that is echoed in Fletcher et al's recent study (2007). Self-promotion is vital in an environment where research has become commodified (De Groot, 1997; Morley, 2003) but is interpreted as 'pushiness' by some women themselves in Bagilhole and Goode's study. Their fear of appearing to embrace such behaviour ensured that they stayed silent and relied on their work to speak for itself. Bagilhole and Goode point out the dual standards at work here where self-promotion of research is seen positively as signs of ambition and leadership qualities in men but negatively as bossiness or pushiness in women whom society dictates should be modest and unassuming.

Discipline and the Gendered Research Culture

Interaction and identity, then, are two levels of the gendered research culture. But although we have talked about the research culture as a single entity, it is of course far from being homogenous throughout academia and one of the key things affecting the nature of the research culture is discipline. Becher and Trowler's (2001) extensive examination of the life and work of academics, identifies many patterns of difference between the disciplines which impact on the research culture. They argued that academic communities were defined to a large extent by the discipline in which they operated rather than the institution they worked in. With regard to research they noted that type and frequency of research output could be closely correlated with discipline and was characterised in the 'urban' disciplines (those where the ratio of researchers to research areas was high) by journal articles, brief position papers and conference papers whose shelf life was relatively short whereas in the 'rural' disciplines (where researchers' fields of interest are many and varied) books were the most prestigious type of research output which might take many years to appear.

Henkel (2000) also found some disciplinary related findings in her study of academic identities and policy change. She interviewed 153 academics in seven disciplines and concluded that discipline played a major part in whether academics saw themselves primarily as researchers or as researcher/teacher/administrator. The scientists in her sample were more likely to state that research was of primary importance to them and to express concern when administration or teaching duties took them away from this. Those in arts based subjects were strongly focussed on research being interconnected with teaching and therefore their identities were bound up with pursuing both aspects of the job.

Lucas (2006) analysed discipline in terms of different types of Bourdieuian capital showing how economic and material capital (external funding) is particularly important in SET departments for example who have heavy requirements for equipment, consumables, technical support staff and research assistants. Social capital, described as collaboration, networking and reputation, was also vital where such activities attracted further work and funding and enhanced standing in the discipline. In English, however, economic capital was less important for conducting research and although reputation was as important, social capital in the form of collaboration was found to be less so. A reading of Lucas and Henkel would suggest that interaction and identity could be more of an issue for women working in SET disciplines than in others and thus any breakthroughs here in terms of Internet use would perhaps benefit SET women the most.

However, Brew (2001) has argued that discipline is not the sharply defining factor that many scholars would have us believe. She has pointed to shifting conceptions of where disciplinary boundaries are, to the growth in interdisciplinary collaboration and the emergence of new areas of study which cross disciplinary boundaries such as cultural studies, gender studies and nationally-based studies. She acknowledged that researchers in different disciplines may do research in different ways and have different ideas about the nature of research and knowledge production, fostered by working in departmental, disciplinary communities. However, she emphasised the fact that at the individual level, discipline does not adequately explain academics' views about research and that moreover, many academics she has studied have difficulty consigning themselves to one particular discipline.

Despite differing academic views on discipline, its importance in any discussion of research seemed clear and it was crucial to represent a range of disciplines in the study to pick up any differences that discipline might make to a consideration of gender and Internet use. Economics, Biology and History were selected since these were fairly universally taught amongst universities in the UK and the gender ratios of academic staff were similar to the staff gender ratio of all subjects at 35:65.

THE STUDY

The main data collection method used to examine the extent to which Internet technologies were proving to be a mechanism for disrupting the

Table 1. Framework for evaluating academic web profiles

Indicators of Self-Projection from web pages	
Category	**Criteria**
Academic credibility measured in the form of achievements noted on the website	Biography/publications/conferences organised/prizes/funding awarded/organisations involved in/collaborators
Reference to different aspects of the role	Research/Teaching/Admin/combinations/none
Self disclosure – personal details/photo	Yes/No Informal/formal
Creativity	Artwork/writing/photos
Formality – 3rd or 1st person used in text	1st/3rd /neither

gendered research culture at the levels of Interaction and Identity, was semi-structured interviews with 25 male and female academics (See Table 2 for a profile of interviewees). The interviews were designed to explore how interviewees were using the Internet for research as well as to hear their views on the use of the Internet in research. In addition, a review of websites was conducted to see if there were any gendered elements in the way academics use their web presence. Websites can be seen as a form of self-promotion (Dominick, 1999; Papacharissi, 2002) and therefore very much tied up with the concept of both identity and interaction. The review provided an opportunity to observe the behaviour of academics in addition to hearing about their experiences and opinions from the interviews.

Web Review

The aim of the web review was to see what sort of pictures male and female academics were painting of themselves in terms of identity projection and also to establish whether there was a gender element to the content of the web pages themselves. In order to establish criteria for analyzing the web pages, previous studies of academic identity on the web were reviewed. Miller and Arnold (2001) carried out interviews with women academics about their use of homepages in the 1990s and found that they felt a need to establish their credentials as academics through prominent display of publications, qualifications, biographies etc but at the same time felt some reluctance to reveal too much about themselves as people, for example by not including or 'hiding' photos of themselves on their pages. They concluded that the vulnerability felt by some women in promoting themselves in public was replicated in the 'virtual' world. If this were still the case in the 21st Century, this might act against the potential of the Internet to provide a 'gender-neutral' environment in which self-promotion could take place. Similar criteria to Miller and Arnold were used in this study to evaluate the extent to which the web pages were being used by men and women to promote themselves and their research and to network with others. To this end, evidence of qualifications, publications, awards and prizes, photos and personal data was sought. Some of the criteria used are listed in Table 1.

Ten universities were selected, ensuring both pre and post 1992[1] universities were included in the sample. All the individuals' personal websites in the three departments History, Economics and Biology, were then reviewed, resulting in a total of 750 sites. These were analysed for the criteria noted above and the results recorded in an Access database. The data were cleaned and reviewed and transferred into the statistics software package, SPSS. Chi-square tests were applied, cross-tabulating both gender and discipline with each criteria.

Table 2. Profile of interviewees

	History			Biology			Economics			*Total*	
	Male	Female	*Total*	Male	Female	*Total*	Male	Female	*Total*	Male	Female
Pre 1992 University	1	2	*3*	2	3	*5*	1	1	*2*	4	6
Post 1992 University	2	3	*5*	1		*1*	3	4	*7*	6	7
Total	3	5	*8*	3	3	*6*	4	5	*9*	10	13

Interviews

A key part of this study was the semi-structured interviews with a cross section of male and female academics in the UK. The interview questions were compiled using the guidelines of my three research strands (gender inequality in UK academe, the research culture and the increasing use of internet technologies, as set out in the Introduction to this chapter):

- *Section 1 Research culture* – these questions focussed on how the interviewee fitted into the research culture they worked in and how the culture affected their work.
- *Section 2 Internet and Research* – Here I wanted to find out how different aspects of the Internet impacted on the interviewees' research work and their attitudes to their degree of dependence on the Internet.
- *Section 3 Gender inequality, research and technology* – these questions focussed on how interviewees and their colleagues coped with increasing research and technological pressures; whether they had experienced any problems that were related to their gender or whether they saw any patterns in research productivity or Internet use.

Interviewees were found using a number of different methods, including self-selected volunteers from the pilot survey, conference networking, 'snowballing' from known contacts and finally purposive emailing to individuals to ensure a spread of disciplines, gender and institution type (Charmaz, 2001). (See Table 2, which shows a brief profile of the interviewees). They came from eleven universities, five of which were from Post 1992 institutions and six from Pre 1992 institutions. Interviews were conducted in interviewees' offices, on neutral territory (at a conference), or in the interviewees' own homes during the academic year 2006/7. In order to preserve anonymity, all names used in the data analysis have been changed.

RESULTS

Web Review

The aim of this part of the study was to consider how men and women project themselves from their websites. What impression were they giving of themselves to the rest of the world? How were they using the Internet to establish themselves in their academic community? Before proceeding, it should be noted that via the widespread use of 'templating', (the provision of a basic format for a web page which the user then populates) university administration departments may, to a large degree, dictate how a personal academic website appears. However, the content of the websites is more likely to be provided by the academics themselves and it was therefore through the content that I examined how the academics paint their own portraits.

The criteria reviewed here were credibility (how much information they included about their status – qualifications, biography, publications, experiences etc), job function, self-disclosure and presence of their photograph (See Table 1).

In terms of credibility, men were more likely than women to include details of their publications on the website (X=19.17, df=1, p<0.001), what funding they had successfully attracted (X=5.998, df=1, p=0.014), what journals they edited (X=3.762, df=1, p=0.05), what consultancies they had provided (X=4.531, df=1, p=0.03) and what awards or prizes they had won (X=4, df=1, p=0.04). Women and men were equally likely to include details of their biography, organisations they belonged to and people they collaborated with.

This seems to reinforce the literature on self-promotion which suggests that men are more likely to make their achievements known than women (Dolgin & Minowa, 1997). All the areas examined where there was a significant difference noted, were ones where there was an element of achievement. The areas where there were no differences between men and women were ones where the individuals were situating themselves in the community – who they worked with, what organisations they belonged to, the details of their career to date. The women in this review seemed reluctant to promote their achievements which might suggest that the constraints of societal boundaries apparent in the 'real' world appear to be being replicated in the 'virtual' one (Miller & Arnold, 2001; Woolgar, 2002).

Another aspect of self-presentation is the very physical one of photo inclusion. I was interested to see how many academic women revealed their physical persona as well as their academic one. The web review produced a highly significant association between being female and not including a photo on the website (X=23.72, df=2, p<0.01). 40% of women academics websites contained no photo of themselves compared to just 22% of men's. Again this has implications for female academics in that their sites may not be so noticeable as male colleagues' to people reviewing their profile if they do not include a photo - another example of women's relative invisibility in academia (Harley, 2003).

Discipline did not play a significant role in these findings. Controlling for gender, there were no statistically significant results when cross-tabulating the web page criteria with discipline. Thus as far as self-promotion is concerned at least, biologists were as likely as historians or economists to include photos or to provide details of funding awarded, prizes, editorships etc. Despite literature to suggest that collaboration, reputation building and self-promotion are perhaps more important in SET disciplines than in others (Lucas, 2006), this does not seem to be reflected in their web presence.

The web review findings seem to suggest that the Internet's potential for disrupting the Interaction and Identity aspects of the research culture are not being fully exploited by women despite the fact that this type of 'absent' self-promotion and remote networking might well suit them better than the more traditional face-to-face methods that were created, and largely practiced, by a male academic community.

Interviews

The interviews were analysed in terms of two of Acker's five levels of the gendered organization, interaction and identity, to see where Internet technologies might be making a difference to the gendered research culture.

Interaction - Networking and Communication

This level is particularly concerned with the networking and communication aspects of the research culture. As we saw above, the problem for women has been that although networking is important for the establishment of research reputation, collaboration and therefore publication opportunities, information gathering and career enhancement (O'Leary & Mitchell, 1990), male networks have been difficult for women to access. In this section I draw from the interview data, some of the ways in which ITs might be assisting women through new ways of networking and communicating.

Reputation Building

Academic reputation is not just about how much is published but about the number of citations of the publications, the number of invitations to speak at conferences, requests to contribute to publications and to act as a potential supervisor. Being known in the academic community and being accepted by the 'core set' as Atkinson and Delamont (1990) term it, enhances reputation and career prospects. Being visible in the community is therefore vital and as we have seen visibility can be a problem for women. Having a web site – a window on the world – should enable women to increase their visibility without having to attend conferences, participate in after-hours socialising and so on that is problematic for those with outside caring responsibilities (Probert, 2005). Results from the web survey have shown that the web pages of women academics contained less self-promotional material than male colleagues. In contrast, many women in the interviews fully recognised the importance of their website for reputation building activities and were experiencing benefits from it:

But I do think it's important to have a good set of webpages that does say about what you do and what you have done. I think that's one of the reasons for doing it that one hopes for potential PhD students and so on to come from that (Professor Bramley, Female, Biology, Pre 1992).

I've certainly been contacted by people to referee and review through them doing a search of the Internet and finding my research interests. 'Cos I have been contacted cold' (Dr Hinchcliffe, Female, History, Post 1992).

This could perhaps echo Miller and Arnold's finding (2001) that there was a tension between the recognition amongst women academics that web presence was important and their feelings of vulnerability around becoming visible on the web.

Collaboration Opportunities

Interview participants, both male and female, saw the many opportunities for collaboration that the Internet produces as being a particular advantage.

One aspect of the grant is to develop links with other researchers [around the world] and that's what my idea is - collaborating with people using the Internet to find out more about overlaps of research areas (Dr Barnett, Female, Biology, Pre 1992)

Many of these collaborations and research relationships are forged with people that have never met face-to-face and this would suggest that from this point of view it is possible to add to ones research capital (Lucas, 2006) through use of the Internet alone. Having said that, a theme amongst the male interviewees emerged which asserted that personal contact was extremely important for establishing research relationships and that Internet-based contact by itself would not suffice.

I might want collaborators so I might email them, I might phone them up. I prefer to talk rather than e-mail (Prof Billings, Male, Biology, Pre 1992)

Well personal contacts are certainly more important as an end result of electronic contact and its all about contact. I don't want to just have people I speak to in cyberspace...I don't place any great store on that sort of contact. It's always a means to an end (Dr Hill, Male, History, Post 1992).

Opportunities for creating and facilitating collaboration via the Internet seem to be being taken advantage of by men and women and this would therefore assist those women who were not well situated in a network of academic contacts. However, it appears from some of the men's comments that the virtual does not create a substitute for the real as previous scholars have suggested (Miller & Arnold, 2001; Woolgar, 2002) and the need for a network of personal contacts remains.

Information Gathering

One of the reasons for networking is to keep up to date in the field, to find out what is going on in other universities and to identify conferences, course and seminars that are forthcoming. Both male and female interviewees remarked that this was a strong point of the Internet.

I think it's probably where it comes into its own... checking out websites as and when, just wondering what's going on in the community and I do find it quite useful and I do it in a very kind of ad hoc way (Dr Hastings, Female, History, Pre 1992).

One of the most useful things you can do is search meeting abstracts online. Is there anything that anyone else is doing that might be relevant to what you're doing? And I do that a lot so if I can't go to a meeting I can still find out the sorts of things people were presenting (Dr Best, Female, Biology, Pre 1992)

This last point shows one way that the Internet is assisting men and women who cannot get to meetings/conferences or other networking events - a problem which has been identified as being of particular pertinence to women who are more likely to be primary carers for children or other relatives (Acker & Armenti, 2004; Deem & Hillyard, 2002). If meetings and conferences are increasingly being relayed online the need for travelling or staying away is lessened and the proceedings may be accessed at any convenient time. The Internet is adding the flexibility to allow more people to benefit from such activities

Thus although both men and women are keen users of the Internet for gathering information about research opportunities, women may well be benefiting slightly more because of the opportunities the Internet provides as an alternative to travelling.

Identity - Self-confidence and Self-promotion

The identity level of the gendered research culture framework is concerned with the way that individuals internalise the norms of the academic world and how that enables them to operate successfully in the academic world. Traditionally, those norms have revolved around socially constructed masculine characteristics such as competitiveness, individual achievement and constant striving for improvement, requiring individuals to display similarly 'masculine' characteristics such as self-promotion, self-assertion and self-confidence (Harris et al., 1998). Whilst both men and women may possess these qualities, it is not always easy for women to display them in a male-dominated culture without attracting criticism and discrimination (Bagilhole & Goode, 2001). As we saw in the web review, self-promotion via the web is not being exploited by women as much as perhaps might have been predicted. In this section I look at the interview data to see whether Internet technologies are enabling women to communicate and promote their research even if they have the self-confidence or self-assertion issues which are often characterized as 'feminine' characteristics.

Nearly all of the women expressed a preference for e-mail as a way of getting in touch with other academics and potential collaborators. Reasons for this varied:

E-mail has helped [my research practice] with the fact that it isn't face to face. It means I could contact people that I would never contact otherwise...

[It's] much easier than walking up to somebody (Dr Hanson, Female, History, Post 1992).

I prefer [e-mail] because it's less invasive in a way and it's less disturbing. Making a phone call can be slightly more awkward. Going to see someone as well - you don't know how you're going to catch them. They might be busy or where can you meet them? (Dr Hastings, Female, History, Pre 1992)

I really like it, it's a sort of contact with no contact. you're in control...I find it so much easier than picking up the phone (Dr Harrison, Female, History, Post 1992).

Throughout these different responses, the theme of 'contact with no contact' comes up. The women here recognize the need to approach other academics but appreciate the fact that e-mail allows an approach to be made with no immediate response being expected in real time.

These responses echo the work of Miller and Arnold (2001) who found that the women academics they interviewed valued the freedom that the Internet gave them to approach people and promote their research without having to promote themselves as individuals – something that might have clashed with a traditional socially-constructed view of acceptable female behaviour. In the voices of the women in my study, we can hear this desire to distance the person (themselves) from the subject matter of their contact.

In contrast a number of men specifically mentioned that they would rather talk to people face to face or by phone, valuing the personal contact more than the remoteness of email

I think email can be too impersonal - you're removing that personal communication and people are sitting in their office doing everything on computers... (Dr Bryson, Male, Biology, Post 1992)

I wonder if there's something about human nature that's amenable to face to face contact and finds it more difficult for stuff that comes through on email? (Mr East, Male, Economics, Post 1992).

Comparing these quotes shows an interesting division in the academics' perspectives. Many of the men in this study seem to be looking for personal networking opportunities whilst a lot of the women seem to want the remoteness of email, using it as a tool to gain information or to ask a question. The Internet appears to enable academics to carry out research tasks such as collaborations, conference organisation, and publication requests which perhaps women in particular might have found less possible in pre-Internet days because of the need to do these things via personal networks. It is interesting to note again, however, that for many of the men interviewed, this sort of contact was insufficient to establish research relationships.

E-mail was not the only Internet technology mentioned that eases the communications process. Academic's personal profile websites were mentioned by several of the female interviewees as being a helpful way of preparing for meetings or conferences by giving a sense of who it was that they were going to meet. It seemed to be important to them that there were personal details on the sites so that they could relate to them as people rather than as anonymous experts (Miller & Arnold, 2001).

Well really I'd use [the website] so I can actually talk to people. I'm not a very confident person so I need to know about these people. I can't question people, bounding up to them - 'Hello, I'm so and so, I do...' I can't do that. So if I know I'm going to meet someone and go and talk to them I like to go and find out who they are. Not just their work but their personal details (Dr Hanson, Female, History, Post 1992).

How the Internet is intervening here is seen in the way that the women academics interviewed readily embraced e-mail as 'contact with no contact'. It is difficult to be seen to be pushy or over-ambitious by sending an unsolicited email to even very eminent field-leaders whereas approaching such a person in person requires the sort of self-assertion that many of the women here and in the literature (Bagilhole & Goode, 2001) seem reluctant to show. The Internet also provides a 'shop front' whereby people can be on display as individuals, particularly if their websites include photos or personal details. This allows those for whom lack of confidence is an issue to meet the person virtually before meeting in person and in this respect the inclusion of personal information is what is important rather than a faceless list of achievements.

However, it is not all good news for female academics. There were signs from the male interviewees that although e-mail was useful as a communications tool, as a way of increasing ones personal network it was less useful. If we were to assume that because academics can approach each other more easily and with less risk of rejection, they can build effective academic networks in this way then we may be making too big an assumption. Thus the use of e-mail and websites perhaps provides a useful adjunct to face-to-face networking but the latter still needs to be done in order to forge academic relationships.

Discipline

The primary analysis of the data concerned how gender impacted on the use made by academics of the Internet for their research and to what degree this disrupted the research culture at each of Acker's levels of the gendered organisation. However, as part of the overall analysis, I also considered how discipline might have impacted on the use of the Internet by academics in their research. From the literature review I anticipated that SET disciplines might be rating research more highly than other disciplines (Henkel, 2000) and that they had a greater incentive to situate themselves in their disciplinary communities through networking and collaboration (Lucas, 2006). What emerged was that discipline did not make as much difference as I had anticipated. Comments concerning the importance of websites for self promotion came from all three disciplines as seen above and there was certainly no less emphasis on the importance of research from the economists and historians than from the biologists. In terms of using the Internet for research, I had anticipated that biologists would carry out much of their research in the laboratory and therefore having the flexibility of the Internet at home would be of less value to them. Contrastingly I expected historians and economists to be less tied to the physical space of the university and therefore home Internet might be of great benefit in increasing productivity. The reality was not so clear-cut. There were of course some individual biologists who worked in the university but several of them noted that they could still carry out a lot of their research from home and in fact preferred to do so. This was because although scientists often work in teams whilst at the university, they need space alone to write, think and plan, and Internet access at home allowed them to do this. Historians tended by and large to be solitary researchers and many did find that they could now do almost all their research from home. However several noted that because of the large amount of documents needed for their research, they needed to spend time in the university printing off or sorting documents and again, others used sources that were not available on the Internet and had to travel to access them. Thus women in biology were just as likely as historians and economists to gain a benefit from the flexibility that having the Internet at home gave them, to do research-related activities and increase productivity.

DISCUSSION

The main aim of this study was to explore the potential of ITs to disrupt the gendered research culture at the two micro levels of Interaction and Identities. Results from the study show that there are areas where this is in fact happening and traditional practices are being undermined or circumvented through IT use. However there are also areas where the results are more ambiguous – the potential is there but a question mark remains over whether it is being fully exploited by women and finally there are areas where the societal influences are exerting too great an influence and ITs are having little effect.

At the Interaction level, ITs seemed to be providing women with new ways of networking that circumvented traditional androcentric networks. They were enthusiastic about web pages as personal 'shop fronts' for their research, about the ease with which potential research collaborators could be contacted and about the increasing practice of publishing conference papers on the Internet. E-mail as a communication medium was valued highly and the necessity of travelling to conferences and meetings to keep up in ones field was becoming perhaps less vital.

However, there was also a strong feeling, coming particularly from male participants in the study that Internet based interaction was useful but certainly not the whole story and they often expressed a preference for the personal interaction of the old fashioned network. This suggests that although the Internet provides new ways of networking which benefit women in particular, the context in which it operates is resistant to this new way of communicating and therefore contacts made in this way may not be as valuable to the male academic as they are to the female. Whilst men dominate the senior positions in the university and in the research community, this situation will only change very slowly since the dominant group's preferred way of working is proving resistant to change.

At the Identity level, the effect of increased Internet use is perhaps even more mixed. The potential for women academics to promote themselves and their research without having to engage in self-assertive or competitive behaviour is certainly recognised and taken up by many. However, there is still a significant trend for women to make themselves less visible than male colleagues through having short or no web pages, displaying no photograph and not including their achievements on their websites. This gap between what women know to be good practice and what they actually do is puzzling and perhaps by highlighting this trend we may go some way towards encouraging women to make the most of their opportunity for what might be termed the 'passive' self-promotion of web profiles.

In terms of circumventing the self-confidence required to approach fellow academics in a research context, women find e-mail to be invaluable as mentioned above. Whether men rate it as highly is in doubt from the results of this survey and since communication is by definition a two-way thing, its value as a disrupting mechanism of the gendered research culture is ambiguous.

A second aim of the study was to consider whether women in SET drew any more or less benefit from the effects of the Internet on their research, than colleagues in other disciplines. Two things can perhaps be observed from the fact that associations between discipline and Internet use were limited and sporadic as far as research was concerned. Firstly there is the recognition that research is a process that involves much more than analysing data. Whereas Biologists may use computers (although not necessarily the Internet) to analyse data more than Economists or Historians, there are many other aspects of research such as communication, collaboration, literature reviewing, and so on, which are common to all three disciplines and which may be carried out on the Internet to the same extent in History, for example, as in Biology. Secondly it reinforces notions that at the individual level, discipline alone cannot adequately explain academics' attitudes towards research (Brew, 2001) because as Brew argues, inter-disciplinary study, collaboration and research make discipline a slippery concept. Moreover although it is sometimes useful to draw overall conclusions about the behaviour of a population from surveys such as these, when viewed at the level of the individual, behaviour may be very different form the overall pattern.

In summary, it appears at these 'micro'-levels of the gendered research culture, the increasing use of Internet technologies is allowing some circumvention of gendered practices and the potential is there for this to increase. However, we have also seen that the many of the values and constraints of the real world are perpetuated into the 'virtual' one rendering some aspects of the gendered research culture little affected by IT use.

FUTURE RESEARCH DIRECTIONS

Research involving the Internet is problematic because of the rapidly changing nature of the Internet itself and also in the ways society is using it. Outcomes from studies such as this shed light on practices at a particular period in time but must be replicated or developed at regular intervals since contexts are rapidly changing. This would be an obvious way to develop the work carried out here. One example might be in the exploration of how the rise of Web 2.0 technologies might change the picture we have found here. The growth of Web 2.0 technologies is a development which may well increase familiarity with making and maintaining contacts online, affecting the way people perceive and value 'virtual' research relationships.

Another area that could be further developed is to consider the longer-term effects of Internet access on the careers of women academics. My research represents a picture of how women and men are using the Internet over a relatively short period of time. The benefits for women of having the Internet that I have explored here have not yet been translated into career success. The logical extension of this would be to carry out studies over a longer period to see if such things as research productivity, speed of promotion or amount of grant funding achieved had improved for women academics.

CONCLUSION

The gendered research culture has proved problematic for women seeking career progression in virtually all disciplines. The increasing use of ITs at all stages of the research process has the potential to disrupt many aspects of the gendered research culture at a number of different levels. What we have seen here is that ITs are providing women with new ways of establishing themselves in the academic community through remote networking and the possibilities of 'virtual' self-promotion via web profiles. These opportunities are not necessarily tied to discipline and women in SET disciplines appear to be no more or less likely to take advantage of them. However, what we have also seen is that the societal constraints that inhibit many women from displaying the 'masculine' characteristics that are important for a successful academic career, appear to be replicated in the 'virtual' world, rendering women academics and their achievements less visible than male colleagues on the Internet. Moreover, although the new ways of networking are well-used and valued by many of the women interviewed, we have seen less enthusiasm for non-contact networking methods amongst male academics suggesting that the traditional face-to-face networks are far from a relic of the past and women cannot rely on ITs alone to establish themselves in the academic research culture.

REFERENCES

Acker, J. (1990). Hierarchies, jobs, bodies: A theory of gendered organizations. *Gender & Society*, *4*(2), 139–158. doi:10.1177/089124390004002002

Acker, J. (1992). From sex roles to gendered institutions. *Contemporary Sociology, 21*(5), 565–569. doi:10.2307/2075528

Acker, S., & Armenti, C. (2004). Sleepless in academia. *Gender and Education, 16*(1), 3–24. doi:10.1080/0954025032000170309

Acker, S., & Feuerverger, G. (1996). Doing good and feeling bad: The work of women university teachers. *Cambridge Journal of Education, 26*(3), 401–422. doi:10.1080/0305764960260309

Alemany, C., & Diaz, C. (2007). *The last bastions: The situation of Spanish women in Spanish universities. A study of two cases.* Paper presented at the Gender, Work and Organization 5th Biennial International Interdisciplinary Conference, 27-29 June 2007, Keele, UK.

Atkinson, P., & Delamont, S. (1990). Professions and powerlessness: Female marginality in the learned occupations. *The Sociological Review, 38*(1), 90–110.

Bagilhole, B., & Goode, J. (2001). The contradiction of individual merit and the reality of a patriarchal support system in academic careers. *European Journal of Women's Studies, 8*(2), 161–180. doi:10.1177/135050680100800203

Britton, D. (2000). The epistemology of the gendered organization. *Gender & Society, 14*(3), 418–434. doi:10.1177/089124300014003004

Brooks, A. (1997). *Academic women*. Buckingham, UK: Open University Press.

Charmaz, K. (2001). Grounded theory and qualitative analysis. In Gubrium, J., & Holstein, J. (Eds.), *Handbook of interview research: Context and method.* Thousand Oaks, CA: Sage.

Coates, J. (1995). Language, gender and career. In Mills, S. (Ed.), *Language and gender: Interdisciplinary perspectives*. London: Longman.

Court, S. (1999). Negotiating the research imperative: The views of UK academics on their career opportunities. *Higher Education Quarterly, 53*(1), 65–87. doi:10.1111/1468-2273.00113

De Groot, J. (1997). After the ivory tower: Gender, commodification and the 'academic'. *Feminist Review, 55*, 130–142.

Deem, R. (2007). Managing a meritocracy or an equitable organisation? Senior managers' and employees' views about equal opportunities policies in uk universities. *Journal of Education Policy, 22*(6), 615–636. doi:10.1080/02680930701625247

Deem, R., & Hillyard, S. (2002). Making time for management - the careers and lives of manager-academics in UK universities. In G. Crow & Heath (Eds.), Social conceptions of time: Structure and process in work and everyday life, (pp. 126-143). Basingstoke, UK: Palgrave.

Deem, R., & Lucas, L. (2007). Research and teaching cultures in two contrasting UK policy contexts: Academic life in education departments in five English and Scottish universities. *Higher Education, 54*, 115–133. doi:10.1007/s10734-006-9010-z

Deem, R., & Morley, L. (2006). Diversity in the academy? Staff perceptions of equality policies in six contemporary higher education institutions. *Policy Futures in Higher Education, 4*(2), 185–202. doi:10.2304/pfie.2006.4.2.185

Dolgin, K., & Minowa, N. (1997). Gender differences in self-presentation: A comparison of the roles of flatteringness and intimacy in self-disclosure to friends. *Sex Roles, 36*(5-6), 371–380. doi:10.1007/BF02766653

Dominick, J. (1999). Who do you think you are? Personal home pages and self-presentation of the world wide web. *Journalism & Mass Communication Quarterly, 76*(4), 646–658.

Donelan, H., Herman, C., Kear, K., & Kirkup, G. (2009). Patterns of online networking for women's career development. *Gender in Management: An International Journal, 24*(2), 92–111. doi:10.1108/17542410910938790

Fletcher, C., Boden, R., Kent, J., & Tinson, J. (2007). Performing women: The gendered dimensions of the UK new research economy. *Gender, Work and Organization, 14*(5), 433–453. doi:10.1111/j.1468-0432.2007.00359.x

Gray, M., & Kurihara, T. (2006). *Networks of exclusion: Job segmentation and gendered social networks in the knowledge economy. Technology and Gender Equalities Seminar*. Cambridge, UK: ESRC Gender Equality Network.

Halsey, A. H. (1992). *The decline of donnish dominion*. Oxford, UK: Clarendon Press.

Harley, S. (2003). Research selectivity and female academics in UK universities: From gentleman's club and barrack yard to smart macho? *Gender and Education, 15*(4), 377–392. doi:10.1080/09540250310001610580

Harris, P., Thiele, B., & Currie, J. (1998). Success, gender and academic voices: Consuming passion or selling the soul? *Gender and Education, 10*(2), 133–148. doi:10.1080/09540259820989

Henkel, M. (2000). *Academic identities and policy change in higher education* (*Vol. 46*). London: Jessica Kingsley.

Heward, C. (1996). Women and careers in higher education: What's the problem? In Morley, L., & Walsh, V. (Eds.), *Breaking boundaries: Women in higher education* (pp. 11–23). London: Taylor and Francis.

Higher Education Statistics Agency. (2008). *Resources of higher education institutions 2006/7*. Cheltenham, UK: HESA.

Ibarra, H. (1992). Homphily and differential returns: Sex differences in network structure and access in an advertising firm. *Administrative Science Quarterly, 37*, 422–447. doi:10.2307/2393451

Knights, D., & Richards, W. (2003). Sex discrimination in UK academia. *Gender, Work and Organization, 10*(2), 213–237. doi:10.1111/1468-0432.t01-1-00012

Lincoln, J., & Miller, J. (1979). Work and friendship ties in organizations: A comparative analysis of relation networks. *Administrative Science Quarterly, 24*(2), 181–199. doi:10.2307/2392493

Lucas, L. (2006). *The research game in academic life*. Buckingham, UK: Open University Press.

Manuel, A. (2008). *From old boys network to virtual network. A study of internet technology use amongst UK academics, and the extent to which it is disrupting the gendered academic research culture*. Bristol, UK: University of Bristol.

Mavin, S., & Bryans, P. (2002). Academic women in the UK: Mainstreaming our experiences and networking for action. *Gender and Education, 14*(3), 235–250. doi:10.1080/0954025022000010703

Miller, H., & Arnold, J. (2001). Breaking away from grounded identity? Women academics on the web. *Cyberpsychology & Behavior, 4*(1), 95–108. doi:10.1089/10949310151088451

Morley, L. (2003). *Quality and power in higher education*. Maidenhead, UK: Society for Research into Higher Education and Open University Press.

O'Leary, V., & Mitchell, J. (1990). Women connecting with women: Networks and mentors. In S. Lie & V. O'Leary (Eds.), Storming the tower: Women in the academic world (pp. 58-74). London: Kogan Page.

Papacharissi, Z. (2002). The presentation of self in virtual life: Characteristics of personal home pages. *Journalism & Mass Communication Quarterly, 79*(3), 643–660.

Priola, V. (2007). Being female doing gender. Narratives of women in education management. *Gender and Education*, *19*(1), 21–40. doi:10.1080/09540250601087728

Probert, B. (2005). 'I just couldn't fit it in': Gender and unequal outcomes in academic careers. *Gender, Work and Organization*, *12*(1), 50–72. doi:10.1111/j.1468-0432.2005.00262.x

Ramsay, K., & Letherby, G. (2006). The experience of non-mothers in the gendered university. *Gender, Work and Organization*, *13*(1), 25–44. doi:10.1111/j.1468-0432.2006.00294.x

Sappleton, N., & Takruri-Rizk, H. (2008). The gender subtext of science, engineering, and technology (SET) organizations: A review and critique. *Women's Studies: An Inter-disciplinary Journal*, *37*(3), 284–316. doi:10.1080/00497870801917242

Saunderson, W. (2002). Women, academia and identity: Constructions of equal opportunities in the 'new managerialism'; a case of 'lipstick on the gorilla'? *Higher Education Quarterly*, *56*(4), 376–406. doi:10.1111/1468-2273.00226

Singh, V., Vinnicombe, S., & Kumra, S. (2006). Women in formal corporate networks: An organizational citizen perspective. *Women in Management Review*, *21*(6), 458–482. doi:10.1108/09649420610683462

Tonge. (2008). Barriers to networking for women in a UK professional service. *Gender in Management: An International Journal, 23*(7), 484-506.

West, C., & Zimmerman, D. (1987). Doing gender. *Gender & Society*, *1*, 125–151. doi:10.1177/0891243287001002002

Woolgar, S. (2002). The five rules of virtuality. In Woolgar, S. (Ed.), *Virtual society? Technology, cyperbole, reality* (pp. 1–22). Oxford, UK: Oxford University Press.

ADDITIONAL READING

Acker, J. (1990). Hierarchies, jobs, bodies: A theory of gendered organizations. *Gender & Society*, *4*(2), 139–158. doi:10.1177/089124390004002002

Acker, J. (1992). From sex roles to gendered institutions. *Contemporary Sociology*, *21*(5), 565–569. doi:10.2307/2075528

Acker, S., & Armenti, C. (2004). Sleepless in academia. *Gender and Education*, *16*(1), 3–24. doi:10.1080/0954025032000170309

Acker, S., & Feuerverger, G. (1996). Doing good and feeling bad: The work of women university teachers. *Cambridge Journal of Education*, *26*(3), 401–422. doi:10.1080/0305764960260309

Alvesson, M., & Due Billing, Y. (1997). *Understanding gender and organizations*. London: Sage.

Bagilhole, B., & Goode, J. (2001). The contradiction of individual merit and the reality of a patriarchal support system in academic careers. *European Journal of Women's Studies*, *8*(2), 161–180. doi:10.1177/135050680100800203

Barry, J., Berg, E., & Chandler, J. (2006). Academic shape shifting: Gender management and identities in Sweden and England. *Organization*, *13*(2), 275–298. doi:10.1177/1350508406061673

Benschop, Y., & Brouns, M. (2003). Crumbling ivory towers: Academic organizing and its gender effects. *Gender, Work and Organization*, *10*(2), 194–212. doi:10.1111/1468-0432.t01-1-00011

Britton, D. (2000). The epistemology of the gendered organization. *Gender & Society*, *14*(3), 418–434. doi:10.1177/089124300014003004

Brooks, A. (1997). *Academic women*. Buckingham: Open University Press.

Brooks, A., & Mackinnon, A. (Eds.). (2001). *Gender and the restructured university: Changing management and culture in higher education*. Buckingham: Open University and Society for Research into Higher Education.

Clegg, S. (2008). Femininities/masculinities and a sense self: Thinking gendered academic identities and the intellectual self. *Gender and Education*, *20*(3), 209–221. doi:10.1080/09540250802000389

Cockburn, C., & Ormerod, S. (1993). *Gender and technology in the making*. London: Sage.

Court, S. (1999). Negotiating the research imperative: The views of UK academics on their career opportunities. *Higher Education Quarterly*, *53*(1), 65–87. doi:10.1111/1468-2273.00113

Currie, J., Harris, P., & Thiele, B. (2000). Sacrifices in greedy universities: Are they gendered? *Gender and Education*, *12*(3), 269–291. doi:10.1080/713668305

De Groot, J. (1997). After the ivory tower: Gender, commodification and the 'academic'. *Feminist Review*, *55*, 130–142.

Deem, R. (2003). Gender, organizational cultures and the practices of manager-academics in UK universities. *Gender, Work and Organization*, *10*(2), 239–259. doi:10.1111/1468-0432.t01-1-00013

Deem, R., & Lucas, L. (2007). Research and teaching cultures in two contrasting UK policy contexts: Academic life in education departments in five English and Scottish universities. *Higher Education*, *54*, 115–133. doi:10.1007/s10734-006-9010-z

Dever, M., Morrison, Z., Dalton, B., & Tayton, S. (2006). *When research works for women*. Melbourne: Monash University.

Ely, R., Foldy, E., & Scully, M.The Centre for Gender in Organizations Simmons School of Management (Eds.). (2003). *Reader in gender, work and organization*. Malden, MA: Blackwell.

Faulkner, W. (2001). The technology question in feminism: A view from feminist technology studies. *Women's Studies International Forum*, *24*(1), 79–95. doi:10.1016/S0277-5395(00)00166-7

Fletcher, C., Boden, R., Kent, J., & Tinson, J. (2007). Performing women: The gendered dimensions of the UK new research economy. *Gender, Work and Organization*, *14*(5), 433–453. doi:10.1111/j.1468-0432.2007.00359.x

Gherardi, S. (1994). The gender we think, the gender we do in our everyday organizational lives. *Human Relations*, *47*(6), 591–610. doi:10.1177/001872679404700602

Halvorsen, E. (2002). Gender audit. In Howie, G., & Tauchert, A. (Eds.), *Gender, teaching and research in higher education. Challenges for the 21st century* (pp. 9–19). Aldershot: Ashgate.

Harley, S. (2003). Research selectivity and female academics in UK universities: From gentleman's club and barrack yard to smart macho? *Gender and Education*, *15*(4), 377–392. doi:10.1080/09540250310001610580

Harris, P., Thiele, B., & Currie, J. (1998). Success, gender and academic voices: Consuming passion or selling the soul? *Gender and Education*, *10*(2), 133–148. doi:10.1080/09540259820989

Henkel, M. (2005). Academic identity and autonomy in a changing policy environment. *Higher Education*, *49*, 155–176. doi:10.1007/s10734-004-2919-1

Hine, C. (2000). Virtual society? The social science of electronic technologies, http://virtualsociety.Sbs.Ox.Ac.Uk/.

Jackson, L. A., Ervin, K. S., Gardner, P. D., & Schmitt, N. (2001). Gender and the internet: Women communicating and men searching. *Sex Roles*, *44*(5-6), 363–379. doi:10.1023/A:1010937901821

Jackson, S. (2002). Transcending boundaries: Women, research and teaching in the academy. In Howie, G., & Tauchert, A. (Eds.), *Gender, teaching and research in higher education. Challenges for the 21st century* (pp. 20–32). Aldershot: Ashgate.

Kantola, J. (2008). 'Why do all the women disappear?' Gendering processes in a political science department. *Gender, Work and Organization, 15*(2), 202–225. doi:10.1111/j.1468-0432.2007.00376.x

Knights, D., & Richards, W. (2003). Sex discrimination in UK academia. *Gender, Work and Organization, 10*(2), 213–237. doi:10.1111/1468-0432.t01-1-00012

Leahey, E. (2007). Not by productivity alone: How visibility and specialization contribute to academic earnings. *American Sociological Review, 72*(4), 533–561. doi:10.1177/000312240707200403

Letherby, G., & Shiels, J. (2001). Isn't he good, but can we take her seriously? Gendered expectations in higher education. In Anderson, P., & Williams, J. (Eds.), *Identity and difference in higher education* (pp. 121–132). London: Ashgate.

Lucas, L. (2006). *The research game in academic life*. Buckingham: Open University Press.

Maddock, S., & Parkin, D. (1994). Gender cultures: How they affect men and women at work. In Davidson, M., & Burke, J. (Eds.), *Women in management current research issues* (pp. 29–40). London: Paul Chapman Publishing.

Mavin, S., & Bryans, P. (2002). Academic women in the UK: Mainstreaming our experiences and networking for action. *Gender and Education, 14*(3), 235–250. doi:10.1080/0954025022000010703

Miller, H., & Arnold, J. (2001). Breaking away from grounded identity? Women academics on the web. *Cyberpsychology & Behavior, 4*(1), 95–108. doi:10.1089/10949310151088451

Morley, L. (1999). *Organising feminisms. The micropolitics of the academy*. Basingstoke: Macmillan Press.

O'Leary, V., & Mitchell, J. (1990). Women connecting with women: Networks and mentors. In S. Lie & V. O'Leary (Eds.), Storming the tower. Women in the academic world (pp. 58-74). London: Kogan Page.

Papacharissi, Z. (2002). The presentation of self in virtual life: Characteristics of personal home pages. *Journalism & Mass Communication Quarterly, 79*(3), 643–660.

Probert, B. (2005). 'I just couldn't fit it in': Gender and unequal outcomes in academic careers. *Gender, Work and Organization, 12*(1), 50–72. doi:10.1111/j.1468-0432.2005.00262.x

Smith, J. (2008). Quality assurance and gender discrimination in English universities: An investigation. *British Journal of Sociology of Education, 29*(6), 623–638. doi:10.1080/01425690802423304

Travers, C., & Pemberton, C. (2000). Think career global but act local: Understanding networking as a culturally differentiated skill. In Davidson, M., & Burke, J. (Eds.), *Women in management: Current research issues ii* (*Vol. 2*, pp. 84–103). London: Sage.

Woolgar, S. (2002a). *Virtual society? Technology, cyberbole and reality*. Oxford: Oxford University Press.

ENDNOTE

1 Post 1992 universities' are derived from polytechnics and other vocationally-focussed institutions, and traditionally have placed less emphasis on research than the older Pre 1992 Universities.

Profile 12
Engineering Career Spotlight:
Investigation of Domestic Water End Use in Toowoomba

© 2008 Queensland Government Office of Women. Used with permission

Nicola Mead, GradIEAust, Toowoomba Regional Council, Toowoomba

Nicola Mead graduated from the University of Southern Queensland in 2008 with a dual Bachelor of Engineering and Business. During her time at University she was President of the USQ Women in Engineering Club and an advocate for the promotion of women in non-traditional areas. Nicola worked as a Vacation student at the Toowoomba City Council (prior to amalgamation) working on a project that investigated Water Demand Management. After graduating, she commenced at the Toowoomba Regional Council as a graduate engineer within Integrated Regional Planning. Currently working within the Projects team, Nicola's professional skills and interests range from Water Sensitive Urban Design to flood mapping. Nicola also facilitates

Copyright © 2010, IGI Global. Copying or distributing in print or electronic forms without written permission of IGI Global is prohibited.

teams of students for a problem based learning course within the Faculty of Engineering at USQ as well as marking assignments for Public Health Engineering. She is also an active committee member of the Engineers Australia Toowoomba Local Group.

Introduction

Recently, Nicola Mead undertook a project to explore domestic water end use and investigate water consumption in community households to determine quantity and scheduling of use. It is a highly innovative and important project for the Toowoomba Region which is currently experiencing a water crisis. The information gained from this project will significantly enhance the current body of knowledge related to end water use in Australia, in particular the South East Corner of Queensland. The project supports the sustainable use of the region's scarcest resource and will have major implications for demand management programs in Toowoomba and will also contribute to similar projects across Australia. The information will be used to educate the community and promote their participation in water conservation activities and behaviours and form a key part of the Local Governments future water management plan.

Background

Toowoomba is still in the grips of the worst drought in history, with dam levels at only 11% and no significant rain forecasted, Toowoomba residents and authorities need to ensure that they do everything in their power to conserve every drop of water. Water is a finite resource but is often considered renewable, in Toowoomba this renewable aspect does not apply, the catchments are simply not receiving the rain and the new pipeline from Wivenhoe will not reach Toowoomba's dams before they run dry, some reports predict this could happen as soon as Nov 2009. The city's bores cannot sustain the current water use in the long term. Demand management is the only option left that will ensure that our dams last as long as possible. With domestic water use accounting for up to 70% of consumption, understanding usage trends/patterns and attitudes of use within homes is especially important. Data that can quantify these variables will enable water authorities to effectively and efficiently implement demand management programs that target water savings to where they will be most effective. Although there are demand management programs in place in Toowoomba, little is known about their effectiveness.

Residential Water Use

This project investigates the end uses of water in residential properties and uses this information to evaluate the effectiveness of the implemented demand management programs along with factors that affect water consumption, demand patterns and peaking factors. This information will be invaluable to water authorities not only in Toowoomba but throughout Queensland. Bruce Gaydon (Water Demand Management Coordinator) commented in a report to Toowoomba Regional Councillors (Water Services 30 May 08) that

Such a study has not been conducted in Toowoomba before and we have previously relied on work done in other areas... The ultimate aim is to analyse water use within the home along with aspects of water use behaviour

The data from this project can also be used for planning future water supply networks and community education. An integral component of this project will be utilising information for education. Volunteer households who have the smart meters installed will gain detailed knowledge about their own water usage and be able to compare that to other study households. Information regarding end use is highly valuable to all households; specific information about domestic water use (e.g. the volume of water used during a shower or the volume of water used by a washing machine) will empower householders, enabling them to make informed decisions about their behaviour regarding shower times or the purchasing of water efficient appliances. Showing residents that they have the power to change their water use habits and using the information from the end water use project to demonstrate just how effective their water conservation efforts can be reinforces that they can and do make a difference to total consumption.

This project is highly innovative for the region, being the second of its kind in Queensland (behind Gold Coast Water) to install smart metering for the purpose of understanding the detailed domestic water end use. Toowoomba is likely to have a significantly different profile to the Gold Coast due to differences in climate, restrictions, water pressures and Toowoomba does not have dual reticulation systems in place which is where most of the Gold Coast study is focused. Local Governments in the South East have been focusing much of their resources and time in securing other/new water supplies like bores, pipelines and recycled water; this project will highlight the savings that can be made by well implemented demand management programs. With local government support of these demand management initiatives, the pressures on current and potential water supplies will be reduced allowing water to be used more sustainably.

Ms Mead is a good example of the innovative student project worker that we aim to develop in our Civil and Environmental Engineering programs. She is developing a very innovative methodology for assessing domestic household water usage. The results could well change the approach of local governments throughout much of eastern Australia. Nicola Mead has linked with the local City Council to undertake the work and has managed the project very professionally. In so doing, she has laid the foundations for a future intensive monitoring program that will have significant community benefits in terms of the intelligent management of domestic water supplies (A/Prof Mark Porter, Head of Environmental, Civil and Agricultural Engineering at USQ).

The initial stage of this project is to establish an effective methodology that can be used for a more extensive study of up to 200 homes.

The end water use study consists of number of steps:

1. Selection of study participants
2. Selection & installation of monitoring equipment
3. Audit & data collection
4. Analysis of results and report writing

Study participants have been chosen based on water consumption, household make up and willingness to participate in the study. A total of 10 households are included in the pilot study, which are scattered throughout Toowoomba. Each of the households have had their residential water meters replaced with more accurate meters that have a high pulse output (72 pulses/litre). The meters have been connected to a data logger that logs the pulses every 10 secs. The loggers then report log traces via email once per week for processing. Each household was audited at the time of installation to get 'signature traces' of the fixtures in place i.e. max flow rates, volume/toilet flush etc. The audit also involved a survey of attitudes and behaviours relating to water use. The data that is collected is processed though a software package (Trace Wizard) that disaggregates flow into specific events, e.g. tap, shower, dishwasher etc. Trace Wizard utilises a user defined set of parameters to categorise the water use events and assign a specific fixture to each event. The parameters are developed for each individual study residence. These parameters can include the volume, duration, peak flow, rate and mode flow rate of each specific fixture. For example, a toilet may be defined as using 9 litres/flush, with refilling taking place within 90 secs. Similar parameters are established for each of the fixtures found in the household.

Once the flow traces have been analysed they are outputted to an Access file and processed in conjunction with results from all the other participating households. The project is ongoing, with data collection to be completed by the end of July 2008and reporting of results to be finalised by the end of October 2008.

This project aims to establish the methodology for a larger, statistically appropriate study of domestic water use for a future PhD student to complete. This project will ensure that all the hardware, software and technical issues are solved before the major project takes place. It will also give some preliminary results that can be investigated or expanded in the major study.

The second aspect of the project is community liaison and education. Nikki had the opportunity to talk to Flagstone Creek Primary School students about her project and water savings in general. Dr Shelley Kinash, leader of the learning project wrote:

We would like to thank you so much for your contribution! Your presentation to the children at the Tea and Talk made a particularly lasting impression.

Future Directions

Through this project Nikki is hoping to inspire other females to consider engineering as a career by demonstrating that engineers do make a huge contribution to society by providing sustainable and appropriate solutions for communities. She currently runs the Women in Engineering Club at USQ which is focussing much of its activities on attracting and retaining females in this male dominated discipline. They are currently implementing a student championed mentoring program for girls in 1st year to provide them with the support and encouragement to complete their engineering degree. She has been organising this program with assistance from the GO WEST project team at USQ which aims to establish links with female staff, students and professionals in sciences, engineering and technology in rural and remote areas. They are also trying to run more collegiate and social activities to increase the opportunity to network, this is especially important to Toowoomba since being a regional centre, it tends to miss out on the Engineers Australia activities.

Chapter 13
Progression of UK Women Engineers:
Aids and Hurdles

Haifa Takruri-Rizk
University of Salford, UK

Natalie Sappleton
Manchester Metropolitan University, UK

Sunrita Dhar-Bhattacharjee
University of Salford, UK

ABSTRACT

Whilst evidence presented in recent scholarship suggests initiatives aimed at increasing female recruitment to the industry may be proving somewhat effective, retention of female engineers is falling, and there has been minimal progress on breaking down vertical segregation in the engineering workplace. In this chapter the authors explore the situation of women engineers in the UK; they examine, using qualitative and quantitative data collection methods, whether women in engineering do indeed suffer from the application of descriptive and prescriptive stereotypes in the workplace, and how they are able to circumvent them. In particular, the authors instigate the factors, such as self-promotion and confidence, which may aid or hinder the progression of women's careers. They urge educational institution and engineering industries and organisations to nurture such qualities. Their findings indicate that confidence and self-promotion should be high on the agenda of women engineers to survive and progress in the engineering industry.

INTRODUCTION

Social psychologists examining organisational gender relations have confirmed the existence of a 'double bind" for professional women (Rudman & Glick 2001). This bind arises because of the application of stereotypes to women in their gender and occupational roles. As women, they are assumed to possess communal traits (nurturance, kindness, consideration for others) (Parsons & Bales, 1955), and at work, because of their shorter working histories and lesser social status they are assumed to be less competent than men (Acker, 1990). Put simply,

DOI: 10.4018/978-1-61520-657-5.ch013

Copyright © 2010, IGI Global. Copying or distributing in print or electronic forms without written permission of IGI Global is prohibited.

because of the prevalence and pervasiveness of stereotypes regarding the proper role of women in the workplace, women who exhibit agentic behaviour may suffer from a backlash effect expressed in negative social repercussions (Glick & Fiske, 1999). Engaging in agentic behaviour might allow them to break down the stereotype of lesser capability, but such behaviour and its correlates (independence, forcefulness, competitiveness, ambition) are incompatible with the prescriptive communal stereotype. Rudman (1998: p. 629) describes succinctly how self-promotion – highlighting and marketing one's skills, qualifications, prior successes and expertise - can be a double-edged sword for women:

Because social influence is a function of both competence and social attraction (e.g., Carli, LaFleur, & Loeber, 1995), women may suffer from backlash effect in which self-promotion may enhance perceptions of their qualifications, but at the cost of social rejection. That is women may be stuck in a catch-22 in which they are damned if they do self-promote and damned if they don't. (Rudman, 1998, p629)

Thus, although a woman can be perceived as competent when assuming "masculine" behaviours, she risks being judged as insufficiently likeable. In contrast, agentic men are acting in accordance with the stereotypic prescription that this is natural behaviour, and although this may mean that they too, such men are not perceived as likable, likeability is not a prescribed stereotype for men. Thus women and men who behave in an identical fashion are judged differently, with women being treated more harshly (Heilman, 2001).

Recent statistics suggests that initiatives aimed at increasing female recruitment to engineering may be proving somewhat effective in the UK[1]. However, the retention of female engineers is falling and there has been minimal progress on breaking down vertical segregation in the engineering workplace (Charles, 2004). In the review and critique of the science, engineering and technology (SET) literature, Sappleton & Takruri-Rizk (2008) consider the persisting low representation of women in SET professions and argue that the path or "pipeline" for female progression through SET professions endures an increasing level of attrition at specific points after the entry stage for varied reasons. This is despite over 30 years of equality legislation and the recent introduction of the gender equality duty. We suggest that the 'double bind' may be a major contributing factor to this stasis. Women in nontraditional fields like engineering suffer from the likeable-but-incompetent impasse to a greater extent that women in more traditional domains of work because these women have what Glick and Fiske (1999) refer to as a negative (competitive) interdependence with their sector's in-group, i.e. men.

In this chapter we examine whether women in engineering careers in the UK do indeed suffer from the application of descriptive and prescriptive stereotypes in the workplace, and how, if at all, they are able to circumvent them. In particular, we instigate the factors, such as self-promotion and confidence, which may aid or hinder the progression of women's careers. After the background literature we present our research methodology. Using focus groups, questionnaires and in-depth interviews with engineers, technical directors and human resources (HR) professionals, we captured the experiences of men and women working in engineering occupations. Next, we present an analysis of the qualitative and quantitative data we collected supported by the research of other scholars in similar fields. We investigate the personality/character that may help female engineers to "get in and on" in the engineering industry. Then, we inspect the perception of the women who do posses the "winning female engineer character" and hence managed to shatter the glass-ceiling and climb up the career ladder. Our findings indicate that confidence and self-promotion should be high

Table 1. The Glick & Fiske (1999) model

		Dyadic Relationship to In-group	
		Co-operative	**Competitive**
Relative Status to the In-Group	**High**	Unlikeable, Competent	Likeable OR Competent (not both)
	Low	Likeable, Incompetent	Unlikeable, Incompetent

on the agenda of women engineers if they are to survive and progress in the engineering industry.

BACKGROUND LITERATURE

Glick and Fiske (1999) argue that positive (co-operative) interdependence with the in-group predicts who will be stereotyped as likeable, whereas negative interdependence predicts who will be respected as competent. Their model is illustrated in the typology in Table 1.

According to this theory, the generic female stereotype – likeable, but incompetent arises from men's dependency on women in close relationships. Thus, the conventional housewife in a low status, but co-operative relationship with a man is nice, but incompetent, in accordance with established gender stereotypic roles. Women who violate these conventions (such as female engineers) challenge the societal power relationship between the sexes. Glick and Fiske (1999) add that the greater the dependency of the higher status group to the subordinates, the more prescriptive are stereotypes, since this serves to maintain the status quo. Where stereotypes are both descriptive and prescriptive (women are likeable; women should be incompetent), counter stereotypical behaviour is negatively evaluated. And, where individuals become serious competitors to the in-group, a double bind arises. The authors (Glick & Fiske, 1999: 216) explain:

Sexist men split women into (at least) two polarized types: the traditional woman who is likable but incompetent and the nontraditional woman who is competent but not likable. The former remains a type on which traditional men remain dependent for intimate relationships; as a result, the interpersonal traits required for the kind of relationships these men desire – in which the woman is supportive, nurturant, and deferent – continue, we think to be prescriptive aspects of the stereotype. In contrast, the stereotypes of nontraditional woman are simply descriptive. Such women are seen as competent because of the considerable (even if still limited) success they have had in the workplace... such woman may be penalized for not fulfilling the prescriptions that are seen characterizing the prototypic woman. It is when such prescriptions about interpersonal traits 'spill over" into the workplace that someone who fits the stereotypes of a career women but therefore violates the prescriptive stereotypes associated with traditional woman is told that she is not 'feminine enough'

While Glick and Fiske use the term "sexist" to describe men who apply prescriptive and descriptive stereotypes to women, Rudman and Glick (2001) have argued that the communality dimension of female stereotypes may not be actively seen as sexist because communality is generally a widely-endorsed and favourable trait. To test this hypothesis, they presented a videotape to 172 students in which an agentic or a communal male or female applicant was being interviewed for a position as a computer lab manager. As expected, agentic female applicants were evaluated as less socially skilled and likable than a man using an identical strategy. Curiously, the agentic female was viewed as less hireable than the male for a job, but only when the job description was presented as feminized (this was done by highlighting that the role required interpersonal skills). Rudman and Glick suggest that this result occurred because the woman's social skills, rather than her competence were devalued. They concluded that it is a subset of agentic traits (competitiveness, aggression and other forms of social dominance) that elicits the backlash effect because such characteristics conflict with female 'niceness'. Thus, agentic women can avoid backlash by being "nice but able" and downplaying the social dominance element of agentism.

Furthermore, confidence building in a female engineer's self and technical abilities begins prior to embarking on an engineering career. Zeldin and Pajares (2000) adopt Bandura's (1986, 1997) self-efficacy theory, "which Bandura defined as people's judgments of their capabilities to produce designated levels of performance", (Zeldin and Pajares, 2000, p 216), to ascertain one of the essential requirements for females to survive in a male-dominated study or career environment. They contend that social encouragement promotes determination in individuals to sustain the persistence necessary to succeed, resulting in perpetual development of skills and of personal efficacy. However, they agree that the lack of such support and encouragement could undermine individuals' self-belief and confidence in their own abilities.

Of course, messages can also work to undermine efficacy beliefs when used to convince people that they lack capabilities. For example, when women receive social messages that they do not belong in a male-dominated field such as mathematics, they may be especially vulnerable to believing that they are not, and cannot be, competent in that area (Zeldin and Pajares, 2000, p 217).

Additionally, many scholars have highlighted the need to create a conducive learning environment for women in science engineering, construction and technology (SECT) to enhance their self esteem and confidence. Williams et al (2002) developed an innovative staff development training offered by the Let's TWIST project, a European funded project aimed at widening participation of women studying, training and working in SECT. The training targeted academics teaching SECT subjects in further and higher education in the UK and in Portugal. Shull and Weiner (2002) investigated techniques to improve the retention of women in engineering studies by employing information technology to enhance the women's self-perception and motivation. Like Zeldin and Pajares (2000), they used the self-efficacy theory to apply teaching methods designed to embrace educational attributes, motivation, effort and persistence. The methods used by Shull and Weiner to teach a course in computer technology created an environment were the women on the course established a greater self-efficacy and associated control beliefs of their abilities and proficiency. Conversely, as stressed by Anna Ravenscroft (2008) in her article in the online ieee.usa today's engineer, negative experiences at college and university exacerbates low self-efficacy amongst women students when studying courses designed to "weed out the less competent, but which mostly weed out the less confident, who are disproportionately female", (Ravensccroft, 2008). Likewise Ridgway (2004) elaborates on the hegemonic gender beliefs in which men are viewed as more competent and status worthy and women are considered as less competent in general but "nicer" and better at communal errands. As she asserts, hegemonic gender beliefs are a "stubborn part of social reality that must be dealt with or accommodated in many contexts, even if they are not personally endorsed" (Ridgway, 2004, p514).

RESEARCH METHODOLOGY

Data Collection Tools

The research team investigated varied methods for the collection and analysis of data to aid the fulfilment of the research aim. The team employed a mix of quantitative and qualitative methods which varied from questionnaire surveys, to individual interviews, and focus groups. To justify the selection of such diverse data collection techniques, the team scrutinised researchers' arguments for the use of mixed methods. Gerald Cupchick (2001) contended that positivist and constructivist ontologies underlie quantitative and qualitative methods and that the two approaches bring distinctive qualities to the research process:

For Udo Kelle (2001), "triangulation", is seen by authors as a concept for method integration. Kelle states that triangulation does not represent a single methodological concept but a metaphor with a broad semantic field[2]. Furthermore, Kelle explains if quantitative and qualitative methods are combined to answer a specific research question, the combination of the two methods may lead to one of the following: convergence of the results where the results lead to the same outcome; the outcome of the quantitative and qualitative methods may be complementary to each other or; the results may be divergent or contradictory. C. C. Ragin (1994) adopted the notion that qualitative studies may capture important facts missed by more general quantitative studies.

Taking into consideration the above arguments the research team combined quantitative and qualitative data collection methods hoping for a fuller picture of the investigated research question. The following sections define the methods employed in achieving various parts of the research.

Questionnaires

The questionnaires were intended to identify good practices, problem areas, and barriers to attracting and retaining engineers. Two sets of questionnaire were designed; the first was aimed at employees in engineering industries and the second was aimed at managerial personnel in engineering industries including, human resources managers and directors. Postal questionnaires were used alongside online questionnaires. 1600 employee questionnaires were distributed amongst engineering companies in the North West and to University of Salford alumni. A total of 169 (10.6% response rate) completed questionnaires were returned. 52.1% (88) of the respondents were men, and 46.2% (78) of the respondents were women. 1.7% (3) of the respondents declined to identify their gender.

140 questionnaires were distributed to technical and human resource managers and directors. 23 (17% response rate) completed questionnaires were returned. 56.6% (13) were men, and 39.1% (9) were women. There was 4.3% (1) of the respondents that declined to declare their gender.

Interviews

Keeping in mind the view that qualitative research should not be considered as an easy substitute for a "statistical" or quantitative study, but should complement and deepen findings from the questionnaires, we employed both data collection methods.

The research team aimed to interview sample groups that represent different layers in engineering companies' workforces, including management groups. However, most of the interviewees were selected from the questionnaire responses and those who indicated their willingness to participate further in the research.

26 individual interviews were carried out:

- 13 interviews with engineering employees (12 female, 1 male)
- 13 interviews with managers/directors including human resources (8 female, 5 male)

Additionally, the team carried out six focus groups (32 participant, 19 male & 13 female) representing six engineering industries.

The interviews were semi-structured and conducted with an interview guide. The main focus was to get the interviewees to share their views and experiences with the researchers. The interview guide simply outlined the areas or themes to cover as demonstrated by the literature, which included recruitment, work practices, support and promotion mechanisms, and equal opportunities and diversity aspects.

Overall Industry Representation

In total 41 UK engineering industries were involved in one form or another of data collection methods, these are questionnaires, interviews or focus groups. Certain firms contributed to two or three of the data collection methods.

Data Coding and Analysis

The questionnaires were analysed with the help of the statistical software program SPSS for Windows. Any further comments added by the participants that could not easily be made to fit the SPSS variables were coded into the software Nvivo. Nvivo was also used to analyse the transcriptions from the interviews and focus groups.

FINDINGS AND ANALYSIS

Engineering Personalities

Our research shows that, overall, the engineering workplace is not always an environment conducive to the effective development of females and their careers. We found that women may be denied opportunities for flexible working, can be subject to subtle and more overt forms of discrimination, can be paid less, and have fewer opportunities for progression. It is obvious why some women exit this industry; but we were equally interested in why others stay and progress.

One suggestion was that certain personalities are better able to survive and adapt to the engineering workplace and its organisational culture. In order to find out the level of truth in this idea, respondents to the questionnaire were asked to rate a series of statements relating to the personality elements deemed necessary to succeed in the industry. Our data, as shown in *Table 2*, emphasises that the most important factor that is perceived to help engineers succeed in the industry is having confidence (27% of females and 23% of males). Receiving support from senior colleagues and peers (23% of females and 13% of males) is the second important factor. We argue that having confidence in own technical skills means that an individual is competent as an engineer which in many ways supports the notions presented in the literature review above. As stressed by Zeldin and Pajares (2000), confidence is an essential requirement for females to survive in a male dominated study or career environment. They also support the second important factor in our data, receiving support from senior colleague, as encouragement promotes determination in individuals to sustain the persistence necessary to succeed, resulting in perpetual development of skills and of personal efficacy.

Many interviewees pointed out that confidence in technical ability can come with age and experience. We argue that confidence in technical abilities could be enhanced at an earlier stage while at university in preparation for entering a career in the relevant industry. As was proposed and revealed by researchers (Williams et al, 2002; Shull & Weiner, 2002; Zeldin & Pajares, 2000) there is a need to develop a teaching and learning environment to enhance female students' self-efficacy and confidence in own technical skills and abilities.

...You will need to be better than your men and you also, I think one of the pieces that I suggest should be incorporated which goes back to my assertiveness, is that women are reluctant in that environment where the fact is you will need to excel the other part of that is you need to tell people that you're better, right?....one piece of advice is that women need to be shown that it is

Table 2. How important are certain factors in helping you succeed in the engineering industry? 102 responses (44 females, 57 males)

Personality element	% Female	% Male
Physical Strength	1	6
Being Confident	27	23
Being "Thick skinned"	15	6
Be part of the organisational culture	5	8
Being part of a diverse work force	6	3
Feeling that you belong amongst peers	13	14
Networking opportunities	12	8
Visibility in the company	9	14
Support from senior colleagues	23	17
Being sensitive to others	5	8

okay to blow your own trumpet which is what men would do (Female, Managing Director, 45+ yrs).

This quote seems to indicate that women may need assertiveness training. Are they culturally pre-programmed to not 'blow their own trumpet'? Does this exacerbate their need to excel more than the men to receive their deserved status and recognition or is something deeper going on to adversely affect their career progression and recognition levels in the industry?

More of our female respondents felt that they were expected to work harder than their male colleagues in order to get the respect at work they felt they deserved:

There have been times in my career... where there was a definite disadvantage to being a woman. I had to fight a lot harder to get technical recognition... When I wanted to move into the engineering side, I had to convince every man and his dog that it was going to be worthwhile for them... they were obviously nervous that they were going to spend all this money on me and then I was going to disappear off and have lots babies (Female, Highways Maintenance Engineer, 31-45yrs)

Theorists in the field of organisational culture and gender seem to concur that there are certain values, philosophies, beliefs, norms, and assumptions embedded within all organisations, and that these values have been shaped by male 'norms' (Mills, 1988; Evetts, 1993, 1994; McIlwee & Robinson, 1992; Bagilhole, 2002; Pretz & Hopkins, 1997, Tienari et al, 2002). Ridgway (2004) examines the hegemonic gender beliefs which consider women to be less competent than men and to prefer to do less challenging tasks and which consider men to be more competent and status worthy seem to shape male dominated work environments. This situation is rooted in the tradition of female exclusion from the public sphere: the male domination of the world of work is possibly one of the most pervasive and tenacious systems of power in history. It follows then, that those industries which are characterised as masculine will be doubly affected by sex role stereotypes, regardless of the numbers of women entering them and hence our assertion that women in male dominated environment need to use self promotion and have confidence in their skills and abilities. Conversely, if women engineers do not posses such qualities and do not feel secure in a male dominated career environment, they are at risk of giving up and leaving the industry. This was strongly expresses by one of our interviewees who felt intimidated in male dominated environment and eventually decided to leave the industry.

...sitting in a room full of lads, you know, all waiting for me to say the wrong thing; I just felt uncomfortable... [if there were women there]... that would scare me as well. They're usually quite determined women aren't they? (Female, Former Civil Engineer, 31-45 yrs)

She further stated:

I just wasn't the right person for the job. I wasn't like a fella really. I wasn't hard enough (Female, Former Civil Engineer, 31-45 yrs)

It is clear that the lack of self-esteem and the feeling of not fitting the bill prevent women from being encouraged or accepted into a 'man's world'. Consequently, the acceptance of femininity for a woman becomes a kind of showpiece - a defence-against overstepping the gender divide. Women reject their own power for fear of threatening masculine rationality. It is possible that some women in SET put on a 'performance of poor performance' to maintain the status quo? There is evidence for this, but our data also shows that there are women who are able and posses the qualities to survive the 'man's world'.

Our quality manager is a woman but she has a very big personality. The women who have got on in the company tend to be women with big personalities (Male, Chemical Engineer, 31-45 yrs)

Getting on in Engineering

The career difficulties experienced by women attempting to scale organisational career ladders have received growing attention in the research literature (Kanter, 1977). Of course, it has been well documented that the majority of senior positions in all organisations, including engineering organisations, are still occupied by men. Several studies have shown that women engineers occupy lower status positions and are assigned less responsibility than their male colleagues (Roberts and Ayre, 2002; Martin, 2001). Tara Ramdin (2005) identified coping strategies that may be adopted by women who face hindering experiences in a male dominated environment like engineering. One of the participants in Ramdin's research explained that she would speak up if excluded from a significant meeting that related to her work but didn't put equivalent emphasis on being excluded from a social event. Our research shows that social events are one form of the networking opportunities that help build contacts which may lead to promotion prospects and job openings. Ramdin (2005) also reported that women would deal differently with hindering experiences as they gain more confidence in their abilities.

Roberts and Ayre (2002) also acknowledged the need for the engineering industry to enjoy the benefits of diversity and recognise that it is important to retain, as well as recruit, women into the profession. We add that it is necessary to facilitate the progression and advancement of women as many will be dissatisfied if they repeatedly encounter organizational glass-ceilings; remaining at the same level and receiving less pay than their male colleagues. Indeed, this view was expressed by one of our male interviewees. It seems then that a very real and present glass ceiling may exist for women in engineering. Employees, both male and female, have certainly recognised this:

I've not really seen any [glass ceilings] but I think there are. You definitely see people leaving this company because they feel that they should be getting on, and they're not; I would sense this (Male, Chemical Engineer, 31-45 yrs).

I would prefer to get the management role... I would quite like that, but I don't think a woman could do it, would be liked to go to a higher level (Female, Building Services Engineer, 18-25 yrs)

We were interested in finding out more and posed some questions. Is it possible for women to reach senior positions in engineering organisations? Do women exit the industry because they have difficulties progressing? Job roles and titles differ across the industry; therefore it would be difficult to compare levels of seniority across gender. However, using a number of indicators of seniority, we were able to build up a picture of female managers in engineering.

Figure 1 shows that the bulk of male and female respondents either agreed or strongly agreed with the statement that "people in relatively senior positions tend to be men". However, twice as many female engineers as male engineers strongly agreed with the statement, indicating that males

Figure 1. People in relatively senior positions tend to be men

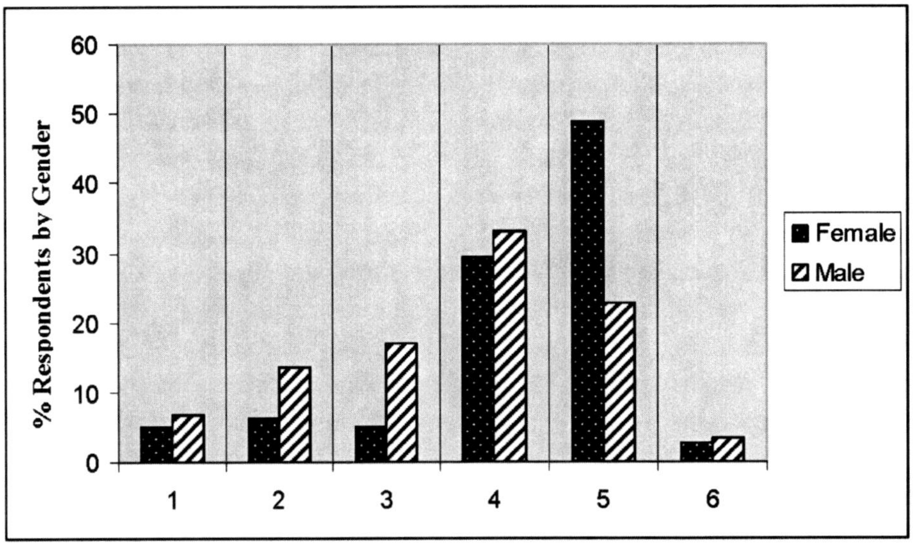

Notes: 166 responses (78 female, 88 male). Respondents were asked to rate the statement on a Likert scale where 1=strongly disagree and 5= strongly agree, 6=not applicable

perceive that any female representation at management level – however small – is evidence that there exists no glass ceiling. The data presented in Figure 1, however, supports the fact that managerial position in engineering organisations are more likely to be held by men.

Evetts (1994) points out that there is a paucity of older women engineers who have spent all their careers in the profession, and have reached senior positions. In a statistical analysis of the Labour Force Survey data, Smeaton et al (1997) show that women with SET qualifications begin to enter the teaching profession in large numbers after the age of 35, and the numbers increase with age. At the same time, they observe that the incidence of managerial positions in SET peaks for women aged 28-34. The literature indicates that older women drop away from management and pursue careers with more flexible working practices in order to combine career and family commitments.

Through an analysis of the qualitative data, the team set about trying to provide possible explanations for the paucity of older women engineers in senior positions. Of course, it may be the case that women are being subject to sex discrimination and employers are simply not passing them over for advancement. Larger employers in particular may be able to conceal such activities given the size of their operations. If women do not question these activities, which may be the case for fear of appearing pushy or aggressive, discrimination may be allowed to continue:

Some of my male colleagues who had come through the training scheme and were therefore just as experienced as I was were promoted months ahead of the time that I was ... I don't know whether that was down to gender or to department budgets or to the work I was doing compared to what they were doing or whether they shouted about it more. I mean obviously in a large organisation where there's a group of people rising through the ranks, some of them are going to be promoted before others. You know, it's impossible to tell what the reasons for that were (9Female, Former Electrical Engineer, 31-45 yrs).

Again, women may be unwilling to fight for the promotions that they deserve, and the lack of self-esteem mentioned earlier may be hampering women's ability to progress in the field:

we're not very good at promoting ourselves I don't think (Female, Managing Director, Industrial Engineering Firm, 31-45 yrs)

I've not had many women work for me, but I've had one who was very clever but didn't put herself across very well... you have to... if you're not tenacious; if you don't put yourself forward; if you don't demonstrate what you've done there are plenty of people out there who don't have the ability but are very good at promoting themselves, who'll get there before you (Male, Chemical Engineer, 31-45 yrs)

Another reason for this vertical segregation may be the fact that women are more likely to take career breaks in order to have children, or prefer to work flexible or reduced hours. Women may also be less able to commit to the requirements deemed necessary to reach senior positions. For example, one senior female engineer interviewed described the difficulties of combining family and work responsibilities with undertaking the Masters degree her employer required. Studying for the degree meant long periods spent away from home on a bimonthly basis; a commitment to which many females would be unable to subscribe to. Indeed, several engineers mentioned colleagues or friends who had left the industry because they felt they were unable to balance work pressures with domestic responsibilities.

Yes, that was another reason for leaving the company I was with previously; it was very, very, not aggressive, quite competitive and you had to put the hours in to be noticed. So basically a lot of the guys that made it to the top didn't have a home life and I don't want that. My personal life is too important to me not to, not to put first but you know, take care of really (Female Highways Design Engineer, 26-30 yrs).

The compatibility of motherhood and management seems to be a perceptual issue. In the literature, theorists have argued that the two cultures are not incompatible. Aaltio-Marjosola & Lehtinen (1998) contend that the act of motherhood actually improves women's capacity to lead; enhancing their communication skills and developing their ability to deal with chaos and respond to several problems simultaneously and efficiently. Bendix (1996) describes the art of leadership as a way of managing organisational resources in such a way as to help the organisation reach its goals: "The major objective of management is to foster co-operative teamwork amongst its employees." (Bendix, cited in Schwartz, 1996, p. 1023)

Many respondents felt that taking a career break or reducing working hours would adversely affect their opportunities for progression. This would have a far greater impact on females as they are more likely to take time out of the industry in order to have children. In fact, several female interviewees provided specific examples of times in which they felt that senior management considered them unsuitable for promotion because of their family commitments:

[Three years ago] Just before I found out I was pregnant I was told, yes you're on the associate promotion board this year, but as soon as I told them I was pregnant, that was the end of that... I only just got my promotion last month... I had to wait nearly two years to get that promotion and I had to go elsewhere and get another job offer and threatened to resign... I really caused a stink in personnel... my career since having the little boy had become just a bit more like having a job and

not a career... I've reached the stage where I'm associate, I will not go any further part time, they will never promote me and I don't think they'll ever promote me any higher... (Female, Mechanical Engineer, 31-45 yrs (4 days a week))

It is clear that this engineer had to pay the price for her choice to have a family. She felt that she is at a stagnant stage of her career and no matter how hard she worked, her company will not provide the incentives or the encouragement for her to progress. We argue that reaching such a stagnant stage in an individual's career could be the start of the exit path from an engineering career.

The Perception of Females who Progressed!

Despite the above, there were several examples of females who had reached board positions and directorships, and were respected for those accomplishments:

There are quite a few lady Directors... and they're Directors because they're better than the other guys in what they're doing; it's as simple as that. (Male, Managing Director, Construction Firm, 45+ yrs)

However, the literature suggests that those few women who are able to reach positions of leadership in organisations must first work harder than male managers, and must be able to demonstrate leadership without slighting or alienating male staff (Roberts and Ayre, 2002). In one way such female leaders managed to combine the competent and likeable character which contradicts Glick and Fiske (1999) typology; if competent, a female leader is unlikable.

Tower and Alkadry (2008) investigated the notion which indicates that women may experience social costs at different levels of their career progression. They surveyed public servants to examine this notion by exploring the characteristics of women who have made it to the top of a professional hierarchy in comparison to other women who are on various rungs of the hierarchy ladder. They urge governments and organisations to pay attention to the social costs of work and progression that affect women more than men.

A number of our interviewees who progressed gave us more insight into what it takes to climb up the ladder;

You will need to be better than your men (Female, Managing Director, 45+ yrs).

An alternative theory posits that women must either be extremely high achievers or exit the industry altogether. However, many interviewees noted that engineering environments can be extremely difficult places in which to survive and thrive. One interviewee pointed out in this scenario, women may actually lose the motivation to remain in the industry:

Half the problem is that clearly these women aren't just sort of very good women. They're top class people... I think that then intimidates other women who want to do what they started doing. So they think, 'my God, if you've got to be like, for example Alison to become a good Project Manager and she's a very very very good Project Manager', they think, 'my God If I've got to be like Alison I'm never going to make it, so I won't bother' (Male, Managing Director, Construction Firm, 45+ yrs).

These women may feel it is necessary to reflect the behaviour and activities of their male colleagues (Greed, 1991; Acker, 1990). As well as shunning domestic commitments in order to accommodate full-time career patterns, this may also mean adopting male modes of dress or styles of leadership. Bagilhole (2002, see also Evets, 1998) and others point out that women who have reached senior positions are frequently childless, whilst it is a common negative gender-role expectation that the prospect for mothers to progress in their chosen field is gloomy.

I'm now…approaching…middle age and I've never considered having children. If I had considered having children, I think it would have been very difficult. I think what I accepted early on in life is that if I wanted to go into a man's world, as it certainly was then, I needed to be more of a man than a woman (Female, Managing Director, Industrial Engineering Firm, 31-45 yrs).

If you had more women in it then there'd be the chance but the problem is in order for the women to get to the top they have to behave like men to get there. Like Margaret Thatcher, Margaret Thatcher was the worse thing for women in power because she behaved like a man (Female, Managing Director, 45+ yrs).

However, many interviewees, particularly females suggested that women should not have to act 'like men' in order to succeed in the industry.

…Girls…think they've got to be really butch and be able to stomp around out on site and scream and holler and swear, and it's not like that (Female, Highways Maintenance Engineer, 31-45 yrs).

"I bring this feminine side to it and I don't feel the need to try and act like one of the blokes, so I feel I've got quite a good balance going on that has actually been commented on by somebody else (Female, Director, Engineering Design Consultancy, 31-45 yrs).

Feminists working in the fields of technology and organisational studies have explained the lack of women in senior positions as being a product of social stereotypes and the patriarchal nature of hierarchy (Kanter, 1997; Acker, 1990, Tienari et al, 2002). Societal stereotypes emphasising the importance and identity of the male as breadwinner; and that of the female as family-oriented has a long-lasting effect in the workplace:

Managers grow up in this society and at the same time they reconstruct its culture. The old images of an ideal father and of modern organizational man share common features such as orderliness, rationality, efficiency, and unemotionality… Therefore, management, which is historically so closely related to men's ways of acting, is faced with the need to change, just as fatherhood does (Aaltio-Marjosola & Lehtinen, 1998, p.126)

This means that, subconsciously, men are expected to make better leaders than women. This would explain why some females feel they must imitate men in order to succeed. The qualitative and quantitative research explored male and female engineers' perception of women in senior positions in the engineering organisations. The team was interested in finding out whether employees and managers believe that women make capable managers, and whether this perception could have an effect on female engineers' opportunities for progression.

It is reassuring to note from Figure 2 that the majority of women (63%) and men (61%) are in agreement with the statement that female managers are as effective as male managers. The overall findings from both genders give a positive message regarding the potential effectiveness of female managers as perceived by both men and women.

The majority of both genders in Figure 3 either agreed (female, 46%, men, 34%) or strongly agreed (female, 28%, male 49%). This is good news (at least in terms of the respondents that answered this question) as it indicates that both genders recognise the potential leadership merits of both men and women.

Certainly amongst the interviewees, there was little indication that engineers believe that women cannot make effective managers.

Figure 2. Female managers are as effective as male managers

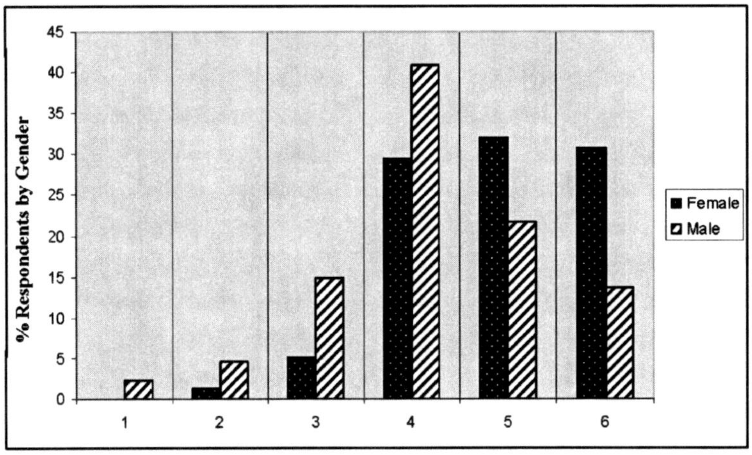

Notes: 163 responses(77 female, 86 male). Respondents were asked to rate the statement on a Likert scale where 1= strongly disagree and 5= strongly agree, 6= Not applicable

...women do the job... they have the ability to, you know, manage people, keep all the balls in the air, manage budgets you know etc (Female, Managing Director, 45+ yrs).

I think that's where women sort of have the edge over men quite often; that we can take on these sort of assertive characteristics as well as being able to keep hold of our feminine side. i.e. understanding and nurturing people and...situations (Female, Managing Director, Industrial Engineering Firm, 31-45 yrs).

My company look to take on people who can perform in their role and I think if it is a man or a woman it makes no difference (Male, Civil Engineer, 26-30 yrs).

However, differences between males and females as managers were recognised. Surprisingly, several of the female interviewees indicated that they preferred to work for male managers. This again takes us back to Glick and Fiske (1999) model of the competent but unlikeable woman:

Sometimes they're [female managers] very aggressive...I know they're trying to be assertive but I think it's a fine line between aggression and assertiveness and they always seem to, they have to prove their worth really more than a man and I think this aggression makes them feel as though they've got power over whoever they're trying to control (Female, Highways Engineer, 45+ yrs).

One of the male interviewees stated:

"must admit, I have found it quite difficult working for a woman boss... I didn't know how to handle it. I've gone through university which is completely male dominated. I've come into a working environment which is completely male dominated and when I had a woman boss it completely threw me (Male, Engineer, 31-45 yrs).

Having a woman boss which is a clear diversion from the "masculine norm" for this male engineer is something he could not handle. The last quotation, in one way, sums it all up and reinforces the masculine nature of engineering studies and careers that is designed by men for men.

Figure 3. Women and men are accepted in positions of leadership in my company

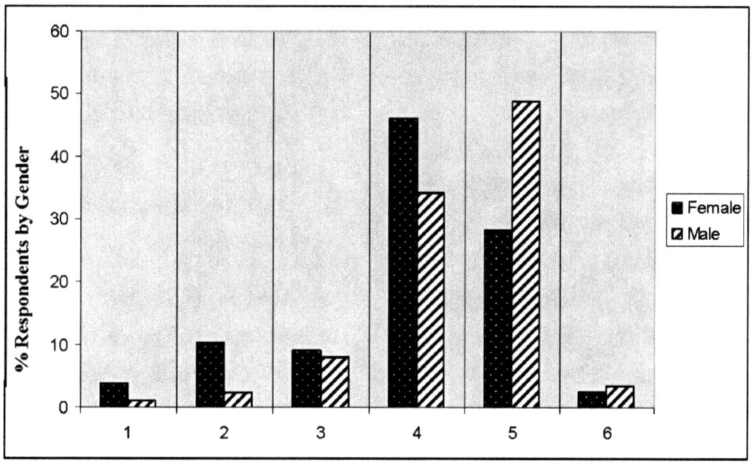

Notes: 164 responses (78 female, 86 male). Respondents were asked to rate the statement on a Likert scale where 1= strongly disagree and 5= strongly agree, 6= Not applicable

CONCLUSION AND RECOMMENDATIONS

Through the literature, we justified the urgent need to construct a learning environment for women in engineering that is conducive to enhancing their self esteem and confidence in their technical abilities. This is crucial in order to guarantee that female engineering graduates would actually pursue a career in engineering, stay and progress in the industry which will give them the necessary career satisfaction. Furthermore, we examined some of the multifaceted causes that may aid or otherwise hinder female engineers' progression in the UK's engineering industry. We investigated the personality/character that female engineers need to possess to "get in and on" in the engineering industry. It is evident that self-promotion, persistence and confidence in technical abilities are imperative to the advancement of women's careers in particular in a male dominated career environment such as engineering. We have also looked at the perception, of colleagues, of the women who made it up the higher rungs of the career ladder.

To foster and develop a culture a self promotion, confidence and self – efficacy amongst women engineers which will aid their progression, we emphasise the responsibility of educational institutions and engineering industries and organisation and hence, we recommend that:

- Engineering departments in higher education institutions should assess and review their structure, culture, practices and the curriculum and adopt the means to better prepare female engineers to enter stay and progress in an engineering career environment.
- Higher education institutions should give students early exposure to engineering concepts, especially in a societal context and by providing practical experience as part of the curriculum are twofold: firstly, it can help to build technical self-confidence in students, and secondly, it will reduce the disparity between learning and working when students make the move to the relevant industry.
- Engineering organisations should ensure that all employees, regardless of sex or

position, are given access to training and development, employers can encourage employees to think about their future career path encouraging them and giving them the opportunity to progress.

- Engineering organisations should build and develop technical skills in female engineers as there is evidence that some females have not had the opportunity to develop the technical skills that are necessary to progress in engineering. Technical ability is a requirement of many Chartership programmes and all staff should be given the opportunity to progress as far as they would like in their chosen field. Technology occupations are destined to be the most lucrative jobs in the future, and building and developing technical skills in women can help to empower them in the workplace.
- There is a need to provide role models and mentors for women at all stages in their career. The development and delivery of mentorship programmes has proven to be an extremely effective way to improve the retention and progression of women in science and engineering roles as well as help to reduce the sense of isolation that women in non-traditional occupations may feel.

REFERENCES

Aaltio-Marjosola, I. (1994). Gender Stereotypes as Cultural Products of the Organization. *Scandinavian Journal of Management*, *10*(2), 147–162. doi:10.1016/0956-5221(94)90017-5

Aaltio-Marjosola, I., & Lehtinen, J. (1998). Male Managers as Fathers? Contrasting Management, Fatherhood and Masculinity. *Human Relations*, *51*(2), 121–136. doi:10.1177/001872679805100201

Acker, J. (1990). Hierarchies, Jobs, Bodies: A Theory of Gendered Organizations. *Gender & Society*, *4*(2), 139–158. doi:10.1177/089124390004002002

Bagilhole, B. (2002). Challenging Equal Opportunities: Changing and Adapting Male Hegemony in Academia. *British Journal of Sociology of Education*, *23*(1), 19–33. doi:10.1080/01425690120102836

Charles, M., & Grusky, D. (2004). *Occupational Ghettos: the Worldwide Segregation of Women and Men*. Stanford, CA: Stanford University Press.

Cialdini, R. B. (1990). Indirect Tactics of Image Management: Beyond Basking. In Giacalone, R. A., & Rosenfeld, P. (Eds.), *Impression Management in the Organisation* (pp. 45–56). Hillsdale, NJ: Lawrence Erlbaum.

Cupchik, G. (2001, February). Constructivist Realism: An Ontology That Encompasses Positivist and Constructivist Approaches to the Social Sciences. *Forum Qualitative Sozialforschung/ Forum: Qualitative Social Research, 2*(1). Retrieved October 4, 2006, from http://qualitative-research.net/fqs/fqs-eng.htm

Evetts, J. (1993). Women and Management in Engineering: the 'Glass Ceiling' for Women's Careers. *Women in Management Review*, *8*(7), 19–25. doi:10.1108/09649429310046391

Evetts, J. (1998). Managing the Technology but not the Organization: Women and Career in Engineering. *Women in Management Review*, *13*(8), 283–290. doi:10.1108/09649429810243144

Glick, P., & Fiske, S. T. (1999). Sexism and Other 'Isms': Interdependence, Status, and the Ambivalent Content of Stereotypes. In Swann, J., William, B., Langlois, J. H., & Gilbert, L. A. (Eds.), *Sexism and Stereotypes in Modern Society*. Washington, DC: American Psychological Association. doi:10.1037/10277-008

Greed, C. (1991). *Surveying Sisters: Women in a Traditional Male Profession*. London: Routledge.

Heilman, M. E. (2001). Description and prescription: How Gender Stereotypes Prevent Women's Ascent Up the Organizational Ladder. *The Journal of Social Issues*, *57*, 657–674. doi:10.1111/0022-4537.00234

Kanter, R. M. (1977). *Men and Women of the Corporation*. New York: Basic Books.

Kelle, U. (2001, February). Sociological Explanations between Micro and Macro and the Integration of Qualitative and Quantitative Methods. *Forum Qualitative Sozialforschung / Forum: Qualitative Social Research, 2*(1). Retrieved October 4, 2006, from http://qualitative-esearch.net/fqs/fqs-eng.htm

Martin, P. Y. (2001). 'Mobilizing Masculinities': Women's Experiences of Men at Work. *Organization*, *8*(4), 587–618. doi:10.1177/135050840184003

Maskell-Pretz, M., & Hopkins, W. E. (1997). Women in Engineering: Toward a Barrier-Free Work Environment. *Journal of Management Engineering*, *13*(1), 32–37. doi:10.1061/(ASCE)0742-597X(1997)13:1(32)

McIlwee, J. S., & Robinson, J. G. (1992). *Women in Engineering: Gender, Power and Workplace Culture*. Albany, NY: State University of New York Press.

Mills, A. J. (1988). Organization, Gender and Culture. *Organization Studies*, *9*(3), 351–369. doi:10.1177/017084068800900304

Parsons, T., & Bales, R. F. (1955). *Family, Socialization and Interaction Process*. New York: The Free Press.

Ragin, C. C. (1994). *Constructing social research*. Thousand Oaks, CA: Pine Forge Press.

Ramdin, T. E. (1998). *Identification of Factors Which Facilitate and Hinder the Career Adjustment of Women Engineers" in Canadian Coalition of Women in Engineering, Science and Technology.* Retrieved September 25, 2006 from http://ccwest.org/Cnu5news/CCWESTfullarticles.asp?dismode=article&artid=174

Ravenscroft, A. M. (2008). What's Keeping Women out of it? *ieee.usa today's engineer.* Retrieved March 20, 2009, from http://www.todaysengineer.org/2008/Mar/wie.asp

Ridgeway, C. L., & Correll, Shelly J. (2004). Unpacking the Gender System: A Theoretical Perspective on Gender Beliefs and Social Relations. *Gender & Society*, *18*(4), 510–531. doi:10.1177/0891243204265269

Roberts, P., & Ayre, M. (2004). *The Careers Review of Engineering Women: An Investigation of Women's Retention in the Australian Engineering Workforce*. Canberra, Australia: Engineers Australia.

Rudman, L. A. (1998). Self Promotion as a Risk Factor for Women: the Costs and Benefits of Counterstereotypical Impression Management. *Journal of Personality and Social Psychology*, *74*(3), 629–645. doi:10.1037/0022-3514.74.3.629

Rudman, L. A., & Glick, P. (2001). Prescriptive Gender Stereotypes and Backlash Toward Agentic Women. *The Journal of Social Issues*, *57*, 743–762. doi:10.1111/0022-4537.00239

Sappleton, N., & Takruri-Rizk, H. (2008). The Gender Subtext of Science, Engineering and Technology (SET) Organizations: A Review and Critique. *Women's Studies*, *37*(3), 284–316. doi:10.1080/00497870801917242

Schwartz, H. S. (1996). The Sin of the Father: Reflections on the Roles of the Corporation Man, the Suburban Housewife, Their Son and Their Daughter in the Deconstruction of the Patriarch. *Human Relations*, *49*(8), 1013–1040. doi:10.1177/001872679604900801

Shull, P. J., & Weiner, M. (2002). Thinking Inside the Box: Self-Efficacy of Women in Engineering. *The International Journal of Engineering Education, 18*(4), 438–446.

Smeaton, D., Glover, J., & Fielding, J. (1997). Recent Trends in Women's Representation in Science, Engineering and Technology (SET) in the UK. *Journal of Women and Minorities in Science and Engineering, 3*, 1–19.

Tienari, J., & Quack, S. & Theobald, Hildegard. (2002). Organizational Reforms, 'Ideal Workers' and Gender Orders: A Cross-Societal Comparison. *Organization Studies, 23*(2), 249–279. doi:10.1177/0170840602232004

Tower, L. E., & Alkadry, M. G. (2008). The Social Costs of Career Success for Women. *Review of Public Personnel Administration, 28*(2), 144–165. doi:10.1177/0734371X08315343

Williams, A., Turrell, P., & Wall, R. (2002). Let's TWIST: Creating a Conducive Learning Environment for Women. *The International Journal of Engineering Education, 18*(4), 447–451.

Zeldin, A. L., & Pajares, F. (2000). Scientific, and Technological Careers Against the Odds: Self-Efficacy Beliefs of Women in Mathematical. *American Educational Research Journal, 37*(1), 215–246.

ENDNOTES

1 Statistics available from http://www.setwomenstats.org.uk/set4statistics/03_analysis.htm

2 Kelle, Udo (2001) discussed three different understandings of the triangulation metaphor: triangulation as mutual validation, triangulation as the integration of different perspectives on the investigated phenomenon and triangulation in its trigonometrical meaning.

Profile 13
Engineering Career Spotlight:
Environmental Engineer

Rachelle Willis, Brisbane, Australia

Rachelle Willis is a Research and Planning Engineer with Griffith University and Gold Coast Water. She is the PhD candidate and project manager of the 'Gold Coast Watersaver End Use' Study, a collaborative ARC-Linkage (Industry) research project between the organisations. Other areas of vocation include sustainability in industry and water management and planning. Rachelle graduated in 2006 from Griffith University with First Class honours and was awarded the 2006 'Smart Women Smart State' in the undergraduate category. She was recently acknowledged at an International Water Association Conference with a 'Best Paper Presented by a Young Water Professional' award. She has also co-authored numerous peer reviewed papers from her End Use research in both journals and conferences (for more information see http://www.griffith.edu.au/engineering-information-technology/centre-infrastructure-engineering-management/research/special-projects/gold-coast-watersaver-end-use-project). Rachelle is actively involved with environmental engineering and sustainability and is a board member of both the Queensland Chapter of the Society of Sustainability and Environmental Engineering, and the National Environmental Engineering College, both elements of Engineers Australia. Rachelle embarked upon environmental engineering as a profession due to a strong interest on improving and maintaining the natural environment while allowing for future development in the built environment.

Copyright © 2010, IGI Global. Copying or distributing in print or electronic forms without written permission of IGI Global is prohibited.

Introduction

In 2006, Rachelle Willis was successful in the Smart Women Smart State Awards. The following is an extract from her successful application.

I am a 4th year Environmental Engineering student at Griffith University, Nathan. I have recently completed my final year thesis which focused on reducing pollutants in stormwater runoff to minimise the adverse impacts on waterways in South East Queensland including Moreton Bay. This thesis focused primarily on improving water quality to ensure that flora and fauna within waterways, such as sea grass, dugongs, turtles, and other endangered species have unpolluted natural environments to thrive in. Poor water quality also has the potential to affect the economic viability of businesses which rely upon water and the flora and fauna which thrives within it, fisheries are an example of such industries.

Research Thesis

My thesis was titled 'An investigation into nitrogen speciation, removal processes and improving removal through stormwater quality improvement devices'. This project is significant as waterways throughout Australia and the world are becoming increasingly polluted and it has been documented that excessive accumulation of pollutants, such as nitrogen, can lead to ecological imbalances. The thesis findings can be applied to aid in the protection of inland and coastal waterways such as Morton Bay and the Great Barrier Reef, not only for environmental benefits but also to ensure the viability of industries such as tourism and fisheries. Nitrogen can cause negative environmental and social problems, for example, the eutrophication of waterways, and the growth of cyanobacteria which can reduce dissolved oxygen in water and have devastating effects on waterway ecosystem diversity. Eutrophication can also affect a municipal water supply and strain water treatment plants, and farmers which rely on the water for irrigation or livestock are also impacted and need to find water from alternative sources.

In order to address these problems it was important to understand various speciations of nitrogen present in the environment in both natural and pollutants forms. Water quality objectives (WQOs) were investigated to determine the state of environmental regulations in Australia and also to determine if these objectives were accomplishable and sustainable. Different stormwater quality improvement devices (SQIDs) were then researched to determine the most appropriate methods for removing various forms of nitrogen from stormwater to protect waterways. Many aspects of SQIDs were considered, some of these included: nitrogen speciation removal potential, the size of device, the size of catchment which the device can treat, cost of installation, life-cycle costs and life-cycle benefits. Other aspects such as the ease of modelling SQIDs were also considered, as well as the possibility of modelling nitrogen speciation removal from stormwater.

My research found that SQIDs are not removing sufficient levels of nitrogen to meet WQOs, which are guidelines set by Councils throughout Australia to protect environmental values of waterways. This means that most existing developments in Australia and around the world are releasing higher nitrogen concentrations into waterways than are environmentally sustainable. The potential pollutant speciations of nitrogen, especially nitrates, are particularly hard to remove due to the environmental conditions which are required and replicating these conditions has proved a challenge for engineers. It was found that the use of SQIDs such as wetlands, street trees with stormwater filtration areas, permeable pavements, and well vegetated bioretention devices removed the highest concentrations of potential pollutant speciations of nitrogen. The use of vegetated soils with anaerobic zones and wetlands proved to be the most efficient removers. Life-cycle cost benefits determined that wetlands were the most cost effective and they also provide significant long term social and environmental benefits.

This research was important as it delved into and identified solutions to reduce the concentrations of nitrogen in stormwater which flows into waterways in South East Queensland and Moreton Bay. This is significant as waterways provide rich, naturally diverse ecosystems which are constantly utilised for recreation, the consumption of water, industrial uses such as fishing, and education. Waterways need to be protected not only to maintain their natural state, which in Moreton Bay is inclusive of dugongs and sea grass forests, but also to maintain and support the numerous human uses such as a productive fisheries or tourism industry.

This thesis was carried out with Brown Consulting (QLD) Pty Ltd as an industry based project. Brown Consulting sponsored the project to improve the stormwater runoff quality from various developments which they work on. The ideas developed through the project have been conveyed to Brown Consulting for consideration and application in upcoming developments. This thesis is of importance to Brown Consulting and other engineering firms, as complying with environmental regulations is paramount, and often developments will not be approved if pollutants in stormwater are over the regulated objectives. Results from this thesis could be used by many engineering firms to improve their nitrogen removal from stormwater, to comply with WQOs, and to protect waterways and Moreton Bay. Cost benefit analysis was aimed at educating stormwater managers to consider all aspects of development, including the environmental and community benefits of SQIDs, rather than just considering the cost. For further information about this project please contact me and I will be happy to provide a copy, subject to approval from Brown Consulting (QLD) Pty Ltd.

Dissemination of Information and Knowledge

The thesis was presented in a poster format at the Industrial Affiliate Program (IAP) Expo at the Brisbane Technology Center in June 2006. I was one of the two (2) students out of approximately 130 students, nominated by my peers to present my thesis to industry representatives, university officials, peers and other students, and other attendees of the day. This was a short presentation carried out in the formal proceedings of the Expo. In this presentation I stated the importance of my research, gave an outline of the topics that I covered in my research, and discussed some of my key findings. Many attendees approached me after the formal proceedings had concluded to further discuss my research and significant findings.

I was also nominated at the IAP Expo by academics as having an exceptionally high quality and interesting thesis. University officials including the Dean of the Faculty of Engineering and the Deputy Vice Chancellor (Academic) and invited business delegates were directed to my project to discuss the overall basis of the thesis and my outcomes. I received the best poster award on the day.

I received a high distinction (HD), and my thesis came top of the environmental engineering stream. I have also been asked by academics for permission for their postgraduate students to read my thesis for reference in their studies.

The 6th Annual Environmental Student Project Award Evening (October 2005)

This award evening allowed Queensland wide final year environmental students to competitively showcase their environmental research (desktop, field and research) topics to peers, university academics, industry representatives, governing bodies, businesses and the wider community. This event was held at Griffith University in

2005 and I was the co-organiser of this annual event which was supported by The Society for Sustainability and Environmental Engineering and the Waste Management Association of Australia (QLD Branch). It was also sponsored by a large number of industries including; GHD, Queensland Government Department of Main Roads, SKM, the University of Queensland, Griffith University, Abigroup, Brown Consulting, Collex and Tarong Energy.

This event allows sharing of information about environmental research being carried out by environmental students at universities in Queensland and New South Wales. I thoroughly enjoyed organising and facilitating fellow environmental students to promote their innovative research to the community and their peers. This event was well attended and I believe that the people present were surprised at the level of research and work that environmental students were producing.

Promoting and Supporting Environmental Engineering

President of the Griffith University Environmental Engineering Society (GUEES) 2005

My role as the president of this society was primarily to promote Environmental Engineering to fellow students and to potential students through events such as career nights and to run social events for students. I was often asked to speak to other students about the role of environmental engineers and to organise information sessions about future employment for environmental engineers.

Student Representative of the Environmental Engineering School Committee 2005

This committee was made up of the Griffith University environmental engineering lecturers. The role that I played on this committee was to give a students perspective on topical university issues which affected environmental engineers. I was also a nominated student representative for the IEAust accreditation of the Environmental Engineering course at Griffith University.

Mythbusters 2005

This is an event organised by Griffith University for high school students to reveal the truth about engineering and science at the university and thereby attract students to enroll in engineering and science degrees. I was asked to speak at this event and attended the event to promote environmental engineering to younger students. I found that many students were grateful to have a student's perspective on engineering and I was approached by many students and parents to discuss my experience as an engineering student. I believe that it is important to provide a positive female role model to potential engineering students.

Griffith University Open Day 2005

I was asked to attend the Open Day as a student representative to share my experiences with younger students and to encourage them to consider a career in engineering. I found that many of the girls that I talked to on the day were skeptical that they could pass the courses or become an engineer. I was happy to be able to share my experiences and encourage them into this field using examples of achievements of female students and engineers in this traditionally male dominated career.

Closing Comments (2009)

As a young woman in engineering I have worked and studied in a male dominated profession. I personally strive for success in my career choice and to promote the engineering profession to other women. I was proud to win the Undergraduate Engineering category in the Smart State awards that promotes engineering to smart women who may also be considering a career in engineering.

Chapter 14
Women in Brazilian CS Research Community:
The State-of-the-Art

Mirella M. Moro
Universidade Federal de Minas Gerais – UFMG, Brazil

Taisy Weber
Universidade Federal do Rio Grande do Sul – UFRGS, Brazil

Carla M.D.S. Freitas
Universidade Federal do Rio Grande do Sul – UFRGS, Brazil

ABSTRACT

Many communities have been concerned with the problem of bringing more girls to technology and science related areas. The authors believe that the first step in order to solve such a problem is to understand the current situation, like to investigate the "state-of-the-art" of the problem. Therefore, in this chapter, they present the first study to identify which areas of Computer Science have more and less feminine participation. In order to do so, they have considered the program committees of the Brazilian conferences in those areas. The authors' study evaluates the 2008 and previous editions of such conferences. They also discuss some Brazilian initiatives to bring more girls to Computer Science as well present what else can be done.

INTRODUCTION

Many communities from different countries have been concerned with the problem of the low number of women studying and working on technology, engineering, and science related areas (Brainard & Carlin, 2001; Ceci & Williams, 2006; Dewandre 2002; Hyde et al 1990; Ivie & Ray, 2005; Patterson 2005; Ramsey & McCorduck, 2005; Ticoll, 2008). Specifically, technological areas are fascinating and provide a vast range of opportunities to their professionals. Still, we wonder why the interest of girls in joining the Engineering and Computer Science fields (among others) is decreasing.

Studies inform that there are usually three points of view to understand the low interest of women

DOI: 10.4018/978-1-61520-657-5.ch014

Copyright © 2010, IGI Global. Copying or distributing in print or electronic forms without written permission of IGI Global is prohibited.

toward science and technology oriented careers: (*i*) the biological, gender differences; (*ii*) the social construction of technological careers as a male domain; and (*iii*) the individual differences among women related to technological oriented work and workplace (Trauth et. al, 2004). Nonetheless, what we can do to reverse such a scenario and bring more girls to those fields is still a big question (Harris & Raskino, 2007; Ramsey & McCorduck, 2005; Simard 2007).

We believe that the first step in order to solve such a problem is to understand the current situation, like to investigate the "state-of-the-art" of the problem. Instead of evaluating all the so-called "hard sciences" in general, we focus in *Computer Science* (CS) because it is our area of expertise. Furthermore, we notice that there is a variety of studies specific for women in CS and Information Technology (IT), such as (Ramsey & McCorduck, 2005; Simard, 2007; Ticoll 2008; Trauth et al 2004). Nonetheless, we focus this work on studying the women participation in *research* on CS related areas (*i.e.*, we do not consider women presence in the CS/IT industry). It is very important to notice that CS per se is a major science composed of many subareas, which are completely different from each other. For example, Databases, Computer Networks, Formal Methods, and Computer Graphics (just to cite a few) are entirely distinct areas with their own specific features, problems, techniques, and applications. Furthermore, each of those has its own research community with distinguished conferences and symposia, such as VLDB (Intl. Conf. on Very Large Data Bases), INFOCOM (IEEE Conf. on Computer Communications), FM (Intl. Symp. on Formal Methods), and SIGGRAPH (Intl. Conf. and Exhibition on Computer Graphics and Interactive Techniques), and journals such as ACM Transactions on Data Base Systems, Computer Networks, ACM Transactions on Graphics, among many others.

Given the unique profile of each subarea, it does not seem accurate to study the number of women in the general CS area. For example, a quick look at those conferences shows that the women participation in their program committees is very distinct from each other. Therefore, in order to have more accurate statistics, it is necessary to investigate each subarea individually. Such investigation is vital for qualifying the state of the art scenario. In particular, we believe that once we have identified the "most feminine subareas", we can investigate why those areas are so successful in their number of women. So, instead of trying to find out what on the technological areas do not attract girls, we will focus on what has *attracted* them.

Nonetheless, there is one problem with this idea. To the best of our knowledge, current statistics about women (and other minorities) participation in CS and other sciences present just the general number (CRA 2007; Ivie & Tay, 2005; Harris & Raskino, 2007; Simard et al, 2008; Simard 2009). In other words, there is no division of the number among the subareas in which women study or do research. Another particular problem is that, in Brazil, statistics on college and graduate programs (over students and faculty members) do not differentiate gender. In this chapter, we attempt to provide a *first* overview of the current presence of women in the Brazilian CS research community through an empirical study. While it would be interesting to have worldwide numbers, we decided to focus on Brazilian numbers. Then, a second step would be to investigate whether such numbers reflect the international situation as well. Once we have identified which of the CS subareas have more women participation, we can investigate those areas and find out which specific features attract the "feminine eyes". Furthermore, we can target the communities with less feminine participation in order to reverse such a situation. Finally, in this chapter, we also overview two Brazilian initiatives for bringing the CS female community together, and comment on what else can be done.

BACKGROUND INFORMATION AND RELATED WORK

Since the world wide revolution for women's right (from 1792 on, year in which Mary Wollstonecraft, often called the first feminist philosopher, published *A Vindication of the Rights of Woman*), women have claimed equality in many fronts, including the right to work and getting fair paycheck. In academic, science and research communities, it is not different, for high level education was a men's privilege until late nineteenth century. In order to provide a historical perspective, some important dates of women achievements during such a period include[1]:

- 1840, Catherine Brewer Benson is the first woman to earn a bachelor's degree, by Georgia Female College (now Wesleyan College);
- 1849, Elizabeth Blackwell is the first American woman to receive a medical degree, by Geneva Medical College (today, the college is part of the State University of New York Upstate Medical University and is called Hobart College);
- 1865, Elizabeth Garrett Anderson is the first British woman to obtain a medical qualification, from the License of Apothecaries' Hall;
- 1873, Ellen Swallow Richards earns her B.S. degree as the first women admitted to the Massachusetts Institute of Technology;
- 1887, Rita Lobato Velho Lopes is the first Brazilian woman (and second Latin American) to receive a medical degree, by Medicine College of Salvador.

With a very controversial access to high level education, woman scientists were also recognized with Nobel prizes. The first one was Marie Curie, who won it for Physics in 1903 and for Chemistry in 1911. Many years later, Gerty Theresa Cori won it for Physiology or Medicine in 1947. Although there is no Nobel Prize for Computer Science, ACM (Association for Computing Machinery) established an award in 1966, called the Turing Award, for recognizing an individual with distinguished contributions of technical nature to the computing community.

The first woman to receive this award is Frances E. Allen (IBM), in 2006, in the 40th edition of the award. The second one is Barbara Liskov (MIT), in 2008. Barbara is also the first woman to obtain a Ph.D. from a Computer Science Department, by Stanford University in 1968.

Given such a historical perspective, this section follows with some statistics about Computer Science and a discussion on related work.

Computer Science Numbers in US

The number of women receiving a Bachelor degree in CS has decreased. Data from the National Center for Educational Statistics show that the number of bachelors graduated peaked in middle 1980s and early 2000s, and has declined since then. Figure 1 illustrates such a trend based on those data. It is very important to notice that such data do not distinguish CS subareas of expertise. Therefore, there is no way to infer the areas in which those bachelors work.

Table 1 shows statistics on women working on computer-related positions. According to it, most women within computer-related areas are Software Engineers. However, considering the research community, there are more areas of expertise beyond the ones in this table. For example, this table does not consider any of the theoretical areas of CS (such as formal methods). Although this table gives some perspective on working women, it is not enough to imply women distribution over CS *research* areas. Again, it in Brazil (and many other countries), statistics present the general number of bachelor degree's recipients, without any separation of gender.

Figure 1. Bachelor's, master's, and doctor's degrees conferred by degree-granting institutions, by sex of student and field of study: 2005-06

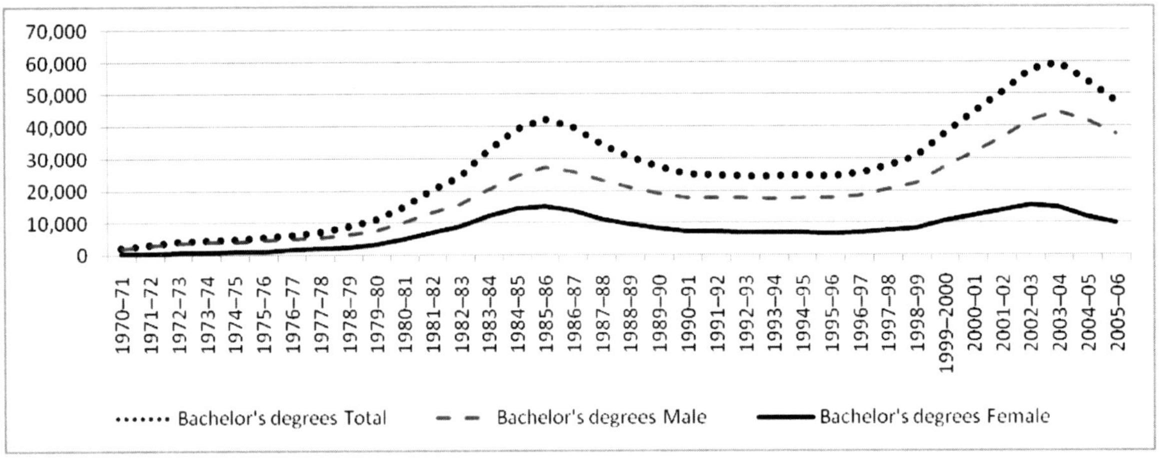

SOURCE: U.S. Department of Education, National Center for Education Statistics, Higher Education General Information Survey (HEGIS), "Degrees and Other Formal Awards Conferred" surveys, 1970–71 through 1985–86; and 1986–87 through 2005–06 Integrated Postsecondary Education Data System, "Completions Survey" (IPEDS-C:87–99), and Fall 2000 through Fall 2006.

Related Work

Computer Science major is in a downward cycle in the U.S., especially for women (Patterson 2005), as many science-related majors are in the U.S. and other countries (Welde et al., 2007). Notice that such a drop occurs while the number of female students in colleges and universities increases (Patterson 2005). In this section, we overview some of the related work that focus on understanding and solving this problem.

Welde, Laursen and Thiry (Welde et al., 2007) present some reasons for the low number of women in science, technology, engineering, and math (STEM). Those reasons were collected from previous work and papers from different persons and organizations. In summary, the low number of women in STEM-related careers may be justified by: classroom climate, lack of role models, poor preparation and lack of encouragement in STEM courses, lack of "critical mass" of women, bias and discrimination in hiring and advancement of women, salary differences and low status, and issues of work-life balance.

Another recent report, from Canada (Ticoll 2008), discusses why the enrollment in ICT (Information and Communications Technology, which includes many computer-related topics) majors has dropped. It argues that the causes for such a change are connected to six perspectives: the dotcom crash and the offshoring phenomenon, computers are everyday tools for young people and have no appealing to them, people advise students that ICT is not for them, girls (in particular) think they lack the necessary abilities to succeed in this area, the K-12 curriculum, and students believe they have more future in other areas. This study also makes a very strong statement: "Women are probably better than men at the competencies needed for the jobs of today [in ICT] but no one is telling them this important fact".

Table 1. Occupations for Women in 2008 within computer-related area

Occupation	Employed Both Sexes	Employed Female	Percent Female
Computer control programmers and operators	68	6	8.7
Computer hardware engineers	69	13	19.4
Network and computer systems administrators	227	49	21.4
Computer, automated teller, and office machine repairers	335	35	10.5
Network systems and data communications analysts	422	100	23.7
Computer programmers	534	119	22.4
Computer software engineers	1,034	216	20.9

Data not shown where base (total employed, both sexes) is less than 50,000. Source: U.S. Department of Labor, Bureau of Labor Statistics, Annual Averages 2008. Women's Bureau, April 2009.

This report also reviews another survey (by Wendy Cukier, ICTC 2007) on the underrepresentation of girls and women in ICT. The reasons for that include: socialization and early education (computers perceived as male domain), systematic barriers to start and keep studying ICT, misalignment of career descriptions with job requirements, impact of influencers, and lack of role models. Finally, the report also cites three ideas for tackling the problem: ICT curricula in high school, delivered by effective and compelling teachers; high school visits by role models; and exposure to uses of ICT that help solve social problems, or are multidisciplinary in nature.

A recent survey report published by the Anita Borg Institute (Simard et al., 2008) discusses the obstacles faced by women in technology, which includes family, success, values, workplace culture and climate. It also presents initiatives for keeping and advancing women in CS by summarizing policies and practices employed in top companies. Among those, we can cite: flexible work schedules, parental leave, childcare benefits, career planning, fair career development, and mentoring.

A different perspective is provided by Harris, Morello and Raskino (Harris et al., 2007). According to this study, there are three key traits in woman that can affect team performance within an IT organization: listening skills, proficiency in language skills, and superior social skills. Moreover, this study recognizes that there is no universal agreement as the influence of gender differences in working matters. After a long discussion, other two gender-related traits are cited in favor of men: special visualization and pattern spotting; and aggression and risk taking. Overall, one important conclusion from this work is that business leaders should improve the organization diversity in order to guarantee better team work, productivity, and satisfaction among its employees.

A complementary view based on the cultural and environmental aspects is presented by Frieze and others (Frieze et al., 2006). The main question the authors try to solve is: gender as a cultural issue (where culture designates the complex and large set of relationships, values, attitudes, and behaviors that bind a community consciously and unconsciously). This paper evaluates different case studies and concludes that there is indeed an impact from culture and environment that determines the women's choices and participation in CS.

Overall, current work on women in sciences and technology focuses on discussing the reasons why women do not feel compelled to study and work in those areas, as discussed in the previous section. Different organizations have presented statistics on the gender distribution of students and professionals in computer-related area (including IT and ICT). Although U.S. and Canada (among other countries) have speculated the existing statistics, Brazil is far behind, since its statistics consider only the general number of students in college and graduate schools (with no gender distinction). Nonetheless, to the best of our knowledge, there is no study on women in the CS *research* community. The best approximation is the data of students in CS graduate schools. However, such numbers do not qualify which CS fields women tend to study or do research on, which is the focus of this chapter.

THE STATE-OF-THE-ART

The study presented in this chapter aims to bring some light to an important issue: in which areas of the Brazilian CS research community has more/less women working. Therefore, in this section we present the methodology and results of an initial study on the distribution of women participation within different CS research subareas. Specifically, we present the methodology employed, the results from 2008, followed by some historical results and complementary numbers.

Methodology

In order to find such "feminine CS subareas", we have investigated the women participation in the program committees (PC) of the most traditional Brazilian research conferences. Note that, since we do not have access to an official census, we have focused on publicly available data. In this case, all research events have either a webpage or published proceedings (or both) with a set of persons participating in their PC. Finally, we are aware of the fact that, usually, PCs consider more senior members than juniors. However, we have also noticed that there is an effort of the conference chairs to include junior researchers for promoting their inclusion into the community as well.

Therefore, we have defined a set of 12 CS subareas. Those areas were chosen based on the tradition of their Brazilian conferences and the size of their communities. Specifically, the areas and events considered (as well as some related keywords) are as illustrated in Figure 2.

Then, we gathered the PCs of their major conferences, and computed their number of women. The number of women is separated in two parts: the total number of women and the number of Brazilian women (since some of the events have an international program committee).

Statistics from 2008 Events

The first step in our study was to compute the women participation in the 2008 events. Table 2 presents those results, showing for each selected Computer Science area, the acronym for its Brazilian conference (with the number of the analyzed edition), the total number of persons in its program committee, the percentage of women, and the percentage of Brazilian women. Data is presented classified in decreasing order of the percentage of Women participation in the committees.

According to this table, Human-Computer Interaction and Databases are the "most feminine" subareas, while Fault Tolerance and Integrated Circuits as the "less feminine" ones. Overall, the rate for women participation is 21% in average, and Brazilian women 16%. Also, there is *no* relation between the size of the PC and the number of women. For example, the largest PC (SBIA) has 21% of women, while the smallest PC (SBMF) has 24% of women. A first analysis of Table 2, considering just the first and last lines (i.e., Databases and Integrated Circuits) could lead us to conclude that areas most related to interaction with users (i.e., closer to social sciences) attract more women as opposed to areas more related to electronic engineering. It is also important to notice that most research done on Databases deal with organizing the data (e.g., modeling, improving its performance) and presenting it

Figure 2. CS research subareas, Brazilian event, and some keywords

~ **Artificial Intelligence**: SBIA (Braz. Symp. on Artificial Intelligence). Some keywords: constraints and search, computer vision and pattern recognition, distributed AI, knowledge representation and reasoning, machine learning and data mining, model-based reasoning, natural language processing, planning and scheduling, robotics.
~ **Computer Architecture**: SBAC-PAD (Intl. Symp. on Computer Architecture and High Performance Computing). Some keywords: processor, cache, memory, storage, and network architecture; operating systems and virtualization; tools for parallel and distributed programming; parallel and distributed systems, algorithms, and applications; grid, cluster, and peer-to-peer systems; reconfigurable and fault-tolerant systems; power and energy-efficient systems;.
~ **Computer Graphics**: SIBGRAPI (Braz. Symp. on Computer Graphics and Image Processing). Some keywords: animation and simulation, augmented reality, coding, compression and transmission of images / videos, computational geometry, games, geometric & solid modeling, global illumination, graphical interfaces and hardware, illumination, color and reflectance, image synthesis and realism, image-based rendering, information visualization, medical imaging, motion analysis / synthesis and optical flow, parallel visualization and graphics clusters, remote sensing and geophysical imaging, segmentation and grouping in images / videos, wavelets.
~ **Computer Networks**: SBRC (Braz. Symp. on Computer Networks and Distributed Systems). Some keywords: active networks and VPN, digital TV, distributed algorithms, fault tolerance, grid and cloud computing, middleware, mobility, P2P, protocols, QoS, security, sensor networks, traffic control, web services.
~ **Databases**: SBBD (Braz. Symp. on Databases). Some keywords: active databases (DB), access methods and indexing, authorization, privacy and security, concurrency control and recovery, data mining and knowledge discovery, data semantics and visualization, data warehousing, DB design, digital libraries, GIS, information retrieval, mobile data, distributed DB, query languages, processing and optimization, spatial DB, textual DB, web DB, web services.
~ **Fault Tolerance**: LADC (Latin American Symp. on Dependable Computing). Some keywords: engineering of dependable systems (DS), dependability of hardware, software, networks, data, and maintenance, security, dependability and human issues, algorithms and methods for DC, dependability and security benchmarking.
~ **Formal Methods**: SBMF (Braz. Symp. on Formal Methods). Some keywords: formal methods (FM), languages formal aspects, logics and semantics of programming and specification languages, reusable domain theories, type systems and category theory in CS, computational complexity methods and models, FM for software/hardware development, FM applied to model-driven engineering, model checking, theorem proving, and software testing.
~ **Human-Computer Interaction**: IHC (Braz. Symp. of Human Factors on Computer Systems). Some keywords: accessibility and therapeutic use of interactive systems, affective computing, context-sensitive interaction, ergonomics, HCI and aging, education in HCI, social aspects of HCI and their application on health systems, public transportation, education, and environment, HCI and Software Engineering, HCI design techniques and tools, formal methods and diagrams for user interface design, human-robot interaction, legal issues on HCI, usability evaluation methods, social interaction, information visualization techniques for interacting with large data sets.
~ **Information Systems**: SBSI (Braz. Symp. on Information Systems). Some keywords: information systems (IS), fundamentals and conceptual modeling for IS, business process, E-business, E-commerce, E-government, enterprise IS, decision support systems, estrategical planning, data warehousing and OLAP, service-oriented architectures.
~ **Integrated Circuits**: SBCCI (Symp. on Integrated Circuits and Systems Design). Some keywords: system level modeling and synthesis, hardware-software co-design and co-verification, high-level and logic synthesis, physical design, verification, simulation, emulation and prototyping techniques, testability issues, reconfigurable architectures and novel applications of FPGAs, analog, digital and mixed signal designs, embedded systems design and industrial applications, low power tools and design techniques, nanoelectronic circuits and nanoarchitectures.
~ **Programming Languages**: SBLP (Braz. Symp. on Programming Languages). Some keywords: programming language (PL) design and implementation, formal semantics of PL, theoretical foundations of PL, scripting languages, PL for mobile, Web and network computing, program analysis and verification.
~ **Software Engineering**: SBES (Braz. Symp. on Software Engineering). Some keywords: agent-oriented software engineering (SE), aspect-oriented SE, component-based SE, CSCW and SE, experimental SE, methods, techniques, languages and tools for SE, model driven software development, requirements engineering, SE for the WWW, social aspects of SE, software dependability, maintenance, metrics and measurement, software processes, product lines, quality and quality models, software reengineering, reuse, verification, validation and testing.

(e.g., query languages, data visualization). Those tasks require constant interaction with the users in order to identify what exactly they need from the data. Such interaction is practically nonexistent in Integrated Circuits. Nonetheless, this conclusion cannot explain why programming languages (an area not directly related to hardware) has less feminine participation than computer architecture

Considering the first four subareas, we notice that all of them are related to the development

Table 2. Women participation major Brazilian Program Committees in 2008

			Percentage	
Computer Science Area	Event in 2008	PC	Women	Braz.W
Human-Computer Interaction	10th IHC	57	42.1	42.1
Information Systems	4th SBSI	56	30.4	30.4
Databases	23rd SBBD	61	47.5	27.9
Software Engineering	22nd SBES	74	36.5	24.3
Formal Methods	11th SBMF	34	23.5	17.6
Computer Networks	26th SBRC	94	14.9	14.9
Artificial Intelligence	19th SBIA	149	21.5	13.4
Computer Architecture	20th SBAC	75	16.0	13.3
Computer Graphics	21st SIBGRAPI	77	14.3	10.4
Programming Languages	12th SBLP	45	8.9	8.9
Fault Tolerance	3rd LADC[1]	89	6.7	6.7
Integrated Circuits	21st SBCCI	70	2.9	0.0
	TOTAL	*881*	*21.2*	*16.1*

[1] Observation: LADC is a bi-annual event. The data on this paper were collected from LADC 2003, 2005, and 2007. Also, LADC is an expansion of the Brazilian Symposium on Fault Tolerance (SCTF, whose first edition was in 1987) to include Latin American researchers.

of software systems, which is farther from the "engineering border" of CS. This would be a consequence of the historical small participation of women in electronic engineering career. The position of Formal Methods (5th place) is explained by the fact that, in this area, the math background is even stronger than in other subareas. Historically, in Brazil, Mathematics has attracted more women along the years due to teaching jobs. Women with mathematics degrees eventually migrate to CS, more specifically to Formal Methods, Theory of Computation, and areas alike. A surprising factor is that the so called "popular" areas (such as computer networks and computer graphics) do not attract many women. Networks can be considered closer to hardware or a more technical career, what bring us back to the first conclusion. As for computer graphics, we believe that there are two main reasons for the larger masculine participation: the first reason is due to the fact that along the first 30 years (1965-1995) graphics were mainly applied in supporting mechanical and structural engineering applications, while in recent years boys came to computer courses because they wanted to work in the games industry.

Overall, those areas that require more social and organizing skills (i.e., for interacting with users) and that make CS more user-friendly seem to attract more women. Such an attraction could be motivated by the inherent ability of women to interact and to view technology from the user's side (while men seem to be more concerned with the technical aspects). Also, most women have a natural talent to make things very organized and uncluttered, skills that are highly necessary in the first four areas (it may sound cliché, but it is no coincidence that, in general, women are in charge of organizing the family house in most different cultures).

Historical Results

In order to complement the results from the previous section, we have also evaluated the PCs from 2006 and 2004. The goal is to evaluate how the feminine participation has evolved in those

Table 3. Women participation in major Brazilian Program Committees in 2004, 2006, and 2008

Year	2004		2006		2008	
Event	PC	Wom	PC	Wom	PC	Wom
SBBD	49	20.41	47	21.28	61	47.54
IHC	23	69.57	31	48.39	57	42.11
SBES	49	26.53	63	36.51	74	36.49
SBSI	36	22.22	86	36.14	56	30.36
SBMF	21	23.81	31	22.58	34	23.53
SBIA	62	37.10	147	27.21	149	21.48
SBAC	56	3.57	61	8.19	75	16.00
SBRC	60	13.33	83	14.46	94	14.89
SIBGRAPI	67	11.94	105	15.24	77	14.29
SBLP	42	11.90	49	12.24	45	8.89
LADC	33	15.15	36	13.89	30	6.7
SBCCI	62	8.06	69	4.35	70	2.86
TOTAL	*560*	*13.1*	*805*	*20.9*	*822*	*22.1*

same areas in recent years[2]. Table 3 presents the numerical results, while Figure 3 presents those same results in a bar chart. The results are ordered in decreasing number of women participation in the 2008 events. The analysis of Figure 1 shows that although the Human- Computer Interaction is the area with the highest percentage of Brazilian women in the PC (see Table 1), it is also the area with the largest decrease in such number (27.46%) from 2004 to 2008, while in the Da-

Figure 3. Women participation in Brazilian PC – 2004, 2006, and 2008

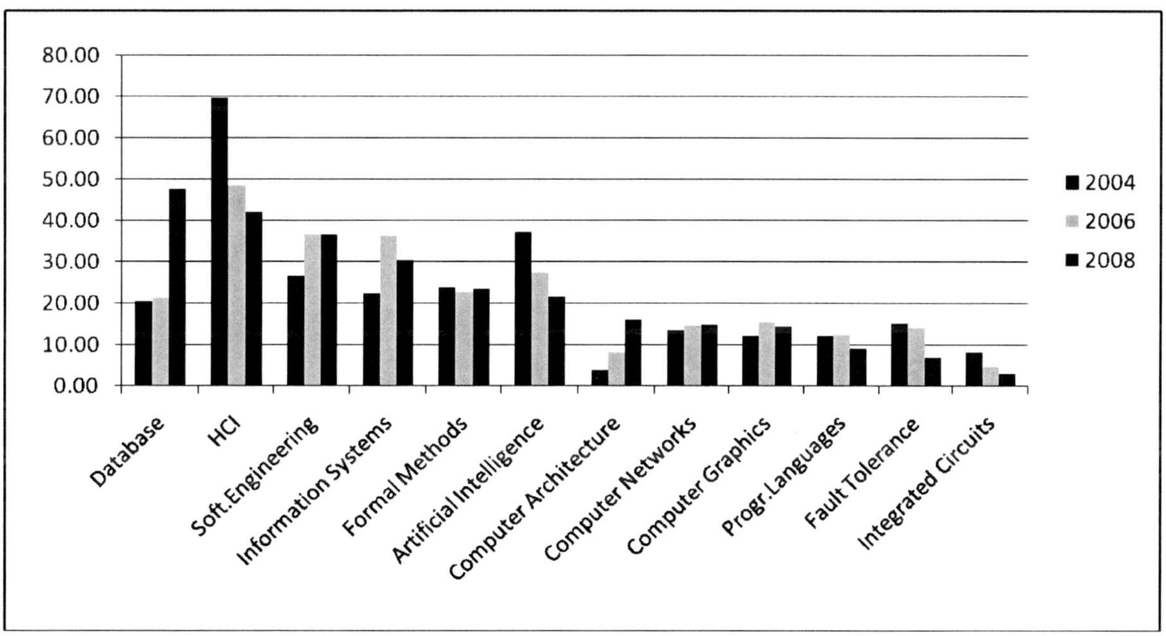

Table 4. Brazilian conferences ordered by feminine participation in their PCs

	2004	2006	2008
More Women	Human Comp. Int.	Human Comp. Int.	Databases
	Artificial Intellig.	Software Eng.	Human Comp. Int.
	Software Eng.	Information Sys.	Software Eng.
	Formal Methods	Artificial Intellig.	Information Sys.
	Information Sys.	Formal Methods	Formal Methods
	Databases	Databases	Artificial Intellig.
	F. Tolerance	Comp. Graphics	Comp. Architecture
	Comp. Networks	Comp. Networks	Comp. Networks
	Comp. Graphics	F. Tolerance	Comp. Graphics
	Program. Lang.	Program. Lang.	Program. Lang.
	Integr. Circuits	Comp. Architecture	F. Tolerance
Less Women	Comp. Architecture	Integr. Circuits	Integr. Circuits

tabase community it was verified an increase of 27.13%. Other areas showing increasing number of women participation are Software Engineering, Computer Architecture, and Computer Networks. Computer Architecture surprisingly shows a steady increase along this time frame. Computer Graphics also showed a small increase, although it can be considered as stable as Formal Methods. On the other hand, Artificial Intelligence, Fault Tolerance, and Integrated Circuits showed a decreasing participation of women in the PCs. Finally, notice that overall, the number of women in CS has actually increased, from 13.1% total in 2004 to 22.1% total in 2008.

An area-migration process could explain the decreasing women participation in Human-Computer Interface, Artificial Intelligence and Fault Tolerance favoring Software Engineering, Computer Architecture and Databases. For example, it is well known that many women working in Artificial Intelligence in Brazil became also active in the education field, mainly in distance learning, due to the increasing resources and governmental programs related to this subject.

Since a 5-year time interval can be considered quite small, for some areas, mainly the stable ones, we cannot establish a trend right now. However, as can be observed in Table 3, the relative position of CS sub-areas regarding women participation in PCs oscillates among those that are more related to software (Human Computer Interaction, Software Engineering, Databases, Information Systems, Formal Methods and Artificial Intelligence) in the first half of table, while in the second half, except of Programming Languages, we have the CS sub-areas closer to the engineering careers, either as the required background or application area.

Complementary Numbers

This section presents some complementary numbers from other sources. Specifically, it describes the feminine participation in: graduate thesis awards, graduate programs faculty advisors, and international program committees.

Graduate Thesis Award

Besides the program committees, we have also analyzed the number of women that have won the most prestigious Brazilian student research award – the *Brazilian Computer Society Contest of MSc*

Table 5. Women participation as faculty in CS graduate programs (June 2009)

Brazil. Region	Number Programs	Programs		Type		Faculty	
		Master	M+PhD	Public	Private	Total	% Women
Center-West	1	1	0	1	0	20	30
Northeast	5	2	3	4	1	123	21
North	1	1	0	1	0	18	6
Southeast	9	2	7	8	1	293	27
South	5	3	2	2	3	131	22
TOTAL	**21**	**9**	**12**	**16**	**5**	**585**	**24**

and PhD Thesis (SBC CTD). We have evaluated the first, second, and third prizes as well as all honorable mentions. The results are as follows.

- In the MSc category, there are 19 women out of 69 students awarded (27%), and
- In the PhD category, there are 10 women out of 33 students awarded (30%).

We believe that those are very promising numbers. They tell us that girls have an active and high quality participation in our CS graduate programs. We certainly hope that those girls will join the different PCs in the Brazilian conferences, helping to keep and making the statistics even better.

Graduate Programs Faculty Advisors

In Brazil, most research in Computer Science is developed within university graduate programs. Recently, some companies, such as IBM and Google, have started developing research in their Brazilian branches as well. In this section, we have considered the top 21 graduate programs in Computer Science (according to CAPES rank). From those, 9 programs have only master degree, and 12 have both master and PhD degrees; 16 programs are within public universities and 5 in private universities. Table 5 presents the results of women participation in those programs separated by Brazilian regions (Brazil is divided in 5 different geographic regions). It presents the name of the region, the number of programs in that region, the master and PhD courses, the public and private universities, the total of professors and the percentage of women. In average, 24% of the professors in the grad programs are women; the north region has less women (6%) while the center-west has more women (30%).

International Program Committees

We have also computed some numbers from the international community. Specifically, we have searched for previous editions of SIGMOD (ACM Intl. Conf. on Management of Data, representing the Databases area), ICCAD (IEEE/ACM Intl. Conf. on Computer-Aided Design, representing the Integrated Circuits area), and SIGCHI (ACM Intl. Conf. on Human Factors in Computing Systems, representing the Human-Computer Interaction area). Table 6 presents the results from feminine participation in varied years. We were not able to recover data from all years. We can see that the international numbers are different from the Brazilian ones. The database conference has between 8 and 17% women participation, against the 20 to 47% in the Brazilian counterpart. It is very important to notice that while the Brazilian conference is the only one in a database-related are within Brazil, SIGMOD is just one international conference on that area. For a more detailed evaluation, we would probably have to consider an average of more conferences, such as VLDB

Table 6. Women participation in selected international conferences

SIGMOD – First edition in 1975											
2009		2008		2006		2004		2000		1998	
PC	WOM	PC	WOM	PC	WOM	PC	WOM	PC	WOM	PC	WOM
100	11.00	98	14.29	86	8.14	85	17.65	37	8.11	41	14.63
ICCAD – First edition in 1975											
2008		2006		2004		1998		1994			
PC	WOM	PC	WOM	PC	WOM	PC	WOM	PC	WOM		
90	12.22	96	11.46	81	6.17	58	8.62	54	1.85		
SIGCHI – First edition in 1982											
2008		2006		2005							
PC	WOM	PC	WOM	PC	WOM						
46	36.96	65	27.69	54	38.89						

(Intl. Conf. on Very Large Data Bases) and ICDE (Intl. Conf. on Data Engineering).

For the more engineering-related area, the feminine participation in ICCAD has grown from almost 2% in 1994, to 6% in 2004, and 12% in 2008. Such an increase contradicts the Brazilian numbers, which show a decrease from 8% in 2004 to almost 3% in 2008. Finally, for the human-computer interaction area, the feminine participation varies from 27 to 39%, showing a reality more similar to the Brazilian numbers.

Ideally, we would like to recover data from more conferences in order to have better statistics. However, there is a major problem between comparing Brazilian numbers to international ones. In Brazil, there is usually one big conference that is specific for one big area. In the international community, there are usually more than one conference related to the same big area. For example, consider Databases. While in Brazil there is only SBBD, in the international community there are SIGMOD, VLDB, and ICDE plus more specific conferences such as SIGKDD, SIGIR, EDBT, ICDT, PODS, SSDBM, among others. All of those deal with different aspects related to Databases. In Brazil, all such aspects are discussed in one single event (SBBD).

TARGETING THE PROBLEM

Now, we describe two of the Brazilian initiatives for bringing more girls to CS and related areas. The first one is the Women in Information Technology workshop, held with the Congress of the Brazilian Computer Society, SBC. The second one is Bits, Bytes & Batom[3], which is a column of the SBC's career-focused magazine SBC Horizontes[4]. Then, we discuss some ideas on new initiatives that can be applied to the Brazilian community.

WIT: Women in Information Technology Workshop

One of the first initiatives that brought attention to the lack of women in CS was the creation of WIT – the workshop for Women in Information Technology (WIT). The workshop is sponsored by SBC and focuses on different aspects of women in IT. Its first edition was held in 2007, during the Congress of the Brazilian Computer Society. Historically, the congress has attracted over 2000 participants, who have access to all its events. Even though not all of those participants may attend WIT (since there are other events in parallel), simply having the workshop in the program helps to promote similar initiatives and raise overall awareness on that issue. The workshop usually has keynote speakers talking about relevant topics followed by an open panel for discussions.

One interesting observation about the workshop is that male professors also participate in the seminars. The feminine audience has warmly referred to such men as "honorary women". This heterogeneous participation is very important to everybody. It shows that bringing more women to CS is not just "women business" because it affects every single person in the area. During the first WIT, in 2007, the participants discussed the image of people who study in computer science and engineer related majors. We verified that TV shows and movies usually represent such a student as a *nerd*, who wears glasses, has no social skills, and is always single. The young generations grew up with easy access to TV and movies (which was not very common before the 70s) and the negative image of "the nerd" is fixed in their brains. Hence, when they choose a major, most of the students associate the nerd to the science related options, and simply discarded them.

One solution for such problem would be to try to develop a new image of the science students. For example, in Brazil, soap operas are very popular (as much as soccer). One idea suggested during the workshop was to have a very handsome, cool actor playing a CS major student. Well, the idea is simple but, in reality, it requires a big support from other entities (TV channels and their sponsors) that go beyond the current society reach.

During the second WIT, in 2008, the panel discussed the participation of women in computer undergrad courses in all Brazilian regions. Different from what was expected, the numbers are approximately equal in industrialized south/southeast and in more conservative, rural north/northeast. Computer engineering counts with approximately 4 to 5% of female students, while computer science counts with the double, from 8 to 9%, in the whole country. This percentage of women participation increases significantly to almost 20% when graduate courses (masters and PhDs) are considered. The main conclusion from such numbers is that geography does not play any important role in the women participation, neither does regional culture.

The participants also discussed the motivation of girls and boys to choose computer related majors. Most of the young participants informed that they were not aware of all the possibilities that a computer careers offers to women. The generalized view was of a career close to machines and with little human contact opportunities. Another conclusion of the workshop was that we need to strongly publicize computer career to young women and girls before they make their career choices.

Bits, Bytes & Batom

Bits, Bytes & Batom is another initiative, a column in the new electronic magazine of the Brazilian Computer Society, called *SBC Horizontes*. The magazine is formed by different sections that target important aspects of the career in Computer Science (career options, graduate studies, industry and research challenges, ethics, and so on). The editorial board is coordinated by professors Mirella M. Moro (UFMG) and Marta Mattoso (from Federal University of Rio de Janeiro, COPPE/UFRJ), and formed by a number of distinguished members of the Brazilian research and industry communities.

All aspects related to women in CS are discussed in the specific column *Bits, Bytes & Batom*. The editors of this column are professors Claudia Bauzer Medeiros (from the State University of

Campinas, UNICAMP, São Paulo) and Sandra Fabbri (from the Federal University of São Carlos, UFSCAR, São Paulo). The magazine is publicly available online, and any student and professional (member of the Society or not) may access its articles. Since its release (in December 2008), the magazine has had over 7500 accesses.

In its first edition, *Bits, Bytes & Batom* presented a photo shoot that happened during SBBD, the Brazilian Symposium on Databases. SBBD brings together the Brazilian community each year in a different city. The Symposium is organized by the Brazilian Computer Society (SBC) and aggregates a set of events, such as the Workshop of Thesis and Dissertations in DB, Demos and Poster Sessions. In 2008, the conference was hosted by UNICAMP, in Campinas, São Paulo. Each year the event has something new. In 2008, the novelty was the realization that practically **50%** of the Brazilians within the SBBD's program committee are women. In order to celebrate such achievement, the women decided to take a picture together, and the pictures were published in the ACM Women Newsletter, Spring 2009[5], as well as in the first edition of SBC Magazine Horizontes[6]. Just this photo shoot in the last day of the conference helped to bring attention to the "increasing women in Computer Science" movement. Most of the participants did not even know this was an issue.

Overtime, we hope that this column will discuss not only issues related to being a woman in CS, but also how to bring more women to technological areas. Furthermore, the column may also be a venue for publicizing similar initiatives.

Ideas for Brazilian CS Community

Besides the current initiatives, we also plan to join the Brazilian Computer Society and organize other initiatives for bringing and keeping women in CS.

Idea 1. To promote CS in high schools. Following the idea of women do not know they are very valuable within ICT area (Ticoll 2007), the idea here is to give talks about CS in high schools, preferably a talk given by a successful role model. The objective is twofold: to advertise computer science (and engineering related areas) to future college students and to bring more girls to the field as well. In these talks, the social relevance of computer science and engineering as well as the various possible careers for women could be explored. SBC can support the elaboration and distribution of the presentation material, but we face the scale problem. The number of female professors is not enough to cover the huge number of schools in Brazil. Also, there is a great amount of schools in places with difficult access. Such places should be reached by a recorded talk which could be sent to the school by mail or via Web.

Idea 2. To promote CS in Math Olympics. SBC organizes the Computing Olympics for children and teenagers. Likewise, Brazil has also the Math Olympics with an impressive participation of girls (which could be justified by the fact that most Math teachers are women). Since last year, SBC offers preparatory lessons in CS as well as the registration in the Computing Olympics for the participants of the Math Olympics that have reached a minimum grade. This action has already brought some results by attracting more girls to the CS Olympics. Currently, we are planning to present to these girls the different computer areas and the career opportunities for women. We hope that those girls distribute this knowledge between their female friends and be motivated to follow a computer career.

Idea 3. To organize women meetings in Brazilian events. Brazilian CS events usually start early morning in a weekday. Also, it is common practice to arrive at the event's city a day prior to the event begin. Most of the times, people arrive in time to have dinner with other participants (who are usually accommodated in a set of hotels). The idea here is to make the dinner prior to the conference start as an official event for women to get together. For example, considering the database community, in 2009 there will be the 1st WomB:

Women in database workshop. In this event, we plan to have a panel to discuss which features of the databases area are more appealing to women, followed by dinner. We hope that WomB will serve as an example to the other Brazilian CS communities to help women in those other areas to deploy similar initiatives. Also, we would like to extend the discussion on why some of the areas are more appealing to women in such meetings.

Idea 4. To organize women meetings in Latin American events. Finally, CS Latin American community also gets together every year in the CLEI congress (Latin American Centre for Informatics Studies). An idea is to promote an event similar to WIT for a bigger, more diverse audience. We have noticed that other countries in Latin America also suffer from the same problem of not having many women students in CS majors. By bringing such discussion to CLEI, we hope to extend our discussions to other countries as well as to make even more people aware of such problems.

CONCLUDING REMARKS

This chapter has presented an overview of the feminine participation in the Brazilian CS research community. We have evaluated the program committees of the Brazilian conferences in CS related areas and computed the percentage of women participation. Our study evaluates the 2008 and some previous editions of such conferences. The main conclusions are as follows:

- Regarding 2008 numbers:
    - Human-Computer Interaction and Databases are the "most feminine" subareas, while Fault Tolerance and Integrated Circuits as the "less feminine" ones.
    - There is no relation between the size of the PC and the number of women.
    - Overall, the rate for women participation is 21% in average, and Brazilian women 16%.
    - Overall, considering the first four subareas, all of them (Human-Computer Interaction, Information Systems, Databases, and Software Engineering) are related to the development of software systems, which is farther from the "engineering border" (hardware based) of CS.
- Regarding historical numbers from 2004, 2006, and 2008:
    - Human-Computer Interaction is the area with the highest percentage of Brazilian women but also the one with largest decrease.
    - Areas with increasing number of women include: Databases, Software Engineering, Computer Architecture, and Computer Networks.
    - Areas with decreasing number of women include: Human-Computer Interaction, Artificial Intelligence, Fault Tolerance, and Integrated Circuits.
    - Overall, the number of women in CS has actually increased, from 13.1% (2004) to 22.1% (2008).
    - Overall, the areas with more women are those more related to software (Human-Computer Interaction, Software Engineering, Databases, Information Systems, Formal Methods, and Artificial Intelligence) while the ones with less women are those more related to engineering (except Programming Languages).

Furthermore, we have also studied some international conferences in order to compare their numbers to the Brazilian ones. The main conclusion is that for a fair comparison, we would have to consider more than one event per area. Finally, we have also discussed some Brazilian initiatives to bring more girls to Computer Science as well presented what else can be done.

ACKNOWLEDGMENT

This work was partially funded by CNPq, Brazil.

REFERENCES

Brainard, S. G., & Carlin, L. (2001). A six-year longitudinal study of undergraduate women in engineering and science. In Lederman, M., & Bartsch, I. (Eds.), *The Gender and Science Reader*. New York: Routledge.

Ceci, S. J., & Williams, W. M. (2006). Why Aren't More Women in Science? Top Researchers Debate the Evidence. Chicago: American Psychological Association (APA).

CRA. (2007). *CRA Taulbee Survey*. Retrieved from http://www.cra.org/statistics/

Dewandre, N. (2002) Women in Science: European Strategies for Promoting Women in Science. *Science, 11*, 295 (5553), 278-279.

Frieze, C., Hazzan, O., Blum, L., & Dias, M. B. (2006). Culture and environment as determinants of women's participation in computing: revealing the "women-CS fit". In *Procs. of ACM SIGCSE – Technical Symposium on Computer Science Education.*

Harris, K., Morello, D., & Raskino, M. (2007). *Women and men in IT: Breaking sexual stereotypes.* Gartner. Retrieved from http://www.gartner.com/DisplayDocument?doc_cd=146320

Hyde, J. S., Fennema, E., & Lamon, S. J. (1990). Gender differences in mathematics performance: A meta-analysis. *Psychological Bulletin, 107*(2), 139–155. doi:10.1037/0033-2909.107.2.139

Ivie, R., & Ray, K. N. (2005). Women in Physics and Astronomy. *AIP Publication Number R-430.02*. American Institute of Physics. Retrieved from http://www.aip.org/statistics/trends/gendertrends.html

Mannix, E. A., & Neale, M. A. (2005). What difference makes a difference? *Psychological Science in the Public Interest, 6*(2), 31–55. doi:10.1111/j.1529-1006.2005.00022.x

Patterson, D. A. (2005). Restoring the Popularity of Computer Science. *Communications of the ACM, 48*(9), 25–28. doi:10.1145/1081992.1082011

Ramsey, N., & McCorduck, P. (2005). Where are the Women in Information Technology? *Anita Borg Institute*. Retrieved from http://anitaborg.org/news/research

Simard, C. (2007). The Business Case for Gender Diversity. *Anita Borg Institute*. Retrieved from http://anitaborg.org/news/research

Simard, C. (2009). Obstacles and Solutions for Underrepresented Minorities in Technology. *Anita Borg Institute*. Retrieved from http://anitaborg.org/news/research

Simard, C., Henderson, A. D., Gilmartin, S. K., Schiebinger, L., & Whitney, T. (2008). Climbing the Technical Ladder: Obstacles and Solutions for Mid-Level Women in Technology. *Anita Borg Institute*. Retrieved from http://anitaborg.org/news/research

Ticoll, D. (2008). IT enrollments and retentions: situation report. *Labor Market Intelligence, ICTC's Report*. Information and Communicattions Technology Council. Retrieved from http://www.ictc-ctic.ca/en/content.aspx?id=1922

Trauth, E. M., Quesenberry, J. L., & Morgan, A. L. (2004). Understanding the under representation of women in IT: toward a theory of individual differences. In *Procs. of ACM SIGMIS – Conference on Computer Personal Research.*

Vegso, J. (2005). Interest in CS as a Major Drops Among Incoming Freshmen. *Computing Research News*, *17*(3). Retrieved from http://www.cra.org/CRN/articles/may05/vegso.

Welde, K. D., Laursen, S., & Thiry, H. (2007). Women in Science, Technology, Engineering and Math (STEM). *Fact Sheets*. Sociologists for Women in Society. Retrieved from http://www.socwomen.org

ENDNOTES

1. http://wikipedia.org
2. Note that we leave more complete historical evaluation, considering all editions of the events, for future work.
3. *Batom* is the Brazilian word for lipstick.
4. *Horizontes* is the Brazilian word for horizons.
5. http://women.acm.org/newsletter_v01.04.pdf
6. http://www.sbc.org.br/horizontes/edicoes/v01n01/22.html

Compilation of References

Aaltio-Marjosola, I. (1994). Gender Stereotypes as Cultural Products of the Organization. Scandinavian Journal ofManagement, 10(2), 147–162. doi:10.1016/0956-5221(94)90017-5

Aaltio-Marjosola, I., & Lehtinen, J. (1998). Male Managers as Fathers? Contrasting Management, Fatherhood and Masculinity. Human Relations, 51(2), 121–136. doi:10.1177/001872679805100201

ACED. (2008). Addressing the supply and quality of engineering graduates for the new century: Report on ALTC funded project. Retrieved 12 May, 2009, from http://www.altc.edu.au/system/files/resources/Grants_DBIprojec_engineeringquality_project%20report_25march08.pdf

Acker, J. (1990). Hierarchies, Jobs, Bodies: a theory of gendered organization. Gender & Society, 4(2), 139–158. doi:10.1177/089124390004002002

Acker, J. (1992). From sex roles to gendered institutions. Contemporary Sociology, 21(5), 565–569. doi:10.2307/2075528

Acker, S., & Armenti, C. (2004). Sleepless in academia. Gender and Education, 16(1), 3–24. doi:10.1080/0954025032000170309

Acker, S., & Feuerverger, G. (1996). Doing good and feeling bad: The work of women university teachers. Cambridge Journal of Education, 26(3), 401–422. doi:10.1080/0305764960260309

Acker, S., & Oatley, K. (1993). Gender issues in education for science and technology: current situation and prospects for change. Canadian Journal of Education, 18(3), 255–272. doi:10.2307/1495386

Ackers, H. L. (2003). The Participation of Women Researchers in the TMR Marie Curie Fellowships. Luxembourg: European Communities.

Ackers, H. L., & Oliver, E. (2005). Gender, Mobility and Progression in Science Careers: MOBISC Summary Report. Leeds, UK: University of Leeds.

Adam, A., & Green, E. (2001). Gender, agency, location and the new information society in Alison Adam & Eileen Green. In Virtual Gender. London: Routledge.

Advanced Engineering Capability Network. (n.d.). Retrieved August 26, 2009 from http://www.engineeringcapability.net/

Adya, M., & Kaiser, K. M. (2005). Early Determinants of Women in the IT Workforce: a Model of Girls' Career Choices. Information Technology & People, 18(3), 230–259. doi:10.1108/09593840510615860

Ahuja, M. K. (2002). Women in the Information Technology Profession: A Literature Review, Synthesis and Research Agenda. European Journal of Information Systems, 11(1), 20–34. doi:10.1057/palgrave/ejis/3000417

Ajzen, I. (1991). The Theory of Planned Behavior. Organizational Behavior and Human Decision Processes, 50, 179–211. doi:10.1016/0749-5978(91)90020-T

Copyright © 2010, IGI Global, distributing in print or electronic forms without written permission of IGI Global is prohibited.

Alemany, C., & Diaz, C. (2007). The last bastions: The situation of Spanish women in Spanish universities. A study of two cases. Paper presented at the Gender, Work and Organization 5th Biennial International Interdisciplinary Conference, 27-29 June 2007, Keele, UK.

Alger, J. (2003). Summary of Supreme Court decisions in admissions cases, U.S. Supreme Court – June 23, 2003. Downloaded October 2, 2007 from http://www.vpcomm.umich.edu/admissions/overview/cases-summary.html

Ali, H. (2005). Permanent mission of Malaysia to the United Nations. Revised Version. Available at http://www.un.int/malaysia/GA/60thGA/6030.PDF

Alper, J. (1993). The pipeline is leaking women all the way along. Science New Series, 260(5106), 409–411.

American Association of University Professors. (2006). AAUP faculty gender equity indicators 2006. Downloaded October 2, 2007 from http://www.aaup.org/AAUP/pubsres/research/geneq2006

American Educational Research Association. Association of American Colleges and Universities, & American Association for Higher Education. (2003). Brief of amici curiae in support of the respondents, No. 02-516, Gratz et al. v Bollinger et al. Downloaded October 2, 2007 from http://www.vpcomm.umich.edu/admissions/legal/gra_amicus-ussc/um/AERA-gra.pdf

Ancis, J. R., & Phillips, S. D. (1996). Academic gender bias and women's behavioral agency self-efficacy. Journal of Counseling and Development, 75(2), 131–137.

Anita Borg Institute. (2007). History of website. Retrieved February 12, 2009, from Anita Borg Institute: http://anitaborg.org/about/history/

Anlezark, A., Lim, P., Semo, R., & Nguyen, N. (2008). From STEM to leaf: Where are Australia's science, mathematics, engineering and technology (STEM) students heading? Retrieved 12 May, 2009, from http://www.dest.gov.au/NR/rdonlyres/53AD328C-5620-4935-980A-FB8309F78F4F/24712/STEMreportOP03890100908.pdf

Armstrong, D. J., Riemenschneider, C. K., Allen, M. W., & Reid, M. R. (2007). Advancement, voluntary turnover and women in IT: A cognitive study of work-family conflict. Information & Management, 44, 142–153. doi:10.1016/j.im.2006.11.005

Assendelft, L. V., Gunther-Canada, W., Dolan, J., Palmer, B., & Swers, M. (2003). Political Science in a Different Voice: Women Faculty Perspectives on the Status of Women in Political Science Departments in the South. PS: Political Science and Politics, 36(2), 311–315.

Association of American Colleges and Universities and University of Maryland. (2007). Diversity web. http://www.diversityweb.org/

Association of Professional Engineers. Scientists and Managers Australia. (2007). Women in the Professions Survey Report. Retrieved 11 February 2009, from http://www.apesma.asn.au/women/survey/women_in_the_professions_2007.pdf.

Association of University Teachers. (2004). The Unequal Academy: UK academic staff 1995–96 to 2002–03. London: Association of University Teachers. Retrieved from http://www.aut.org.uk/index.cfm?articleid=916

Atkinson, P., & Delamont, S. (1990). Professions and powerlessness: Female marginality in the learned occupations. The Sociological Review, 38(1), 90–110.

Australian Bureau of Statistics. (2006). Calculations based on data in: Average weekly Ordinary time earnings (AWOTE) - Full time adults, 6302, Australian Bureau of Statistics, 1984-2006. Retrieved 30 January 2009 from http://www.abs.gov.au/AUSSTATS/abs@.nsf/DetailsPage/6302.0Nov%202006?OpenDocument

Australian Bureau of Statistics. (2008). Average Weekly Earnings Australia, August 2008, 'Table 13C Average Weekly Earnings, Queensland (Dollars) - Original', data cube: Excel spreadsheet, Cat. no. 6302.0. Retrieved 30 January 2009 from http://www.abs.gov.au/AUSSTATS/abs@.nsf/DetailsPage/6302.0Aug%202008?OpenDocument

Australian Bureau of Statistics. (2009). Employed Persons by Sex, Industry, Occupation, State, 6291.0.55.003 E09_aug86, 1986-1996 and 1996-2007. Retrieved 2 March 2009, from http://www.abs.gov.au/AUSSTATS/abs@.nsf/DetailsPage/6291.0.55.003Feb%202009?OpenDocument

Australian Computer Society – Women. (n.d.). Retrieved 26 August 2009 from http://www.acs.org.au/acswomen/

Bagilhole, B. (1993). Survivor in a male preserve: a study of British women academics' experiences and perceptions of discrimination in a UK University. Higher Education, 26, 431–447. doi:10.1007/BF01383737

Bagilhole, B. (2002). Challenging Equal Opportunities: Changing and Adapting Male Hegemony in Academia. British Journal of Sociology of Education, 23(1), 19–33. doi:10.1080/01425690120102836

Bagilhole, B., & Goode, J. (2001). The Contradiction of the Myth of Individual Merit, and the Reality of a Patriarchal Support System in Academic Careers, A Feminist Investigation. European Journal of Women's Studies, 8(2), 161–180. doi:10.1177/135050680100800203

Bagozzi, R. P. (2007). The legacy of the technology acceptance model and a proposal for a paradigm shift. Journal of the Association for Information Systems, 8(4), 244–254.

Baker, D. P. (1988). School charter and parental management in West Germany. Sociology of Education, 255–265.

Balasingam, M. (n.a.). Female involvement in information technology in developing countries. Retrieved from http://trincoshanmuga.sch.lk/article/Female_Involvement.doc/

Balka, E. (2005). Women's access to on-line discussion about feminisme. Retrieved from http://feminisme.eserver.org/gender/cyberspace/feminist-use-of-cyberspace.txt

Bandura, A. (1977). Self-efficacy: toward a unifying theory of behavioral change. Psychological Review, 84(2), 191–215. doi:10.1037/0033-295X.84.2.191

Bandura, A. (1986). Social Foundations Of Thought and Action: A Social Cognitive Theory. Englewood Cliffs, NJ: Prentice- Hall.

Bandura, A. (1994). Self Efficacy. In V.S. Ramachaudran (Ed.), Encyclopedia of Human Behavior, (Vol. 4, pp.71-81). New York: Academic Press.

Bandura, A. (1995). Self-Efficacy in Changing Societies. New York: Cambridge University Press.

Bandura, A. (1997). Self-efficacy: the exercise of control. New York: W.H. Freeman.

Barber, L. A. (1995). U.S. Women in Science and Engineering, 1960-1990: Progress Toward Equity? The Journal of Higher Education, 66(21), 213–234. doi:10.2307/2943912

Barker, L., & Aspray, W. (2006). The state of research on girls and IT. In J. M. Aspray, Women and information technology: Research on underrepresentation. Cambridge: MIT Press.

Baz, J., & Oxford, P. (2008). The University of Sydney's Women in Science (WiSci) project, WISENET Journal, 79. Retrieved 25 April, 2009, from http://www.wisenet-australia.org/issue79/Women%20in%20Science-WiSci.htm

Beasley, M., Thompson, T., & Davidson, J. (2003). Resilience in response to life stress: the effects of coping style and cognitive hardness. Personality and Individual Differences, 34, 77–95. doi:10.1016/S0191-8869(02)00027-2

Becher, A. (1990). Academic Tribes and Territories, Milton Keynes, UK: Open University Press.

Becker, H. J. (1982). Parent involvement: A survey of teacher-practices. The Elementary School Journal, 83, 85–102. doi:10.1086/461297

Berg, H. M., & Ferber, M. A. (1983). Men and Women Graduate Students: Who Succeeds and Why? The Journal of Higher Education, 54(6), 629–648. doi:10.2307/1981934

Beteta, H. C. (2007). What is missing in measures of women's empowerment? Journal of Human Development, 7(2), 221–241. doi:10.1080/14649880600768553

Betz, N. E., & Hackett, G. (1981). The relationship of career-related self-efficacy expectations to perceived career options in college women and men. Journal of Counseling Psychology, 28(5), 399–410. doi:10.1037/0022-0167.28.5.399

Betz, N. E., & Hackett, G. (1997). Applications of self-efficacy theory to the career assessment of women. Journal of Career Assessment.Special Issue: Career assessment for women. Theory into Practice, 5(4), 383–402.

Betz, N. E., & Schifano, R. S. (2000). Evaluation of an intervention to increase realistic self-efficacy and itnerests in college women. Journal of Vocational Behavior, 56, 35–52. doi:10.1006/jvbe.1999.1690

Beutel, A. M., & Nelson, D. J. (2005). The gender and race-ethnicity of faculty in top science and engineering research departments. Journal of Women and Minorities in Science and Engineering, 11(4). doi:10.1615/JWomenMinorScienEng.v11.i4.50

Birbaumer, A., Lebano, A., Ponzellini, A., Tolar, M., & Wagner, I. (2007). From the margins to a field of opportunities: Life story patterns of women in ICT. Women's Studies International Forum, 30, 486–498. doi:10.1016/j.wsif.2007.09.001

Bird, G. W., & Schnurman-Crook, A. (2005). Professional identity and coping behaviors in dual-career couples. Family Relations, 54(1), 145–160. doi:10.1111/j.0197-6664.2005.00012.x

Black Data Processing Association. (2009). About BDPA. Retrieved March 3, 2009, from Black Data Processing Association: http://www.bdpa.org/portal/index.php?option=com_content&task=view&id=16&Itemid=43

Blackwell, L., & Glover, J. (2008). Women's Scientific Employment and Family Formation: a longitudinal perspective. Gender, Work and Organization, 15(6), 579–599. doi:10.1111/j.1468-0432.2007.00385.x

Blickenstaff, J. C. (2005). Women and science careers: leaky pipeline or gender filter? Gender and Education, 17(4), 369–386. doi:10.1080/09540250500145072

Blossfeld, H.-P., & Drobnic, S. (2001). Careers of couples in contemporary societies. Oxford, UK: Oxford University Press.

Blumenthal, K. (2005, July 6). Title IX's next hurdle. The Wall Street Journal, p. B1.

Bogg, J. (2007). Dr Jekyll and Ms. Hide: Where are the women in science? And what would attract them from other sectors? Nature, 447.

Brainard, S. G., & Carlin, L. (1998). A six-year longitudinal study of undergraduate women in engineering and science. Journal of Engineering Education, 87(4), 17–27.

Brainard, S. G., & Carlin, L. (2001). A six-year longitudinal study of undergraduate women in engineering and science. In Lederman, M., & Bartsch, I. (Eds.), The Gender and Science Reader. New York: Routledge.

Brainard, S., & Carlin, L. (1997). A longitudinal study of undergraduate women in engineering and science. In Proceedings of 1997 ASEE/IEEE Frontiers in Education Conference.

Britton, D. (2000). The epistemology of the gendered organization. Gender & Society, 14(3), 418–434. doi:10.1177/089124300014003004

Brooks, A. (1997). Academic women. Buckingham, UK: Open University Press.

Brunner, C., Bennett, D., & Honey, M. (1998). Girl Games and Technological Desire. In J. Cassell & H. Jenkins (Eds.), From Barbie to Mortal Kombat: Gender and Computer Games, (pp. 72-88). Cambridge, MA: MIT Press.

Bryce, J., & Rutter, J. (2003). The gendering of computer gaming: experience and space. In S. Fleming, & I. Jones (Eds.), Leisure cultures: investigations in sport, media and technology, (pp. 3-22). Eastbourne, UK: Leisure Studies Association.

Buck, G. A., Clark, V. L. P., Leslie-Pelecky, D., Lu, Y., & Cerda-Lizarraga, P. (2008). Examining the cognitive processes used by adolescent girls and women scientists in identifying science role models: A feminist approach. Science Education, 92(4), 688–707. doi:10.1002/sce.20257

Building Engineering and Science Talent. (2002). The quiet crisis: Falling short in producing American scientific and technical talent. Downloaded January 9, 2008 from http://www.bestworkforce.org/PDFdocs/Quiet_Crisis.pdf

Building Engineering and Science Talent. (2004). A bridge for all: Higher education design principles to broaden participation in science, technology, engineering and mathematics. Downloaded January 9, 2008 from http://www.bestworkforce.org/PDFdocs/BEST_BridgeforAll_HighEdFINAL.pdf

Building Engineering and Science Talent. (2004). The talent imperative: Diversifying America's science and engineering workforce. Downloaded January 9, 2008 from http://www.bestworkforce.org/PDFdocs/BEST-TalentImperative_FullReport.pdf

Bujaki, M. L., & McKeen, C. A. (1998). Hours spent on household tasks by business school graduates. Women in Management Review, 13(3), 105–113. doi:10.1108/09649429810215848

Bureau of Labor Statistics. (2007a). Changes in men's and women's labor force participation rates. Retrieved on March 15, 2009 from http://www.bls.gov/opub/ted/2007/jan/wk2/art03.htm

Bureau of Labor Statistics. (2007b). Employed Persons by Detailed Occupation, Sex, Race, and Hispanic or Latino Ethnicity. Retrieved on January 25, 2009 from http://www.bls.gov/cps/cpsaat11.pdf

Bureau of Labor Statistics. (n.d.). Occupational Outlook Handbook (OOH), 2008-09 Edition. Retrieved on March 15, 2009 from www.bls.gov/OCO

Burns, B., Griffiths, M., Moore, K., & Richardson, H. (2007). Disappearing women: North West ICT final report. Manchester, UK: Salford University.

Cancro, R., & Slotnik, D. (1970). Computer graphics and resistance to technology. American Journal of Psychotherapy, 24, 461–469.

Carlone, H. B., & Johnson, A. (2007). Understanding the science experiences of successful women of color: Science identity as an analytic lens. Journal of Research in Science Teaching, 44(8), 1187–1218. doi:10.1002/tea.20237

Catalyst, Inc. (2000a). Cracking the glass ceiling: Catalyst's research on women in corporate management #16133. Downloaded September 10, 2007 from http://www.catalyst.org/xcart/product.php?productid=16133

Catalyst, Inc. (2000b). Women and the MBA: Gateway to opportunity. New York.

Catalyst, Inc. (2007). Retrieved from http://www.catalyst.org/

Ceci, S. J., & Williams, W. M. (2006). Why Aren't More Women in Science? Top Researchers Debate the Evidence. Chicago: American Psychological Association (APA).

Ceci, S. J., Williams, W. M., & Barnett, S. M. (2009). Women's Underrepresentation in Science: Sociocultural and Biological Considerations. Psychological Bulletin, 135(2), 218–261. doi:10.1037/a0014412

Charles, M., & Grusky, D. (2004). Occupational Ghettos: the Worldwide Segregation of Women and Men. Stanford, CA: Stanford University Press.

Charmaz, K. (2001). Grounded theory and qualitative analysis. In Gubrium, J., & Holstein, J. (Eds.), Handbook of interview research: Context and method. Thousand Oaks, CA: Sage.

Chen, T. T., & Farr, J. L. (2007). An empirical test of the glass ceiling effect for Asian Americans in science and engineering. In R.J. Burke & M.C. Mattis (Eds.), Women and minorities in science, technology, engineering and mathematics: Upping the numbers (pp. 128-156). Cheltenham, UK: Edward Elgar Publishing Lmtd.

Cialdini, R. B. (1990). Indirect Tactics of Image Management: Beyond Basking. In Giacalone, R. A., & Rosenfeld, P. (Eds.), Impression Management in the Organisation (pp. 45–56). Hillsdale, NJ: Lawrence Erlbaum.

Clark, L. (2003). SET for Success in the Smart State: Increasing the participation of women in Science, Engineering and Technology in Queensland. Retrieved 12 May, 2009, from http://www.learningplace.com.au/uploads/documents/store/resources/res_11094_setforsuccess.doc

Coates, J. (1995). Language, gender and career. In Mills, S. (Ed.), Language and gender: Interdisciplinary perspectives. London: Longman.

Cockburn, C. (1983). Brothers: Male Dominance and Technological Change. London: Pluto Press.

Cockburn, C. (1991). In The Way of Women: Men's Resistance to Sexual Equality in Organisations. London: Macmillan.

Collins, R., & Palmer, A. (2007). Perceptions of Rewarding Excellence in Teaching: Expectancy Theory Makes a Comeback? Presented at the International policies and practices for academic enquiry International Colloquium, Southampton Solent University.

Commission on Professionals in Science and Technology. (2007). Workforce data. Retrieved from http://www.cpst.org/

Commission on the Advancement of Women and Minorities in in Science. Engineering, and Technology Development. (2000). Land of plenty: Diversity as America's competitive edge in science, engineering and technology. Downloaded January 9, 2008 from http://www.nsf.gov/publications/pub_summ.jsp?ods_key=cawmset0409

Committee on Maximizing the Potential of Women in Academic Science and Engineering. (2007). Beyond bias and barriers: Fulfilling the potential of women in academic science and engineering. Committee on Maximizing the Potential of Women in Academic Science and Engineering / Committee on Science, Engineering, and Public Policy.

Consalvo, M. (2008). Crunched by passion: women game developers and workplace challenges. In H. Kafai, Denner & Sun (Eds.), Beyond barbie and mortal kombat: new perspectives on gender and gaming (pp. 177-192). Cambridge, MA: The MIT Press.

Coontz, S. (2007). Motherhood stalls when women can't work. Hartford Courant, May 13, 2007. Copyright Council on Contemporary Families. Downloaded September 10, 2007 from http://www.contemporaryfamilies.org/subtemplate.php?t=pressReleases&ext=motherhoodstalls

Costa, P. T., Terracciano, A., & McCrae, R. R. (2001). Gender Differences in Personality Traits Across Cultures: Robust and Surprising Findings. Journal of Personality and Social Psychology, 81(2), 322–331. doi:10.1037/0022-3514.81.2.322

Court, S. (1999). Negotiating the research imperative: The views of UK academics on their career opportunities. Higher Education Quarterly, 53(1), 65–87. doi:10.1111/1468-2273.00113

CRA. (2007). CRA Taulbee Survey. Retrieved from http://www.cra.org/statistics/

Crittenden, A. (2001). The price of motherhood. New York: Henry Holt & Co.

Cronin, C., & Roger, A. (1999). Theorizing progress: women in science, engineering, and technology in higher education. Journal of Research in Science Teaching, 36(6), 637–661. doi:10.1002/(SICI)1098-2736(199908)36:6<637::AID-TEA4>3.0.CO;2-9

Crowley, K. C. (2001). Parents explain more often to boys than to girls during shared scientific thinking. Psychological Science, 12, 258–261. doi:10.1111/1467-9280.00347

Cupchik, G. (2001, February). Constructivist Realism: An Ontology That Encompasses Positivist and Constructivist Approaches to the Social Sciences. Forum Qualitative Sozialforschung/ Forum: Qualitative Social Research, 2(1). Retrieved October 4, 2006, from http://qualitative-research.net/fqs/fqs-eng.htm

Current Population Survey (CPS). (2007). US Labor force statistics. Retrieved 20/02/09 from http://www.bls.gov/cps/wlf-databook2008.htm

Davis, F. D. (1989). Perceived usefulness, perceived ease of use, and user acceptance of information technology. MIS Quarterly, 13(3), 319–339. doi:10.2307/249008

Davis, K. S. (2002). Advocating for equitable science-learning opportunities for girls in an urban city youth club and the roadblocks faced by women science educators. Journal of Research in Science Teaching, 39(2), 151–163. doi:10.1002/tea.10014

De Cohen, C. C., & Clewell, B. C. (2006). External evaluation of the ADVANCE program Georgia Institute of Technology: Year four report. Downloaded October 2, 2007 from http://www.advance.gatech.edu/files/GT-YearFourReport.pdf

De Groot, J. (1997). After the ivory tower: Gender, commodification and the 'academic'. Feminist Review, 55, 130–142.

Dean, C. (2007, April 17). Computer science takes steps to bring women to the fold. Retrieved February 27, 2009, from The New York Times: http://www.nytimes.com/2007/04/17/science/17comp.html

Deem, R. (2007). Managing a meritocracy or an equitable organisation? Senior managers' and employees' views about equal opportunities policies in uk universities. Journal of Education Policy, 22(6), 615–636. doi:10.1080/02680930701625247

Deem, R., & Hillyard, S. (2002). Making time for management - the careers and lives of manager-academics in UK universities. In G. Crow & Heath (Eds.), Social conceptions of time: Structure and process in work and everyday life, (pp. 126-143). Basingstoke, UK: Palgrave.

Deem, R., & Lucas, L. (2007). Research and teaching cultures in two contrasting UK policy contexts: Academic life in education departments in five English and Scottish universities. Higher Education, 54, 115–133. doi:10.1007/s10734-006-9010-z

Deem, R., & Morley, L. (2006). Diversity in the Academy? Staff Perceptions of Equality Policies in Six Contemporary Higher Education Institutions. Policy Futures in Education, 4(2). doi:10.2304/pfie.2006.4.2.185

Deuze, M., Martin, C.B., & Alen, C. (2007). The professional identity of game-workers. Convergence: The international journal of research into new media technologies, 13(4), 335-353.

Dewandre, N. (2002). Women in Science: European Strategies for Promoting Women in Science. Science, 295(5553), 278–279. doi:10.1126/science.1063487

Diamond, C., & Whitehouse, G. (2007). Gender, computing and the organization of working time: Public/private comparisons in the Australian context. Information Communication and Society, 10(3), 320–337. doi:10.1080/13691180701409879

Dillman, D. (2007). Mail and Internet Surveys, The Tailored Design Method. Hoboken, NJ: John Wiley & Sons.

Dolgin, K., & Minowa, N. (1997). Gender differences in self-presentation: A comparison of the roles of flatteringness and intimacy in self-disclosure to friends. Sex Roles, 36(5-6), 371–380. doi:10.1007/BF02766653

Dominick, J. (1999). Who do you think you are? Personal home pages and self-presentation of the world wide web. Journalism & Mass Communication Quarterly, 76(4), 646–658.

Donelan, H., Herman, C., Kear, K., & Kirkup, G. (2009). Patterns of online networking for women's career development. Gender in Management: An International Journal, 24(2), 92–111. doi:10.1108/17542410910938790

Douvan, E., & Walker, A. M. (1956). The sense of effectiveness in public affairs. Washington, DC: American Psychological Association, Psychological Monographs.

Downing, R. A., Crosby, F. J., & Blake-Beard, S. (2005). The perceived importance of developmental relationships on women undergraduates' pursuit of science. Psychology of Women Quarterly, 29(4), 419–426. doi:10.1111/j.1471-6402.2005.00242.x

Durbin, S. (2006). Gender, Skills and Careers in UK Call Centres. In Burgess, J., & Connell, J. (Eds.), Developments in the Call Centre Industry: analysis, changes and challenges (pp. 117–135). Oxon, UK: Routledge.

Durbin, S. (forthcoming). Creating Knowledge through Networks: a gender perspective. Gender, Work and Organization.

Dutta-Bergman, M. J. (2005). Access to the internet in the context of community participation and community satisfaction. New Media & Society, 7(1), 89–109. doi:10.1177/1461444805049146

Dyer, S. K. (2004). Under the Microscope: A Decade of Gender Equity Projects in the Sciences. Washington, DC: American Association of University Women Educational Foundation.

Eagly, A., & Carau, S. (2002). Role congruity theory of prejudice toward female leaders. Psychological Review, 109(3), 573–598. doi:10.1037/0033-295X.109.3.573

Educational Competencies Consortium, Ltd. (2007). About HERA. Retrieved from http://www.hera.ac.uk/about/page.php?page=1

Ellemers, N., van den Heuvel, H., de Gilder, D., Maass, A., & Bonvini, A. (2004). The Underrepresentation of Women in Science: differential commitment or the queen bee syndrome? The British Journal of Social Psychology, 43, 315–338. doi:10.1348/0144666042037999

EOWA. (2009). Employer of Choice for Women. Retrieved 12 May, 2009, from http://www.eowa.gov.au/EOWA_Employer_Of_Choice_For_Women.asp

Epstein, J. L. (1982). Teachers' reported practices of parent involvement: problems and possibilities. The Elementary School Journal, 83, 103–114. doi:10.1086/461298

Equal Opportunity for Women in the Workplace Agency. (2008). Australian Census of Women in Leadership, Australian Government. Retrieved 2 March 2009 from http://www.eowa.gov.au/Australian_Women_In_Leadership_Census/2008_Australian_Women_In_Leadership_Census/Media_Kit/EOWA_Census_2008_Publication.pdf

Erickson, F. (1982). The counselor as gatekeeper. New York: Academic Press.

ESCAP. (2008). Internet penetration in Malaysia. Available at http://malaysiacrunch.blogspot.com/2008/11/internet-penetration-in-malaysia.html

Etzkowitz, H., Kemelgor, C., Neuschatz, M., Uzzi, B., & Alonzo, J. (1994). The paradox of critical mass for women in science. Science, 266, 51–54. doi:10.1126/science.7939644

European Commission. (2004). Towards a European Research Area: science, technology and innovation, Key figures 2004 Eurostat. Brussels, Belgium: Author.

European Commission. (2004). Widening Women's Work in Information and Communication Technology. WWW-ICT. Brussels, Belgium: Author.

European Commission. (2006). Women and Science Statistics and Indicators: SHE Figures 2006. Brussels: European Commission.

Evetts, J. (1993). Women and Management in Engineering: the 'Glass Ceiling' for Women's Careers. Women in Management Review, 8(7), 19–25. doi:10.1108/09649429310046391

Evetts, J. (1998). Managing the Technology but not the Organization: Women and Career in Engineering. Women in Management Review, 13(8), 283–290. doi:10.1108/09649429810243144

Evetts, J. (2000). Analysing Change in Women's Careers: culture, structure and action dimensions. Gender, Work and Organization, 7(1), 57–67. doi:10.1111/1468-0432.00093

Fain, J. (2007a). Occupational segregation home page, 20 leading occupations of employed women, 1997 annual averages. Downloaded September 10, 2007 from http://spears.okstate.edu/home/jrfain/legacy/occseg/20_Leading%20Occupations.html

Fain, J. (2007b). Occupational segregation home page, Non-traditional occupations for employed women in 1997. Downloaded September 10, 2007 from http://spears.okstate.edu/home/jrfain/legacy/occseg/NontraditionalOcc.html

Ferber, M., & Teiman, M. (1980). Are Women at a Disadvantage in Publishing Journal Articles? Eastern Economic Journal, 9(1), 189–194.

Ferreira, M. (2002). The research lab: A chilly place for graduate women. Journal of Women and Minorities in Science and Engineering, 8(85), 85–98.

Ferry, T. R., Fouad, N. A., & Smith, P. L. (2000). The role of family context in a social cognitive model for career-related choice behavior: a math and science perspective. Journal of Vocational Behavior, 57, 348–364. doi:10.1006/jvbe.1999.1743

Fishbein, M., & Ajzen, I. (1975). Belief, attitude, intention, and behavior: An introduction to theory and research. Reading, MA: Addison-Wesley.

Fisher, A. (1992, September). When will women get to the top? Fortune, 44–56.

Flanagan, M. (2005). Troubling 'games for girls': notes from the edge of game design. In DiGRA 2005 Conference: Changing views-worlds in play.

Fletcher, C., Boden, R., Kent, J., & Tinson, J. (2007). Performing women: The gendered dimensions of the UK new research economy. Gender, Work and Organization, 14(5), 433–453. doi:10.1111/j.1468-0432.2007.00359.x

Forster, N. (2000). A case study of women academics' views on equal opportunities, career prospects and work-family conflicts in a British university. Women in Management Review, 15(7), 316–330. doi:10.1108/09649420010378124

Fortenberry, N. (2009). Achieving the Engineer of 2020. Paper presented at the Faculty of Engineering and Surveying Research Seminar.

Fox, M. F. (1998). Women in science and engineering: theory, practice, and policy in programs. Signs, 24(1), 201–223. doi:10.1086/495325

Fox, M. F., & Fonseca, C. (2006). Gender and mentoring of faculty in science and engineering: individual and organizational factors. International Journal of Learning and Change, 1(4), 460–483. doi:10.1504/IJLC.2006.013912

Fox, M. F., & Stephan, P. E. (2001). Careers of Young Scientists: preferences, prospects and realities by gender and field. Social Studies of Science, 31(1), 109–122. doi:10.1177/030631201031001006

Fox, M. F., Sonnert, G., & Nikiforova, I. (2009). Successful programs for undergraduate women in science and engineering: "adapting" versus "adopting" the institutional environment. Research in Higher Education, 50(4), 333–353. doi:10.1007/s11162-009-9120-4

Frear, E. (2009, February 25th). The Career Crunch. Develop Magazine.

Freeman, C. E. (2004). Trends in Educational Equity of Girls and Women. Washington, DC: US Government Printing Office.

Frehill, L. M., DiFabio, N., Layne, P., Turner-Johnson, A., & Hood, S. (2007). Women in Engineering. Society of Women Engineers, 53(3), 1–20.

Frieze, C., Hazzan, O., Blum, L., & Dias, M. B. (2006). Culture and environment as determinants of women's participation in computing: revealing the "women-CS fit". In Procs. of ACM SIGCSE – Technical Symposium on Computer Science Education.

Fullerton, T., Fron, J., Pearce, C., & Morie, J. (2007). Getting girls into the games: Towards a 'virtuous cycle'. In H. Kafai, Denner & Sun, (Eds.), Beyond Barbie and Mortal Kombat: New perspectives on gender and computer games. Cambridge, MA: MIT.

Gaither, M. (2008). Loh-Ludher on Homeschooled Malaysian Women. Available at http://gaither.wordpress.com/2008/

Gardiner, M., Tiggemann, M., Kearns, K., & Marshall, K. (2007). Show me the money! An empirical analysis of mentoring outcomes for women in academia. Higher Education Research & Development, 26(4), 425–442. doi:10.1080/07294360701658633

Gardner, E., Young, P., & Ruth, S. (1989). Evolution of attitudes toward computers: A retrospective review. Behaviour & Information Technology, 8, 89–98. doi:10.1080/01449298908914542

Gateway Community and Technical College. (2007). Student climate survey.: Gateway community and technical College. Covington: Gateway Community and Technical College.

Gavel, D. (2000, April 14). Study Reveals New Priorities for New Generation of Faculty. Harvard Gazette. Retrieved from http://www.news.harvard.edu/gazette/.

Gibb, H. (2001). Gender mainstreaming: good practices from the Asia Pacific Region. The North-South Institute, Canada, and APEC.

Gibbs, G., & Habershaw, T. (2002). Recognising and Rewarding Good Teaching – a guide to good practice. Milton Keynes, UK: Centre for Higher Education Practice, Open University Press.

Gill, R. (2002). Cool, creative and egalitarian? exploring gender in project-based new media work in Europe. Information Communication and Society, 5(1), 70–89. doi:10.1080/13691180110117668

Glick, P., & Fiske, S. T. (1999). Sexism and Other 'Isms': Interdependence, Status, and the Ambivalent Content of Stereotypes. In Swann, J., William, B., Langlois, J. H., & Gilbert, L. A. (Eds.), Sexism and Stereotypes in Modern Society. Washington, DC: American Psychological Association. doi:10.1037/10277-008

Go, W. E. S. T. (2009). Go Women in Engineering Science and Technology: About Us. Retrieved 09 May, 2009, from http://www.usq.edu.au/gowest/

Goh Beng, L. (2002). Modern dreams: An enquiry into power, cultural production and the cityscape in contemporary urban Penang, Malaysia. New York: Cornell University Press.

Goldin, C. (1995). Career and family: Women look to the past (Working Paper No. 5188). Cambridge MA: National Bureau of Economic Research.

Good, C., Aronson, J., & Harder, J. A. (2008). Problems in the pipeline: Stereotype threat and women's achievement in high-level math courses. Journal of Applied Developmental Psychology, 29, 17–28. doi:10.1016/j.appdev.2007.10.004

Goodman, I. F., & Cunningham, M. L. (2002). Final report of the Women's Experiences in College Engineering (WECE) Project. Downloaded September 10, 2007 from http://www.grginc.com/WECE_FINAL_REPORT.pdf

Graner Ray, S. (2004). Gender inclusive game design: expanding the market. Hingham, MA: Charles River Media Inc.

Gray, M., & Kurihara, T. (2006). Networks of exclusion: Job segmentation and gendered social networks in the knowledge economy. Technology and Gender Equalities Seminar. Cambridge, UK: ESRC Gender Equality Network.

Greed, C. (1991). Surveying Sisters: Women in a Traditional Male Profession. London: Routledge.

Greenfield, S. (2002). SET Fair: a report on women in science, engineering and technology. London: Department for Trade and Industry.

Greenfield, S., J. Peters, N. Lane, T. Lees and G.I Samuels. (2002). SET Fair: A Report on Women in Science, Engineering, and Technology. Retrieved February 12, 2009 from http://www.ukrc4setwomen.org/downloads/Government/setfair.pdf

Griffiths, M., Moore, K., & Richardson, H. (2007). Celebrating heterogeneity?: a survey of female ICT professionals in England. Information Communication and Society, 10(3), 338–357. doi:10.1080/13691180701409945

Groysberg, B. (2008). How Star Women Build Portable Skills. Harvard Business Review, 86(2), 74–81.

Gunter, R., & Stambach, A. (2003). As balancing act and as game: how women and men science faculty experience the promotion process. Gender Issues, 21(1), 24–42. doi:10.1007/s12147-003-0020-1

Gupta, A. F. (1997). The Internet & the English Language. First Conference on Postcolonial Theory.

Hackett, G. (1997). Self-efficacy in career choice and development. In A. Bandura (Ed.), Self-efficacy in changing societies (pp. 232-258). Cambridge, UK: Cambridge University Press.

Haines, L. (2004, September). Why are there so few women in games? Research for Media Training North West.

Halford, S., Savage, M., & Witz, A. (1997). Gender, Careers and Organisations. London: Macmillan.

Hall, R. M., & Sandler, B. R. (1982). The Classroom Climate: A Chilly One for Women? (Report). Washington, DC: Project on the Status and Education of Women, Association of American Colleges.

Halsey, A. H. (1992). The decline of donnish dominion. Oxford, UK: Clarendon Press.

Handley, J. (1994). Women, Decision Making and Academia: An Unholy Alliance. Women in Management Review, 9(3), 11–16. doi:10.1108/09649429410056290

Harding, S. (1986). The Science Question in Feminism. Milton Keynes. UK: Open University Press.

Harley, S. (2003). Research selectivity and female academics in UK universities: From gentleman's club and barrack yard to smart macho? Gender and Education, 15(4), 377–392. doi:10.1080/09540250310001610580

Harris, K., Morello, D., & Raskino, M. (2007). Women and men in IT: Breaking sexual stereotypes. Gartner. Retrieved from http://www.gartner.com/DisplayDocument?doc_cd=146320

Harris, P., Thiele, B., & Currie, J. (1998). Success, gender and academic voices: Consuming passion or selling the soul? Gender and Education, 10(2), 133–148. doi:10.1080/09540259820989

Harris, R. W., Yogeesvaran, K., & Lee, L. W. Y. (2007). Telecentres for national e-inclusion in Malaysia. Available at http://14donline.net/ATF/2007/fullpapers/Roger_ATF07ABS113.pdf

Hartman, H. & Hartman, M. (2006). Leaving engineering: Lessons from Rowan University's College of Engineering. Journal of Engineering Education, January.

Hartman, H., & Hartman, M. (2008). How undergraduate engineering students perceive women's (and men's) problems in science, math and engineering. Sex Roles, 58, 251–265. doi:10.1007/s11199-007-9327-9

Harvard University. (2007). Project Implicit. Retrieved from https://implicit.harvard.edu

Hashim, J. (2008). Learning barriers in adopting ICT among selected working women in Malaysia. Gender in Management: An International Journal, 23(5), 317–336. doi:10.1108/17542410810887356

Hassoun, S., & Bana, S. (2001). Practices for Recruiting and Retaining Graduate Women Students in Computer Science and Engineering. Paper presented at the meeting of the International Conference on Microelectronic Systems Education, Las Vegas, NV.

Hax, C. (2007, July 29). Tell me about it. Washington Post, Sunday Source.

Heckman, J. J. (2006, June 30). Skill formation and the economics of investing in disadvantaged children. Science, 312, 1900–1901. doi:10.1126/science.1128898

Heilman, M. E. (2001). Description and prescription: How Gender Stereotypes Prevent Women's Ascent Up the Organizational Ladder. The Journal of Social Issues, 57, 657–674. doi:10.1111/0022-4537.00234

Helsinki Group on Women and Science. (2002). Women in Science: review of the situation in the United Kingdom. London: Department for Trade and Industry.

Hemenway, K. (1995). Human Nature and the Glass Ceiling in Industry. Communications of the ACM, 38(1), 55–62. doi:10.1145/204865.204878

Henderson, A. (1981). Parent participation-student achievement: The evidence grows. Columbia: National Committee for Citizens in Education.

Henkel, M. (2000). Academic identities and policy change in higher education (Vol. 46). London: Jessica Kingsley.

Henwood, F. (1996). WISE choices? Understanding occupational decision-making in a climate of equal opportunities for women in science and technology. Gender and Education, 8(2), 199–214. doi:10.1080/09540259650038860

Herman, C., & Kirkup, G. (2008). Learners in Transition: the use of ePortfolios for women returners to science, engineering and technology. Innovations in Education and Teaching International, 45(1), 67–76. doi:10.1080/14703290701757468

Heward, C. (1996). Women and careers in higher education: What's the problem? In Morley, L., & Walsh, V. (Eds.), Breaking boundaries: Women in higher education (pp. 11–23). London: Taylor and Francis.

Heward, C., & Taylor, P. (1994). Women at the Top in Higher Education. Policy and Politics, 20(1), 111–121.

Heward, C., Taylor, P., & Vickers, R. (1997). Gender, Race and Career Success in the Academic Profession. Journal of Further and Higher Education, 21(2), 205–218. doi:10.1080/0309877970210206

Hewitt, P. (2003). A Strategy for Women in Science, Engineering and Technology. London: Department for Trade and Industry.

Hewlett, S. A., Luce, C. B., & Servon, L. J. (2008, June). Stopping the exodus of women in science. Harvard Business Review. Retrieved from http://hbr.harvardbusiness.org

Higher Education Statistics Agency. (2008). Resources of higher education institutions 2006/7. Cheltenham, UK: HESA.

Hodgson, B. (2000). Women in science--or are they? Physics Education, 35(6), 451–453. doi:10.1088/0031-9120/35/6/313

Hofstede, G. (1980). Culture's Consequences: International Differences in Work Related Values. London: Sage.

Hollinger, C. L. (1983). Self-Perception and the career aspirations of mathematically talented female adolescents. Journal of Vocational Behavior, 22, 49–62. doi:10.1016/0001-8791(83)90005-2

Hooker, C. (2004). Irresistible Forces: Australian Women in Science. Carlton, Victoria: Melbourne University Press.

Hoover-Dempsey, K. V., Bassler, O. C., & Brissie, J. S. (1987). Parent involvement: Contributions of teacher efficacy, school socioeconomic status, and other school characteristics. American Educational Research Journal, 24(3), 417–435.

Hughes, E. L., & Parkes, K. R. (2007). Work hours and well-being: the roles of work-time control and work-family interference. Work and Stress, 21(3), 264–278. doi:10.1080/02678370701667242

Hussar, W., & Bailey, T. (2008). The name assigned to the document by the author. This field may also contain sub-titles, series names, and report numbers.Projections of Education Statistics to 2017.Washington, DC: Publisher name and contact information, as provided by the publisher; updated only if notified by the publisher. National Center for Education Statistics.

Hyde, J. (2005, September). The gender similarities hypothesis. The American Psychologist, 60(6), 581–592. doi:10.1037/0003-066X.60.6.581

Hyde, J. S., Fennema, E., & Lamon, S. J. (1990). Gender differences in mathematics performance: A meta-analysis. Psychological Bulletin, 107(2), 139–155. doi:10.1037/0033-2909.107.2.139

Ibarra, H. (1992). Homphily and differential returns: Sex differences in network structure and access in an advertising firm. Administrative Science Quarterly, 37, 422–447. doi:10.2307/2393451

IBISWorld press release. (2008). The new American players: baby boomers and women take on video gaming. Retrieved 18/08/2008 from http://www.ibisworld.com/pressrelease/pressrelease.aspx?prid=133

ICT Policy & Regulations Researches, February 3, 2009,Malaysia's broadband plan – stimulating the private sector. Available at: http://ictpolicynregulation.blogspot.com/2009/02/malaysias-broadband-plan-stimulating.html

IGDA. (2004). Quality of life white paper. International Game Developers Association.

IGDA. (2005). Game developers demographics: An exploration of workforce diversity. International Game Developers Association.

Indepdendent. (2008). Women and Science: how to achieve the right balance, (September).

Inspiring girls now in technology education. (2009). Retrieved June 30, 2009, from IGNITE: http://www.ignite-us.org/

Inzer, L. D., & Crawford, C. B. (2005). A Review of Formal and Informal Mentoring: Processes, Problems, and Design. Association of Leadership Educators, 4(1), 31–50.

Ivie, R., & Ray, K. N. (2005). Women in Physics and Astronomy. AIP Publication Number R-430.02. American Institute of Physics. Retrieved from http://www.aip.org/statistics/trends/gendertrends.html

Jaafar, N. I., & Sulaiman, A. (2005). Domestic computer usage and activities in West Coast Malaysia: age and income differences. Information Development, 21(2), 128–137. doi:10.1177/0266666905054507

Jacobs, J. A., & Winslow, S. E. (2004). Overworked Faculty: Job Stresses and Family Demands. The Annals of the American Academy of Political and Social Science, 596(1), 104–129. doi:10.1177/0002716204268185

Joanna Goode, R. E. (2004). Lost in translation: Gender in high school computer science. Representation of Women in Information Technology.

Johnson, R. D., & Stone, D., L. & Phillips, T. N. (2008). Relations Among Ethnicity, Gender, Beliefs, Attitudes, and Intention to Pursue a Career in Information Technology. Journal of Applied Social Psychology, 38(4), 999–1022. doi:10.1111/j.1559-1816.2008.00336.x

Johnson, R. D., Veltri, N. F., & Hornik, S. (2008). Attributions of Responsibility Toward Computing Technology: The Role of Interface Social Cues and User Gender. International Journal of Human-Computer Interaction, 24(6), 595–612. doi:10.1080/10447310802205784

Joiner, T., Bartram, T., & Garretta, T. (2004). The effects of mentoring on perceived career success, commitment and turnover intentions. The Journal of American Academy of Business, 5(1/2), 163–170.

Jones, M. G., Howe, A., & Rua, M. (2000). Gender Differences in Students' Experiences, Interests, and Attitudes toward Science and Scientists. Science Education, 84(2), 180–192. doi:10.1002/(SICI)1098-237X(200003)84:2<180::AID-SCE3>3.0.CO;2-X

Joshi, I. (Ed.). (1997). Asian women in the information age: new communication technology, democracy, and women. Singapore: Asian Media Information and Communication Centre.

Joy, L., & Carter, N. M. Wagner, H. M. & Narayanan, S. (2007). The Bottom Line: Corporate Performance and Women's Representation on Boards. Retrieved March 5 2009 from http://www.catalyst.org/publication/200/the-bottom-line-corporate-performance-and-womens-representation-on-boards

Kanter, R. M. (1977). Men and Women of the Corporation. New York: Basic Books.

Kass, R. B., Souba, W. W., & Thorndyke, L. E. (2006). Challenges confronting female surgical leaders: overcoming the barriers. The Journal of Surgical Research, 132, 179–187. doi:10.1016/j.jss.2006.02.009

Kelan, E. K. (2008). Emotions in a rational profession: the gendering of skills in ICT work. Gender, Work and Organization, 15(1), 49–71.

Kelle, U. (2001, February). Sociological Explanations between Micro and Macro and the Integration of Qualitative and Quantitative Methods. Forum Qualitative Sozialforschung / Forum: Qualitative Social Research, 2(1). Retrieved October 4, 2006, from http://qualitative-esearch.net/fqs/fqs-eng.htm

Keogh, C., Tattersall, K., & Richardson, H. (2007). The directing equal pay in ICT (DEPICT) final report. European Social Fund and University of Salford.

Kinzie, J. (2007). Women's paths in science: a critical feminist analysis. New Directions for Institutional Research, 2007(133), 81–93. doi:10.1002/ir.206

Klawe, M., & Levenson, N. (1995). Women in Computing: Where are We Now? Communications of the ACM, 38(1), 29–35. doi:10.1145/204865.204874

Klawe, M., Whitney, T., & Simard, C. (2009). Women in Computing - Take 2. Communications of the ACM, 52(2), 68–76. doi:10.1145/1461928.1461947

Klein, S. (2005). Handbook for achivening gender equality through education. Routledge.

Knapp, L. G., & Kelly-Reid, J. E. (2006, February). Postsecondary institutions in the United States: Fall 2003 and degrees and other awards conferred: 2. Education Statistics Quarterly .

Knights, D., & Richards, W. (2003). Sex discrimination in UK academia. Gender, Work and Organization, 10(2), 213–237. doi:10.1111/1468-0432.t01-1-00012

Kobasa, S. C. (1979). Stressful life events, personality and health: an inquiry into hardiness. Journal of Personality and Social Psychology, 37, 1–11. doi:10.1037/0022-3514.37.1.1

Krotoski, A. (2004). Chicks and joysticks: An exploration of women and gaming. Entertainment and Leisure Software Publishers Association (ELSPA) White paper.

Kulis, S., Sicotte, D., & Collins, S. (2002). More Than a Pipeline Problem: Labor Supply Constraints and Gender Stratification Across Academic Science Disciplines. Research in Higher Education, 43(6), 657–691. doi:10.1023/A:1020988531713

Kvasny, A. H., & Lynette, T. (2004). Recruitment is never enough. SIGMIS '04 .

Langham, B. (2009). Horizontal Leadership. Hay Group: People Torque, May (3), 1.

Lapointe Terosky, A., Phifer, T., & Neumann, A. (2008). Shattering the Plexiglas: Continuing Challenges for Women Professors in Research Universities. In J. Glazer-Raymo (Ed.), Unfinished Agendas. Baltimore, MD: John Hopkins University Press.

Lazar, J., & Lidtke, D. (2002). Service-Learning Partnerships in the Information Systems Curriculum. In J. Lazar (Ed.), Managing IT/Community Partnerships in the 21st Century (pp. 1-16). Hershey, PA: Idea Group Publishing.

Lee, R. (1970). Social attitudes and the computer revolution. Public Opinion Quarterly, 34(1), 53–59. doi:10.1086/267772

Levinson, W., & Lurie, N. (2004). When most doctors are women: what lies ahead? Annals of Internal Medicine, 21, 471–474.

Lewin, T. (2008, July 25). Math scores show no gap for girls, study finds. Retrieved 28 2009, February, from The New York Times: http://www.nytimes.com/2008/07/25/education/25math.html?_r=1

Lincoln, J., & Miller, J. (1979). Work and friendship ties in organizations: A comparative analysis of relation networks. Administrative Science Quarterly, 24(2), 181–199. doi:10.2307/2392493

Loh-Ludher, L. L. (2007). The Socioeconomic Context of Home-Based Learning by Women in Malaysia. Distance Education, 28(2), 179–193. doi:10.1080/01587910701439233

Lord, L., Preston, A., & Crosbie, R. (2006). Unearthing New Resources: Attracting and Retaining Women in the Australian Minerals Industry. Canberra, Australia: Australian Government Office for Women and Minerals Council of Australia.

Lucas, L. (2006). The research game in academic life. Buckingham, UK: Open University Press.

Lynn, R. (2008). Review of Why aren't more women in science? Intelligence, 36(4), 380–382. doi:10.1016/j.intell.2007.06.005

Maddux, J. (2002). Self-Efficacy: The Power of Believing You Can. In C. Snyder & S. Lopez (Ed.), Handbook of Positive Psychology (pp. 277-287). New York: Oxford University Press.

Madill, H. M., Ciccocioppo, A.-L., Stewin, L. L., Armour, M.-A., & Montgomerie, T. C. (2004). The potential to develop a career in science: young women's issues and their implications for careers guidance initiatives. International Journal for the Advancement of Counseling, 26(1), 1–19. doi:10.1023/B:ADCO.0000021546.17402.5b

Malhotra, A., Schuler, S. R., & Boender, C. (2002). Measuring women's empowerment as a variable in international development. Prepared for the World Bank Workshop on Poverty and Gender. Downloaded on October 4 from http://www.aed.org/LeadershipandDemocracy/upload/MeasuringWomen.pdf

Mannix, E. A., & Neale, M. A. (2005). What difference makes a difference? Psychological Science in the Public Interest, 6(2), 31–55. doi:10.1111/j.1529-1006.2005.00022.x

Manuel, A. (2008). From old boys network to virtual network. A study of internet technology use amongst UK academics, and the extent to which it is disrupting the gendered academic research culture. Bristol, UK: University of Bristol.

Marakas, G. M., Yi, M. Y., & Johnson, R. D. (1998). The multilevel and multifaceted character of computer self-efficacy: Toward clarification of the construct and an integrative framework for research. Information Systems Research, 9(2), 126–163. doi:10.1287/isre.9.2.126

Maree, M., Maree, D. J. F., Botha, C., & Gcabo, R. (2008). Changing perceptions of women in the science, engineering and technology industry: evaluating the career histories of role models in South Africa. Pretoria, South Africa: Institute for Women and Gender Studies, University of Pretoria.

Markert, L. R. (1996). Gender Related to Success in Science and Technology. Journal of Technology Studies, 22(9), 21–29.

Martell, R. F., Lane, D. M., & Emrich, C. (1996). Male-female differences: A computer simulation. The American Psychologist, 51, 157–158. doi:10.1037/0003-066X.51.2.157

Martin, C. L., & Halverson, C. F. (1981). A schematic processing model of sex typing and stereotyping in children. Child Development, 52, 1119–1134. doi:10.2307/1129498

Martin, L. (1994). Equity and General Performance Indicators in Higher Education.

Martin, P. Y. (2001). 'Mobilizing Masculinities': Women's Experiences of Men at Work. Organization, 8(4), 587–618. doi:10.1177/135050840184003

Martínez Alemán, A. M. (2008). Faculty Productivity and the Gender Question. In J. Glazer-Raymo (Ed.), Unfinished Agendas. Baltimore, MD: John Hopkins University Press.

Martz, B., & Tetua, C. (2008). Students' perception of IS academic programs, IS careers, and outsourcing. Journal of Education for Business, 84, 118–125. doi:10.3200/JOEB.84.2.118-125

Maskell-Pretz, M., & Hopkins, W. E. (1997). Women in Engineering: Toward a Barrier-Free Work Environment. Journal of Management Engineering, 13(1), 32–37. doi:10.1061/(ASCE)0742-597X(1997)13:1(32)

Mason, M. A., & Ekman, E. M. (2007). Mothers on the Fast Track: How a New Generation can Balance Family and Careers. New York: Oxford University Press.

Mason, M. A., & Goulden, M. (2004). Marriage and Baby Blues: Redefining Gender Equity in the Academy. The Annals of the American Academy of Political and Social Science, 596(1), 86–103. doi:10.1177/0002716204268744

Massachusetts Institute of Technology. (1999). A study of the women faculty in science at MIT. Downloaded January 7, 2008 from http://web.mit.edu/fnl/women/Birgfnl.html

Massachusetts Institute of Technology. et al. (2003). Brief of amici curiae in support of the respondents No. 02-241, 02-516. Downloaded January 9, 2008 from http://www.vpcomm.umich.edu/admissions/legal/gru_amicus-ussc/um/MITfinal-both.doc

Mavin, S., & Bryans, P. (2002). Academic women in the UK: Mainstreaming our experiences and networking for action. Gender and Education, 14(3), 235–250. doi:10.1080/0954025022000010703

Mayer-Smith, J., Pedretti, E., & Woodrow, J. (2000). Closing of the gender gap in technology enriched science education: a case study. Computers & Education, 35, 51–63. doi:10.1016/S0360-1315(00)00018-X

McCorduck, N., & Ramsey, P. (2005). Where are the women in information technology? Boulder: Anita Borg Institute for Women and Technology. MentorNet. (n.d.). Retrieved March 1, 2009, from MentorNet: http://www.mentornet.net/

McDowell, J. M., & Smith, J. K. (1992). The Effect of Gender Sorting on Propensity to Coauthor: Implications for Academic Promotion. Economic Inquiry, 30(1), 68–82. doi:10.1111/j.1465-7295.1992.tb01536.x

McIlwee, J. S., & Robinson, J. G. (1992). Women in Engineering: Gender, Power and Workplace Culture. Albany, NY: State University of New York Press.

McKinney, V. R., Wilson, D. D., Brooks, N., O'Leary-Kelly, A., & Hardgrave, B. (2008). Women and Men in the IT Profession. Communications of the ACM, 51(2), 81–84. doi:10.1145/1314215.1340919

MCV. (2008). Recruitment special: Industry salary survey. Retrieved 26th January 2008 from http://www.mcvuk.com/news/29399/Industry-salary-survey

MCV. (2009). Women earn more than men in the UK games industry. Retrieved 03/03/2009 from http://www.mcvuk.com/news/32964/Women-earn-more-than-men-in-the-UK-games-industry

Merit Systems Protection Board. (1992). A question of equity: Women and the glass ceiling: A report to the President and the Congress, October 1992

Metcalf, H., Rolfe, H., Stevens, P., & Weale, M. (2005). Recruitment and Retention of Academic Staff in Higher Education. Department for Education and Skills, Research Report RR658.

Miller, G. (2004). Frontier Masculinity in the Oil Industry: the experience of women engineers. Gender, Work and Organization, 11(1), 47–84. doi:10.1111/j.1468-0432.2004.00220.x

Miller, H., & Arnold, J. (2001). Breaking away from grounded identity? Women academics on the web. Cyberpsychology & Behavior, 4(1), 95–108. doi:10.1089/10949310151088451

Mills, A. J. (1988). Organization, Gender and Culture. Organization Studies, 9(3), 351–369. doi:10.1177/017084068800900304

Morgan, C. S. (1992). College students' perceptions of barriers to women in science and engineering. Youth & Society, 24(2), 228–236. doi:10.1177/0044118X92024002006

Morley, L. (2003). Quality and power in higher education. Maidenhead, UK: Society for Research into Higher Education and Open University Press.

Morris, L. K., & Daniel, L. G. (2008). Perceptions of a Chilly Climate: Differences in Traditional and Non-Traditional Majors for Women. Research in Higher Education, 49(3), 256–273. doi:10.1007/s11162-007-9078-z

Murrell, A. J., & Zagenczyk, T. J. (2006). The gendered nature of role model status: an empirical study. Career Development International, 11(6), 560–578. doi:10.1108/13620430610692953

Mustafa, M.A. (2003, March). Internet access. The World Bank Group Private Sector and Infrastructure Network, (pp. 256).

MyPersonality. Info (2009). Frequency of Personality Types. Retrieved on April 1, 2009 from http://www.mypersonality.info/personality-types/population-gender/

Mysyk, F. (2008). Women Becoming Mentors: Reflection and Mentor Identity Formation as a Process of Lifelong Learning. The International Journal of Diversity in Organisations. Communities and Nations, 8(5), 207–218.

National Advisory Council on Innovation (NACI). (2004). Facing the facts: Women's participation in science, engineering and technology. Pretoria, South Africa: National Advisory Council on Innovation (NACI) and the Department of Science and Technology.

National Alliance for Partnerships in Equity. (2006). Non-traditional occupations SOC/CIP/Career Clusters/DOE Pathways crosswalks. Downloaded September10, 2007 from http://www.napequity.org/nape_publications.htm

National Coalition for Women and Girls in Education. (2002). Title IX at 30: Report card on gender equity. Retrieved September 10, 2007 from http://www.ncwge.org/pubs-reports.html

National Science Board. (2004). Broadening participation in science and engineering faculty (NSF 04-41). Downloaded October 2, 2007 from http://www.nsf.gov/pubs/2004/nsb0441/nsb0441.pdf

National Science Foundation (NSF). (2009). ADVANCE: Increasing the Participation and Advancement of Women in Academic Science and Engineering Careers (ADVANCE). Retrieved March 30, 2009 from http://www.nsf.gov/funding/pgm_summ.jsp?pims_id=5383

National Science Foundation, Division of Science Resources Statistics. (2003). Characteristics of doctoral scientists and engineers in the United States: 2003 (NSF 06-320). Arlington, VA: National Science Foundation. Project Officer, John Tsapogas. Table 3 downloaded on September 10, 2007 from http://www.nsf.gov/statistics/nsf06320/

National Science Foundation, Division of Science Resources Statistics. (2007). Science and engineering statistics. Retrieved from http://www.nsf.gov/statistics/

National Science Foundation. (2007a). ADVANCE: Increasing the participation and advancement of women in academic science and engineering careers. Downloaded from http://www.nsf.gov/funding/pgm_summ.jsp?pims_id=5383&from=fund

National Science Foundation. (2007b). Merit review Broader Impacts criterion: Representative activities. Downloaded January 7, 2008 from http://www.nsf.gov/pubs/gpg/broaderimpacts.pdf

National Science Foundation. (2007c). Division of Human Resource Development. Retrieved from http://www.nsf.gov/div/index.jsp?div=HRD

National Science Foundation. (2007d). New formulas for America's workforce: Girls in science and engineering (NSF 03-207). Downloaded January 9, 2008 from http://www.nsf.gov/ehr/hrd/Newformulas/newformulas.jsp

National Science Foundation. (2007e). New formulas for America's workforce 2: Girls in science and engineering (NSF 06-60). Downloaded January 9, 2008 from http://www.nsf.gov/ehr/hrd/Newformulas/newformulas.jsp

National Science Foundation. (2007f). New tools for America's workforce: Girls in science and engineering (NSF 06-59). Downloaded January 9, 2008 from http://www.nsf.gov/ehr/hrd/Newformulas/newformulas.jsp

National, I. T. Council (NITC). (2004). Multimedia Super Corridor. Available at http://www.nitc.org.my/initiatives/msc.html

NCWIT. (2007). NCWIT Scorecard 2007: A Report on the Status of Women in Information Technology: National Center for Women & Information Technology.

NCWIT. (2009), Our Alliances. National Center for Women & Information Technology. Retrieved on March 29, 2009http://www.ncwit.org/alliance.what.html

Ndubisi, N. O. (2005). Effect of perception and personal traits on computer technology adoption by women entrepreneurs in Malaysia. Journal of Asia Entrepreneurship and Sustainability. Retrieved from www.asiaentrepreneuralshipjournal.com

Nelson, D. J., & Rogers, D. C. (2005). A national analysis of diversity in science and engineering faculties at research universities. Downloaded October 2, 2007 from http://cheminfo.ou.edu/~djn/diversity/briefings/Diversity%20Report%20Final.pdf and later updates in http://cheminfo.chem.ou.edu/faculty/djn/diversity/Faculty_Tables_FY07/07Report.pdf

Newell, H., & Dopson, S. (1996). Muddle in the middle: organizational restructuring and middle management careers. Personnel Review, 25(4), 4–20. doi:10.1108/00483489610123191

Ngini, C. U., Furnell, S. M., & Ghita, B. V. (2002). Assessing the global accessibility of the Internet. Internet Research: Electronic Networking Applications and Policy, 12(4), 329–338. doi:10.1108/10662240210438399

Nielson, T. R., Carlson, D. S., & Lankau, M. J. (2001). The Supportive Mentor as a Means of Reducing Work–Family Conflict. Journal of Vocational Behavior, 59(3), 364–381. doi:10.1006/jvbe.2001.1806

Noonan, M. C., Estes, S. B., & Glass, J. L. (2007). Do workplace flexibility policies influence time spent in domestic labour? Journal of Family Issues, 28(2), 263–288. doi:10.1177/0192513X06292703

Norris, P. (2001). Digital divide: civic engagement, information poverty, and the Internet worldwide. New York: Cambridge University Press.

O'Leary, V., & Mitchell, J. (1990). Women connecting with women: Networks and mentors. In S. Lie & V. O'Leary (Eds.), Storming the tower: Women in the academic world (pp. 58-74). London: Kogan Page.

Oakley, J. G. (2000). Gender-based Barriers to Senior Management Positions: understanding the scarcity of female CEOs. Journal of Business Ethics, 27, 321–334. doi:10.1023/A:1006226129868

Office for National Statistics. (2007). Labour Force Survey: 2007. London: Office for National Statistics.

Office for National Statistics. (2008, September 26th). More women in work but half in part time jobs. London: Office for National Statistics.

Office for Women. (2006a). Smart Women – Smart State Concept Paper: Supporting Women in the Smart State. Brisbane: Queensland Government. Retrieved from http://www.women.qld.gov.au/work-and-life/smart-state-strategy/documents/set-concept-paper.pdf

Office for Women. (2006b). Smart Women – Smart State Strategy Science, Engineering and Technology Action Plan (2006 – 2009). Brisbane: Queensland Government. Retrieved from http://www.women.qld.gov.au/work-and-life/smart-state-strategy/documents/set-action-plan.pdf

Office for Women. (2007). Women's Leadership in Science, Engineering and Technology: The Smart Women – Smart State Taskforce's Board Shadowing Program. Brisbane: Queensland Government. Retrieved from http://www.women.qld.gov.au/leadership-and-community/women-on-boards/documents/board-shadowing-case-study.pdf

Orser, B., Riding, A., Dathan, M., & Stanley, J. (2007, November 2). Gender Challenges of Women in the Canadian Advanced Technology Sector. Paper presented at the Canadian Council for Small Business and Entrepreneurship Conference, Kelwona, B.C., Canada.

Osborne, M., Rees, T., Bosch, M., Ebeling, H., Hermann, C., Hilden, J., et al. (2000). Science Policies in the European Union, Promoting excellence through mainstreaming gender equality. Brussels, Belgium: European Commission. PRAGES. (n.d.). PRActising Gender Equality in Science research project. Retrieved 8 May, 2009, from http://www.retepariopportunita.it/defaultdesktop.aspx?page=2749

Oxford Economics. (2008). The economic contribution of the UK games industry: final report. Oxford, UK: Oxford Economics.

Ozbilgin, M. F., & Woodward, D. (2004). Belonging and Otherness: sex equality in banking in Turkey and Britain. Gender, Work and Organization, 11(6), 668–688. doi:10.1111/j.1468-0432.2004.00254.x

Özbilgin, M., & Healy, G. (2004). The gendered nature of career development of university professors: the case of Turkey. Journal of Vocational Behavior, 64, 358–371. doi:10.1016/j.jvb.2002.09.001

Paglis, L., Green, S., & Bauer, T. (2006). Does Advisor Mentoring Add Value? A Longitudinal Study of Mentoring and Doctoral Student Outcomes. Research in Higher Education, 47(4), 451–456. doi:10.1007/s11162-005-9003-2

Pajares, F. (1996). Self-Efficacy Beliefs in Academic Settings. Review of Educational Research, 66(4), 543–578.

Pajares, F. (2002). Overview of Social Cognitive Theory and Self-Efficacy. Retrieved February 8, 2008 from http://www.emory.edu/EDUCATION/efp/eff.html

Panteli, N. (2006). Returning to IT. Labour & Industry, 16(3).

Papacharissi, Z. (2002). The presentation of self in virtual life: Characteristics of personal home pages. Journalism & Mass Communication Quarterly, 79(3), 643–660.

Papastergiou, M. (2008). Are computer science and information technology stil masculine fields? High school students' perceptions and career choices. Computers & Education, 51(2), 594–608. doi:10.1016/j.compedu.2007.06.009

Papastergiou, M., & Solomonidou, C. (2005). Gender issues in Internet access and favourite Internet activities among Greek high school pupils inside and outside school. Computers & Education, 44(4), 377–393. doi:10.1016/j.compedu.2004.04.002

Parker, J. (2008). Comparing Research and Teaching in University Promotion Criteria. Higher Education Quarterly, 62(3), 237–251. doi:10.1111/j.1468-2273.2008.00393.x

Parsons, T., & Bales, R. F. (1955). Family, Socialization and Interaction Process. New York: The Free Press.

Patterson, D. A. (2005). Restoring the Popularity of Computer Science. Communications of the ACM, 48(9), 25–28. doi:10.1145/1081992.1082011

Pearlin, L., Lieberman, M., Menaghan, E., & Mullan, J. (1981). Mastery scale. Measures of personality and social psychological attitudes. San Diego, CA: Academic Press.

Pell, A. N. (1996). Fixing the leaky pipeline: women scientists in academia. Journal of Animal Science, 74, 2843–2848.

Penner, A. M., & Paret, M. (2008). Gender differences in mathematics achievement: Exploring the early grades and the extremes. Social Science Research, 37, 239–253. doi:10.1016/j.ssresearch.2007.06.012

Perrons, D. (2003). The new economy and the work-life balance: conceptual explorations and a case study of new media. Gender, Work and Organization, 10(1), 65–93. doi:10.1111/1468-0432.00004

Phillips-Jones, L. (2003). The Mentor's Guide. Grass Valley, CA: Coalition of Counseling Centers (CCC)/The Mentoring Group.

Phillis, E., & Pugh, D. (2000). How to get a PhD: A Handbook for Students and their Supervisors (3rd ed.) Buckingham, UK: Open University Press.

Phipphitkul, W. (n.a.). Gender justice: digitally empowered women through information technology. Available at www.wsisasia.org/materials/wil.doc

Phipps, A. (2006). I Can't do With Whinging Women! Feminism and the habitus of 'women in science' activists. Women's Studies International Forum, 29, 125–135. doi:10.1016/j.wsif.2006.03.010

Plumm, K. M. (2008). Technology in the classroom: burning the bridges to the gaps in gender-biased education? Computers & Education, 50, 1052–1068. doi:10.1016/j.compedu.2006.10.005

Poel, E., Wolfe, E., Arroyos-Jurado, E., & Coppola, B. (2006). Women, Mentoring, and a Border University. Journal of Public Affairs Education, 12(4), 501–513.

Powell, A., Bagilhole, B., & Dainty, A. (2007). The good, the bad and the ugly: Women engineering students' experiences of UK higher education. In R.J. Burke & M.C. Mattis, (Eds.), Women and Minorities in Science, Technology, Engineering and Mathematics: Upping the numbers (pp. 47-70). Cheltenham, UK: Edward Elgar Publishing, Ltd.

Powell, A., Baglihole, B., Dainty, A., & Neale, R. (2004). Does the Engineering Culture in UK Higher Education Advance Women's Careers. Equal Opportunities International, 23(7/8), 21–38. doi:10.1108/02610150410787882

Prescott, J. (forthcoming). Women in the computer games industry. Unpublished Ph.D thesis, University of Liverpool, UK.

Priola, V. (2007). Being female doing gender. Narratives of women in education management. Gender and Education, 19(1), 21–40. doi:10.1080/09540250601087728

Probert, B. (2005). 'I just couldn't fit it in': Gender and unequal outcomes in academic careers. Gender, Work and Organization, 12(1), 50–72. doi:10.1111/j.1468-0432.2005.00262.x

Qld Government. (2007). Towards a 10-year plan for science, technology, engineering and mathematics (STEM) education and skills in Queensland. Discussion paper released by the Queensland Government. Retrieved 12 May, 2009, from http://education.qld.gov.au/projects/stemplan/docs/stem-brochure.pdf.

Quimby, J. L., & O'Brien, K. M. (2004). Predictors of student and career decision-making self-efficacy among nontraditional college women. The Career Development Quarterly, 52(4), 323–339.

Ragin, C. C. (1994). Constructing social research. Thousand Oaks, CA: Pine Forge Press.

Ragins, B. R., Cotton, J. L., & Miller, J. S. (2000). Marginal mentoring: The effects of type of mentor, quality of relationship and program design on work and career attitudes. Academy of Management Journal, 43(6), 1177–1194. doi:10.2307/1556344

Ramdin, T. E. (1998). Identification of Factors Which Facilitate and Hinder the Career Adjustment of Women Engineers" in Canadian Coalition of Women in Engineering, Science and Technology. Retrieved September 25, 2006 from http://ccwest.org/Cnu5news/CCWESTfullarticles. asp?dismode=article&artid=174

Ramsay, K., & Letherby, G. (2006). The experience of non-mothers in the gendered university. Gender, Work and Organization, 13(1), 25–44. doi:10.1111/j.1468-0432.2006.00294.x

Ramsey, N., & McCorduck, P. (2005). Where are the Women in Information Technology? Anita Borg Institute. Retrieved from http://anitaborg.org/news/research

Rao, R. (2009). Digital Divide: Issues Facing Adult Learners. Computer and Information Science., 2(1), 132–136.

Raskin, P. M. (2006). Women, Work and Family: Three studies of roles and identity among working mothers. The American Behavioral Scientist, 49(10), 1354–1381. doi:10.1177/0002764206286560

Ravenscroft, A. M. (2008). What's Keeping Women out of it? ieee.usa today's engineer. Retrieved March 20, 2009, from http://www.todaysengineer.org/2008/Mar/wie.asp

Ray, R. M., Sormunen, C., & Harris, T. M. (1999). Men's and women's attitudes toward computer technology: a comparison. Office Systems Research Journal, 17(1), 1–8.

Retrieved 04/03/2009 from http://www.skillset.org/uploads/pdf/asset_9920.pdf?6

Ridgeway, C. L., & Correll, Shelly J. (2004). Unpacking the Gender System: A Theoretical Perspective on Gender Beliefs and Social Relations. Gender & Society, 18(4), 510–531. doi:10.1177/0891243204265269

Rivera, L. M., Chen, E. C., Flores, L. Y., Blumberg, F., & Ponterotto, J. G. (2007). The effects of perceived barriers, role models, and acculturation on the career self-efficacy and career consideration of Hispanic women. The Career Development Quarterly, 56(1), 47–61.

Roberts, C. G., & Sprague, R. (1995). To Compete or Educate: Mentoring and the Research Climate. Professional Ethics (Gainesville, Fla.), 8(4), 1.

Roberts, P., & Ayre, M. (2004). The Careers Review of Engineering Women: An Investigation of Women's Retention in the Australian Engineering Workforce. Canberra, Australia: Engineers Australia.

Roger, A., & Duffield, J. (2000). Factors underlying persistent gendered option choices in school science and technology in Scotland. Gender and Education, 12(3), 367–383. doi:10.1080/713668300

Rose, G. (2005). Group Differences in Graduate Students' Concepts of the Ideal Mentor. Research in Higher Education, 46(1), 53–80. doi:10.1007/s11162-004-6289-4

Rosenbloom, J. L., Ash, R. A., Dupont, B., & Coder, L. (2008). Why Are There So Few Women in Information Technology? Assessing the Role of Personality in Career Choices. Journal of Economic Psychology, 29(4), 543–554. doi:10.1016/j.joep.2007.09.005

Rosenthal, R., & Jacobson, L. (1968). Pygmalion in the classroom. New York: Holt, Rinehart & Winston.

Rosser, S. (2004). Using POWRE to ADVANCE: Institutional Barriers Identified by Women Scientists and Engineers. NWSA Journal, 16(1), 50–78. doi:10.2979/NWS.2004.16.1.50

Rosser, S. V. (2003). Attracting and retaining women in science and engineering. Academe, 89(4), 24–28.

Rosser, S., & Lane, E. O. (2002). Key Barriers for Academic Institutions Seeking To Retain Female Scientists and Engineers: Family-Unfriendly Policies, Low Numbers, Stereotypes, and Harassment. Journal of Women and Minorities in Science and Engineering, 8(2), 161–189.

Ruberg, B. (2009 February 20th). The Gamasutra 20: Top Game Writers. Retrieved 27/02/2009 from www.gamasutra.com

Rudman, L. A. (1998). Self Promotion as a Risk Factor for Women: the Costs and Benefits of Counterstereotypical Impression Management. Journal of Personality and Social Psychology, 74(3), 629–645. doi:10.1037/0022-3514.74.3.629

Rudman, L. A., & Glick, P. (2001). Prescriptive Gender Stereotypes and Backlash Toward Agentic Women. The Journal of Social Issues, 57, 743–762. doi:10.1111/0022-4537.00239

Rutgers University Center for Women and Work. (2002). Fact sheet #1 Facts about working women. Downloaded September 10, 2007 from http://www.cww.rutgers.edu/dataPages/FactSheet1.pdf

Sands, R., Parson, L., & Duane, J. (1991). Faculty Mentoring Faculty in a Public University. The Journal of Higher Education, 62(2), 174–193. doi:10.2307/1982144

Sappleton, N., & Takruri-Rizk, H. (2008). The Gender Subtext of Science, Engineering and Technology (SET) Organizations: A Review and Critique. Women's Studies, 37(3), 284–316. doi:10.1080/00497870801917242

Saunderson, W. (2002). Women, academia and identity: Constructions of equal opportunities in the 'new managerialism'; a case of 'lipstick on the gorilla'? Higher Education Quarterly, 56(4), 376–406. doi:10.1111/1468-2273.00226

Schein, V. E. (2007). Women in Management: reflections and projections. Women in Management Review, 22(1), 6–18. doi:10.1108/09649420710726193

Schwartz, H. S. (1996). The Sin of the Father: Reflections on the Roles of the Corporation Man, the Suburban Housewife, Their Son and Their Daughter in the Deconstruction of the Patriarch. Human Relations, 49(8), 1013–1040. doi:10.1177/001872679604900801

Scott, A. B., & Mallinckrodt, B. (2005). Parental emotional support, science self-efficacy, and choice of science major in undergraduate women. The Career Development Quarterly, 53(3), 263–273.

Sealy, R., Vinnicombe, S., & Singh, V. (2008). The Female FTSE Report 2008. Cranfield University, School of Management.

SET. (2009). UK resource centre for women in science, engineering and technology, requested dataset. Retrieved from www.ukrc4setwomen.org.uk

Settles, I. H., Cortina, L. M., Malley, J., & Stewart, A. J. (2006). The climate for women in academic science: the good, the bad, and the changeable. Psychology of Women Quarterly, 30(1), 47–58. doi:10.1111/j.1471-6402.2006.00261.x

Seymour, E., & Hewitt, N. (1997). Talking about leaving: Why undergraduates leave the sciences. Boulder, CO: Westview Press.

Shah alam, Syed., Abdullah, Zaini, & Ahsan Nilufar. (2009). Cyber Café Usage in Malaysia: An Exploratory Study. Journal of Internet Banking and Commerce, 14(1).

Shah, F. H. (2004). Mainstreaming potential women exporters in international markets through ICT: Malaysia. Melaka, Malaysia: APEC Committee on Trade and Investment.

Shore, L. (1996). Developing gender equity at home: The influence of the exploratorium science at home project on parental attitudes, perceptions, and behaviors. St. Louis.

Shull, P. J., & Weiner, M. (2002). Thinking Inside the Box: Self-Efficacy of Women in Engineering. The International Journal of Engineering Education, 18(4), 438–446.

Siann, G., & Callaghan, M. (2001). Choices and barriers: factors influencing women's choice of higher education in science, engineering and technology. Journal of Further and Higher Education, 25(1), 85–95. doi:10.1080/03098770020030524

Siegel, D. (2008). Building a Pipeline for Diversity Through Intersectoral Collaboration. Higher Education, 55(5), 519–535. doi:10.1007/s10734-007-9072-6

Simard, C. (2007). The Business Case for Gender Diversity. Anita Borg Institute. Retrieved from http://anitaborg.org/news/research

Simard, C. (2009). Obstacles and Solutions for Underrepresented Minorities in Technology. Anita Borg Institute. Retrieved from http://anitaborg.org/news/research

Simard, C., Henderson, A. D., Gilmartin, S. K., Schiebinger, L., & Whitney, T. (2008). Climbing the Technical Ladder: Obstacles and Solutions for Mid-Level Women in Technology. Anita Borg Institute. Retrieved from http://anitaborg.org/news/research

Simpson, R. (1998). Presenteeism, power and organisational change: long hours as a career barrier and the impact on the working lives of women managers. British Journal of Management communication quarterly, 9(September), 37-50.

Singh, V., Vinnicombe, S., & Kumra, S. (2006). Women in formal corporate networks: An organizational citizen perspective. Women in Management Review, 21(6), 458–482. doi:10.1108/09649420610683462

Skillset. (2006). Census 2006: the results of the sixth census of the audio visual industries.

Skillset. (2006). Skillset: Workforce Survey 2006. London: The Sector Skills Council for the Audio Visual Industries.

Skillset. (2007, October 10th). Game boss calls for more ethnic minorities and women to combat skills shortages. Press release. Retrieved 15/02/09 from http://www.skillset.org/skillset/press/releases/article_6286_1.asp

Sloan Work and Family Research Network. (2007). Work-family encyclopedia. Retrieved from http://wfnetwork.bc.edu

Smeaton, D., Glover, J., & Fielding, J. (1997). Recent Trends in Women's Representation in Science, Engineering and Technology (SET) in the UK. Journal of Women and Minorities in Science and Engineering, 3, 1–19.

Smith, D. G., et al. (1997). Diversity works: The emerging picture of how students benefit. Washington, DC: Association of American Colleges and Universities.

Smyth, F. L., & McArdle, J. J. (2004). Ethnic and Gender Differences in Science Graduation at Selective Colleges with Implications for Admission Policy and College Choice. Research in Higher Education, 45(4), 353–381. doi:10.1023/B:RIHE.0000027391.05986.79

Snow, C. P. (1995). Women in Computing. Communications of the ACM, 38(1), 26–28. doi:10.1145/204865.204873

Snyder, C., & Lopez, S. (2007). Positive Psychology: The Scientific and Practical Explorations of Human Strengths. Thousand Oaks, CA: Sage Publications.

Sonnert, G., Fox, M. F., & Adkins, K. (2007). Undergraduate women in science and engineering: effects of faculty, fields, and institutions over time. Social Science Quarterly, 88(5), 1333–1356. doi:10.1111/j.1540-6237.2007.00505.x

Spennemann, D. H. R. (1996). Gender imbalances in computer access among environmental science students. Electronic Journal of Instructional Science and Technology, 1(2). Retrieved from http://www.usq.edu.au/electpub/e-jist/docs/old/vol1no2/contents.htm

Srinivasan, S. (2004, January 12). Computer access unequal in most Asian nations. Available at http://www.asiaitsummit.com

Stake, J. E. (2003). Understanding male bias against girls and women in science. Journal of Applied Social Psychology, 33(4), 667–682. doi:10.1111/j.1559-1816.2003.tb01918.x

Steele, C. M., & Aronson, J. (1995). Stereotype threat and the intellectual test performance of African Americans. Journal of Personality and Social Psychology, 69, 797–811. doi:10.1037/0022-3514.69.5.797

Steele, J. R., Reisz, L., Williams, A., & Kawakami, K. (2007). Women in mathematics: examining the hidden barriers that gender stereotypes can impose. In R.J. Burke & M.C.Mattis (Eds.), Women and minorities in science, technology, engineering and mathematics: Upping the numbers (pp. 159-182). Cheltenham, UK: Edward Elgar Publishing Ltd.

Steinpreis, R. E., Anders, K. A., & Ritzke, D. (1999). The impact of gender on the review of curricula vitae of job applicants and tenure candidates: A national empirical study. Sex Roles, 41(7-8), 509–528. doi:10.1023/A:1018839203698

Stewart, A. J., Malley, J. E., & Lavaque-Manty, D. (2007). Faculty Recruitment: mobilizing Science and Engineering Faculty. Ann Arbor, MI: University of Michigan Press.

Stewart, A., Malley, J., & Lavaque-Manty, D. (2007). Faculty Recruitment: mobilizing Science and Engineering Faculty. Ann Arbor, MI: University of Michigan Press.

Stewart, M. (1998). Gender Issues in physics education. Educational Research, 40(3), 283–293.

Straehley, C. J., & Longo, P. (2006). Family issues affecting women in medicine, particularly women surgeons. American Journal of Surgery, 192, 695–698. doi:10.1016/j.amjsurg.2006.04.005

Sullivan, K. R., & Mahalik, J. R. (2000). Increasing career self-efficacy for women: Evaluating a group intervention. Journal of Counseling and Development, 78(1), 54–62.

Swanberg, J. E. (2004). Illuminating Gender Organization Assumptions: an important step in creating a family-friendly organisation: a case study approach. Community Work & Family, 7(1), 3–28. doi:10.1080/1366880042000200271

Tang, J. (1997). The glass ceiling in science and engineering. Journal of Socio-Economics, 26(4), 383–406. doi:10.1016/S1053-5357(97)90003-2

Tang, J. (2003). Women Succeeding in Science in the Twentieth Century. Sociological Forum, 18(2), 325–342. doi:10.1023/A:1024099831993

Tannen, D. (1990). You Just Don't Understand: Women and Men in Conversation. New York: Ballantine.

Tattersall, A., Keogh, C., & Richardson, H. (2007). The gender pay gap in the ICT industry. University of Salford, UK.

Tavris, C. & Aronson, E. (2007). Mistakes were made (but not by me). AARP Bulletin, June 2007.

Taylor, J. (2006). Report of the Transition and Retention Working Party. USQ.

TBTLA. (2009). Welcome. Tampa Bay Technology Leadership Association. Retrieved on March 29, 2009 http://www.tb-tla.org/

Telterscher, S. (2002). Gender, ICT and Development. UNCTAD Electronic Commerce and Development Report. The presentation given at the World Civil Society Forum Geneva, 14-19 July.

Tenenbaum, H., & Leaper, C. (2003). Parent-child conversations about science: The socialization of gender inequities? Developmental Psychology, 39, 34–47. doi:10.1037/0012-1649.39.1.34

Teoh, M-Y, W. & Chong, S-C. (2008). Improving women entrepreneurs in small and medium enterprises in Malaysia: policy recommendations. Communications of the IBIMA, 2, 31-38.

Thanacoody, P. R., Bartram, T., Barker, M., & Jacobs, K. (2006). Career progression among female academics, A comparative study of Australia and Mauritius. Women in Management Review, 21(7), 536–553. doi:10.1108/09649420610692499

The Associated Press. (2007, May 22). U.S. workplace not family-oriented. Downloaded January 8, 2008 from http://www.msnbc.msn.com/id/16907584

The National Center for Women in Information Technology. (2006). The future of IT. Boulder: The National Center for Women in Information Technology.

Ticoll, D. (2008). IT enrollments and retentions: situation report. Labor Market Intelligence, ICTC's Report. Information and Communicattions Technology Council. Retrieved from http://www.ictc-ctic.ca/en/content.aspx?id=1922

Tienari, J., & Quack, S. & Theobald, Hildegard. (2002). Organizational Reforms, 'Ideal Workers' and Gender Orders: A Cross-Societal Comparison. Organization Studies, 23(2), 249–279. doi:10.1177/0170840602232004

Tierney, J. (2008, July 15). A New Frontier for Title IX Science. NY Times. Retrieved from http://www.nytimes.com

Tinto, V. (1993). Leaving College: Rethinking the Causes and Cures of Student Attrition. (2nd ed.) Chicago, IL: University of Chicago Press.

Tonge. (2008). Barriers to networking for women in a UK professional service. Gender in Management: An International Journal, 23(7), 484-506.

Tower, L. E., & Alkadry, M. G. (2008). The Social Costs of Career Success for Women. Review of Public Personnel Administration, 28(2), 144–165. doi:10.1177/0734371X08315343

Townsend, G. C. (2002). People who make a difference: Mentors and role models. Women and Computing, 34(2), 57–61.

Tracey, C., & Nicholl, H. (2007). The multifaceted influence of gender in career progress in nursing. Journal of Nursing Management, 15(7), 677–682. doi:10.1111/j.1365-2934.2006.00677.x

Tracy, K. (2002). Everyday talk: Building and reflecting identities. New York: Guilford Press.

Trauth, E. M. (2002). Odd girl out: an individual differences perspective on women in the IT profession. Information Technology & People, 15(2), 98–118. doi:10.1108/09593840210430552

Trauth, E. M., Quesenberry, J. L., & Huang, H. (2008). A Multicultural Analysis of Factors Influencing Career Choice for Women in the Information Technology Workforce. Journal of Global Information Management, 16(4), 1–23.

Trauth, E. M., Quesenberry, J. L., & Morgan, A. L. (2004). Understanding the under representation of women in IT: toward a theory of individual differences. In Procs. of ACM SIGMIS – Conference on Computer Personal Research.

Trauth, E. M., Quesenberry, J. L., & Yeo, B. (2008). Environmental Influences on Gender in the IT Workforce. The Data Base for Advances in Information Systems, 39(1), 8–32.

Trix, F., & Psenka, C. (2003). Exploring the color of glass: Letters of recommendation for female and male medical faculty. Discourse & Society, 14(2), 191–220. doi:10.1177/0957926503014002277

Trower, C. (2002, April 12). Women without Tenure, Part IV, Why it Matters, What to do? Next Wave. Retrieved January 19, 2008 from http://www.sciencecareers.sciencemag.org

Trower, C. (2003). Leveling the Field. [NERCHE]. New England Resource Center for Higher Education, 14, 6.

Trower, C. (2007). Collaborative On Academic Careers in Higher Education. Paper Presented at Harvard University, Boston, MA. Retrieved from www.oced.org

Trower, C. (2008). The name assigned to the document by the author. This field may also contain sub-titles, series names, and report numbers.Amending Higher Education's Constitution. The entity from which ERIC acquires the content, including journal, organization, and conference names, or by means of online submission from the author. Academe, 94(5), 16–18.

U.S. Bureau of Labor Statistics. (2007). Occupational employment statistics. Downloaded October 2, 2007 from http://www.bls.gov/oes/home.htm

U.S. Bureau of Labor Statistics. (2007). Standard Occupational Classification (SOC) System. Downloaded September 10, 2007 from http://www.bls.gov/soc/home.htm

U.S. Bureau of Labor Statistics. (2007, November). "Industry output and employment projections to 2016,'. Retrieved March 21, 2009, from U.S. Bureau of Labor Statistics: http://www.census.gov/compendia/statab/tables/09s0601.pdf

U.S. Department of Education, Institute for Education Studies, National Center for Education Statistics. (2006). Table 9. Number persons age 18 and over, by highest level of education attained, age, sex, and race/ethnicity: 2006. Downloaded September 10, 2007 from http://nces.ed.gov/programs/digest/d06/tables/dt06_009.asp?referrer=list

U.S. Department of Education, Institute of Education Sciences, National Center for Education Statistics. (2004). Trends in educational equity of girls & women. Downloaded October 2, 2007 from http://nces.ed.gov/pubsearch/pubsinfo.asp?pubid=2005016

U.S. Department of Education. (2005, March). Postsecondary institutions in the United States: Fall 2003 degrees and other awards conferred. Retrieved March 12, 2009, from U.S. Department of Education: http://nces.ed.gov/pubs2005/2005154.pdf

U.S. Department of Health and Human Services, National Institutes for Health. (2007). Recruitment and retention plan to enhance diversity. Retrieved from http://grants1.nih.gov/training/faq_diversity.htm

U.S. Government Accountability Office. (2004). Gender issues: Women's participation in the sciences has increased, but agencies need to do more to ensure compliance with Title IX. Downloaded January 9, 2008 from http://www.gao.gov/new.items/d04639.pdf

U.S. Glass Ceiling Commission. (1995). A solid investment: Making full use of the nation's human capital, Final report of the Commission. Washington, DC: U.S. Government Printing Office. Downloaded September 10, 2007 from http://digitalcommons.ilr.cornell.edu/key_workplace/120/

UCAS. (2007). Online datasets. Retrieved 04/03/2009 from http://www.ucas.ac.uk/about_us/stat_services/stats_online/annual_datasets_to_download/

UK Resource Centre for Women in Science. Engineering and Technology. (2004). Work Life Balance A SET Perspective. Retrieved on 12 February 2009 from http://www.ukrc4setwomen.org/html/resources/ukrc-publications/

UK Resource Centre for Women in Science. Engineering and Technology. (n.d.). Retrieved on 26 August 2009 from http://www.ukrc4setwomen.org/

UK Resource Centre for Women in Science. Engineering and Technology. (n.d.). Retrieved 30.3.2009 from http://www.ukrc4setwomen.org/html/research-and-statistics/statistics/

United Nations Economic Commission for Europe. (2007). Glossary. Downloaded on October 4, 2007 at http://www.unece.org/stats/gender/glossary/g.html

University of Bradford. (2008). Promotions Exercise 2008. Bradford, UK: University of Bradford.

University of Michigan. (2003). Admissions lawsuits. Downloaded October 2, 2007 from http://www.vpcomm.umich.edu/admissions/overview/

USQ. (2005). Quality Audit: Progress Report. Retrieved 8 May, 2009, from http://www.usq.edu.au/resources/postauditprogressreport2005.doc

USQ. (2008). Equity in Education Policy. Retrieved 1 May, 2009, from http://www.usq.edu.au/resources/122.pdf

Valenduc, G., Vendramin, P., Guffens, C., Ponzellini, A.M., Lebano, A., D'Ouville L., et al. (2004, July). Widening Women's Work in Information and Communication Technologies, final synthesis report. European Commission (IST-2001-34520).

Valian, V. (1998). Why so slow? The advancement of women. Cambridge, MA: MIT Press.

Valian, V. (1999). Why So Slow: The Advancement of Women. London: MIT Press.

Van Deusen, F., Ladge, J., James, J., & Harrington, B. (2008). Building the Business Case for Work-Life Programs. Retrieved on 12 February 2009 from http://www.bc.edu/centers/cwf/research/meta-elements/pdf/BCCWF Business_Case_EBS.pdf

Vancouver Film School. (2009) The game industry: now and in the future. VFS Report.

Vegso, J. (2005). Interest in CS as a Major Drops Among Incoming Freshmen. Computing Research News, 17(3). Retrieved from http://www.cra.org/CRN/articles/may05/vegso.

Venkatesh, M., & Small, R. V. (2002). Active Learning in Higher Education: A Model and Roadmap. In J. Lazar (Ed.), Managing IT/Community Partnerships in the 21st Century (pp. 17-51). Hershey, PA: Idea Group Publishing.

Venkatesh, V., & Brown, S. (2001). A longitudinal investigation of personal computers in homes: adoption determinants and emerging challenges. MIS Quarterly, 25(1), 71–102. doi:10.2307/3250959

Venkatesh, V., & Davis, F. D. (2000). A theoretical extension of the technology acceptance model: four longitudinal field studies. Management Science, 46(2), 186–204. doi:10.1287/mnsc.46.2.186.11926

Venkatesh, V., Morris, M. G., & Ackerman, P. L. (2000). A Longitudinal Field Investigation of Gender Differences in Individual Technology Adoption Decision-Making Processes. Organizational Behavior and Human Decision Processes, 83, 33–60. doi:10.1006/obhd.2000.2896

Viadero, D. (2007, October 24). Experiments aim to ease effects of 'stereotype threat'. Retrieved February 10, 2009, from Education Week: http://www.edweek.org/login.html?source=http://www.edweek.org/ew/articles/2007/10/24/09gap.h27.html&destination=http://www.edweek.org/ew/articles/2007/10/24/09gap.h27.html&levelId=2100

Vinnicombe, S., Singh, V., & Kumra, S. (2004). Making Good Connections: Best Practice for Women's Corporate Networks. Cranfield University.

Von Hellens, A. L., & Nielson, S. H. (2001). Australian women in IT. Communications of the ACM, 44(7), 46–52. doi:10.1145/379300.379310

Wajcman, J. (1998). Managing Like a Man: Women and Men in Corporate Management. Oxford, UK: Blackwell.

Wajcman, J., & Martin, B. (2002). Narratives of Identity in Modern Management: The Corrosion of Gender Difference? Social Compass, 36(4), 985–1002.

Wajcman, J., & Pham Lobb, L. A. (2007). The gender relations of software work in Vietnam. Gender, Technology and Development, 11(1), 1–26. doi:10.1177/097185240601100101

Walby, S. (2004). The European Union and Gender Equality: Emergent Varieties of Gender Regime. Social Politics, 11(1), 4–29.

Walker, J. P., & Swiss, D. J. (1993). Women and the work/family dilemma. New York: John Wiley & Sons.

Weinburgh, M. (1995). Gender differences in student attitudes toward science: a meta-analysis of the literature from 1970 to 1991. Journal of Research in Science Teaching, 32(4), 387–398. doi:10.1002/tea.3660320407

Welde, K. D., Laursen, S., & Thiry, H. (2007). Women in Science, Technology, Engineering and Math (STEM). Fact Sheets. Sociologists for Women in Society. Retrieved from http://www.socwomen.org

Wenneras, C., & Wold, A. (1997). Nepotism and sexism in peer-review. Nature, 387, 341–343. doi:10.1038/387341a0

West, C., & Zimmerman, D. (1987). Doing gender. Gender & Society, 1, 125–151. doi:10.1177/0891243287001002002

White, K. (2003). Women and leadership in higher education in Australia. Tertiary Education and Management, 9(1), 45–60. doi:10.1023/A:1022218403113

Whitten, B. (2003). What works? Increasing the participation of women in undergraduate physics. Journal of Women and Minorities in Science and Engineering, 9(3&4).

Whitten, B. (2004). 'Like a family': What works to create friendly and respectful student-faculty interactions. Journal of Women and Minorities in Science and Engineering, 10(3). doi:10.1615/JWomenMinorScienEng.v10.i3.30

Whitten, B. (2007). What works for women in undergraduate physics and what we can learn from women's colleges. Journal of Women and Minorities in Science and Engineering, 13(1). doi:10.1615/JWomenMinorScienEng.v13.i1.30

Williams, A., Turrell, P., & Wall, R. (2002). Let's TWIST: Creating a Conducive Learning Environment for Women. The International Journal of Engineering Education, 18(4), 447–451.

Williams, J. (2005). Measuring gender and women's empowerment using confirmatory factor analysis. (Research Program on Population Processes POP2005-01 and Population Aging Center PAC2005-01). Downloaded October 2, 2007 from http://www.colorado.edu/ibs/pubs/pop/pop2005-0001.pdf

Wilson, E. M. (1998). Gendered Career Paths. Personnel Review, 27(5), 396–411. doi:10.1108/00483489810230334

Winkler, J. (2000). Faculty Reappointment, Tenure, and Promotion: Barriers for Women. The Professional Geographer, 52(4), 737–750.

Wirt, J., Choy, S., Rooney, P., Provasnik, S., Sen, A., & Tobin, R. (2004). The Condition of Education 2004. NCES 2004-077. Washington, DC: US Department of Education.

Wolcott, B. (2001, April). Role models needed. Mechanical Engineering (New York, N.Y.), 1–9.

Wolffensberger, J. (1993). 'Science is truly a male world': The interconnectedness of knowledge, gender and power within university education. Gender and Education, 5, 37–54. doi:10.1080/0954025930050103

Women in Science Enquiry Network (WISENet). (n.d.). Retrieved on 26 August from http://www.wisenet-australia.org/

Women in Technology (WiT). (n.d.). Retrieved on 26 August from http://www.wit.org.au/

Woo, D. (1994). The glass ceiling and Asian Americans: A research monograph. Washington, DC: U.S. Department of Labor, Glass Ceiling Commission. Downloaded September 10, 2007 from http://digitalcommons.ilr.cornell.edu/key_workplace/129/

Wood, W., & Rhodes, N. D. (1992). Sex Differences in Interaction Style in Task Groups. In C. Ridgeway (Ed.), Gender, Interaction and Inequality (pp. 97-121). New York: Springer-Verlag.

Woolgar, S. (2002). The five rules of virtuality. In Woolgar, S. (Ed.), Virtual society? Technology, cyperbole, reality (pp. 1–22). Oxford, UK: Oxford University Press.

Wynarczyk, P., & Renner, C. (2006). The 'Gender Gap' in the Scientific Labour Market: the case of science, engineering and technology-based SMEs in the UK. Equal Opportunities International, 25(8), 660–673. doi:10.1108/02610150610719128

Sisters in Science. (n.d.). Sisters in science. Retrieved March 13, 2009, from Sisters in Science, LLC: http://www.sistersinscience.org/

Yoder, J. (1994). Looking Beyond Numbers: The Effects of Gender Status, Job Prestige, and Occupational Gender-Typing on Tokenism Processes. Social Psychology Quarterly, 57(2), 150–159. doi:10.2307/2786708

Yost, E. (2008). Why have all the women gone? Exploring Gender Differences in STEM Postdoctoral Education. M.A. Thesis, University of Alabama at Birmingham, Birmingham.

Young, P. (2006). Out of Balance: Lecturers' Perceptions of Differential Status and Rewards in Relation to Teaching and Research. Teaching in Higher Education, 11(2), 191–202. doi:10.1080/13562510500527727

Zarrett, N., Malanchuk, O., Davis-Kean, P., & Eccles, J. (2006). Examining the gender gap in IT by race: Young adults' decisions to pursue an IT career. In W. Aspray, & J. McGrath, Women and information technology: Research on the reasons for under-representation. Cambridge. MIT Press.

Zeldin, A. L., & Pajares, F. (2000). Scientific, and Technological Careers Against the Odds: Self-Efficacy Beliefs of Women in Mathematical. American Educational Research Journal, 37(1), 215–246.

Zeldin, A. L., Britner, S. L., & Pajares, F. (2008). A Comparative Study of the Self-Efficacy Beliefs of Successful Men and Women in Mathematics, Science, and Technology Careers. Journal of Research in Science Teaching, 45(9), 1036–1058. doi:10.1002/tea.20195

Zeldin, A., & Pajares, F. (2000). Against the odds: Self-efficacy beliefs of women in mathematical, scientific, and technological careers. American Educational Research Journal, 37(1), 215–246.

Zumeta, W., & Raveling, J. (2002). The best and brightest for science: Is there a problem here? Commission on Professionals in Science and Technology, Washington, DC.

About the Contributors

Aileen Cater-Steel is Associate Professor in Information Systems and Chair of the IS Research Committee in the Faculty of Business at the University of Southern Queensland, Toowoomba, Queensland Australia. She is an executive member of USQ's GO WEST team – a group providing support and mentoring to female students and staff in disciplines traditionally dominated by men. Dr Cater-Steel has previously published two edited books, one on IT Governance and Service Management and the other on Information Systems Research. Her work has been published in many international journals including the Communications of the ACM, Information and Software Technology, International Journal of IT Standards and Standardization Research and she has presented at many international conferences. She is the Chair of the local chapter of the Australian Computer Society. Her research interests include IT Governance, IT Service Management, e-Commerce adoption, Software Process Improvement and gender issues related to the participation of women in engineering, science and technology disciplines.

Emily Cater has a B.Bus degree (Marketing) from Queensland University of Technology, but has drifted into finance and works in the insurance industry in Bristol, UK. Her interest in gender issues has always been evident, and was (optimistically) voted 'most likely to achieve equal rights for women' in high school, which she successfully petitioned to have changed to 'equal gender rights for all'. Ten years on, she is astounded that the situation has changed so little in her first decade of professional life. She plans on being more than just optimistic in her second decade.

* * *

Jan Bogg is Academic Sub Dean and Research director of the Breaking Barriers program, both positions at the University of Liverpool, UK. Breaking Barriers is a significant program of research that began in 2004. Breaking Barriers is an ongoing program of research that addresses career progression, barriers and drivers for women in science and equity and diversity issues in the workplace. Jan is supervising Julie Prescott's, PhD thesis, developed as a stem emerging from the Breaking Barriers project. Jan is duel qualified as an organizational and health psychologist, her research areas are in the area of workforce issues and psychometrics and global testing.

Daryl E. Chubin is Founding Director, beginning August 2004, of the AAAS Center for Advancing Science & Engineering Capacity at the American Association for the Advancement of Science. His career spans more than 35 years in the academic, federal, and nonprofit sectors. With a Ph.D. in sociology from Loyola University (Chicago, 1973), he is the author of eight books and numerous policy reports

Copyright © 2010, IGI Global, distributing in print or electronic forms without written permission of IGI Global is prohibited.

and articles on science and education policy, program evaluation, careers, and workforce development in science and engineering. He is a AAAS Fellow, a Fellow of the Association for Women in Science, a 2006 QEM Giant of Science, a Sigma Xi Distinguished Lecturer 2007-2009, recipient of the Washington Academy of Sciences' 2008 Social and Behavioral Sciences Award, an alumnus or member of three nonprofit boards, an editorial advisor for three journals, a long-time consultant to corporate and philanthropic foundations, a member of various committees of The National Academies, and has been an adjunct professor in the Cornell in Washington Program since 1991.

Shelia R. Cotten is an Associate Professor in the Department of Sociology at the University of Alabama, Birmingham. In addition to studying gender issues in STEM, Dr. Cotten's research examines technology usage across the life course and the social and health impacts of this usage. Her research is currently funded by the National Science Foundation and the National Institute on Aging.

Carla Dal Sasso Freitas received her PhD degree in Computer Science, in 1994, from Universidade Federal do Rio Grande do Sul, Brazil, where she works since 1980. She is currently associate professor, involved in teaching Computer Graphics, Information Visualization and Human-Computer Interaction, both at undergraduate and graduate levels. During the last 6 years, she was part of the board of directors of the Brazilian Computer Society, SBC. Her research efforts are mainly Data and Information Visualization, and 2D and 3D Interaction, subjects on which she has actively published journal and conference papers and book chapters, as well as coordinated projects including international cooperation projects with institutions from France and Belgium.

Sunrita Dhar-Bhattacharjee (MA) is an experienced Research Fellow with over eight years experience of working in knowledge-based industries in India, Cyprus, Germany and U.K . Principal research and teaching interests lie in the fields of equality and diversity, industry engagement with education, community development, project management, and gender studies in SECT (science, engineering, construction and technology). She is currently involved in a doctoral research on 'Comparative gender segregation: UK and India' and has been working with the construction industry in various equality and diversity issues for the past few years. She is co managing the NWDA Equality and Diversity Guidance Note for Sustainable Buildings and Women and Work project (2009-present). Sunrita has worked on a range of Equality and Diversity research projects funded by different bodies including Women in Engineering (2004-2005), Women in Audio Visual Engineering (2005-2006), training and services delivery for the Housing Sector (2007-2008) and the Construction Knowledge Exchange (2009).

Susan Durbin is a Senior Lecturer and Researcher at the Centre for Employment Studies Research, Business School, University of the West of England, Bristol, UK. Her current research interests and publications are in the areas of senior women's careers, networking and work-life balance; the career trajectories and networking behaviours of female part-time managers; the gendering of knowledge and knowledge work; equality and diversity; and employee involvement and participation in high performance workplaces.

Jessica Guth has an LLB law degree from Leicester University and completed the legal practice course at Nottingham Trent Law School before working in general high street legal practice for 2 years. She then completed a Masters Degree in Social Research at the University of Leeds. Following 3 years as an empirical socio-legal researcher at the Centre for the Study of Law and Policy in Europe, University of Leeds, she took up a lectureship in law at the Bradford University Law School. Jessica has a post-graduate certificate in Higher Education Practice and is a fellow of the Higher Education Academy. She teaches Employment Law and European law and continues to work on socio-legal empirical research projects in both those fields.

Donna M. Handley is an Assistant Professor of Government at the University of Alabama at Birmingham Master of Public Administration program. She is chair-elect for the Southeastern Conference of Public Administration. She also serves on the board of the American Society for Public Administration Section for Women in Public Administration. Her research focuses on intergovernmental fiscal relations and federal/local performance measurements, as well as women in the STEM fields, in particular Public Administration.

Zuraini Ismail started her tertiary education with Diploma in Computer Science (1981) at University Technology MARA. She obtained her Bachelor of Science in Computer Management Systems (1984) at Southern New Hampshire, United States of America and a Masters in Computer Management Science (1987) from University of New Haven, Connecticut, United States of America. She further awarded PhD in Management Information System (2007) at International Islamic University Malaysia. Currently, she is a Senior Lecturer at College of Science and Technology (CST), University Technology Malaysia International Campus Kuala Lumpur. In addition, she holds administrative post as CST's IT Manager. Her present research interest includes Information Security Policy, Information Security - Awareness and Ethics, Physical Security Operations and Management, Cyber Security, Knowledge Management, Information Systems and IT Outsourcing. Her research publications have appeared, among others, in Journal of US-China Public Administration, and ACM SIGCAS Computers and Society.

Birgit Loch is a computational mathematician and Senior Lecturer in the Department of Mathematics and Computing at the University of Southern Queensland. Apart from her mathematical research area of surface fitting, her current research interests are in technology enhanced learning and teaching. She is leading technology focused projects at USQ, such as an inter-faculty trial of tablet PCs and testing and implementation of lecture recording software. She has a secondment with the Division of ICT Services as Principal Advisor Learning & Teaching.

Anne Manuel is the Librarian and Archivist at Somerville College University of Oxford and teaches Legal and Government Information Management at the University of the West of England. Her professional career spans the fields of accountancy, academic administration, digital library development and traditional library management. She holds an LLB from Reading University and Master's degrees in Medieval Studies, Research Methods and Information and Library Management from the University of Bristol. Her PhD in Gender and Higher Education is also from the University of Bristol. Research interests include gender inequality in higher education, gender and technology, information management and technology use and Web 2.0 applications in libraries

David Maree is professor in Psychology at the University of Pretoria. He is responsible for training research psychologists and lectures postgraduate students on research methods, test development and cognitive psychology. His fields of interest include philosophy of science especially the application of Critical Realism to Psychology, and the utilisation of Item Response Theory in psychological test construction. Recently he and his co-author, Marinda, published research on assessing hope and goal-orientation as a psychological construct.

Marinda Maree is working at the Institute for Women's and Gender Studies at the University of Pretoria where she is amongst others responsible for a number of gender-related projects and training. She was involved in projects on women in science as project leader and researcher. Recently she co-authored published research on sexual myths amongst communities in South Africa. She has an MA-degree in Medical Social Work and an honours degree in Psychology. She is an experienced group therapist and life-skills educator.

Maslin Masrom is a Senior Lecturer in College of Science and Technology at University Technology Malaysia International Campus Kuala Lumpur. She received her Bachelor of Science in Computer Science (1989) at University Technology Malaysia. She further received her Master of Science in Operations Research (1992) at Western Michigan University, United States of America. She pursued her Ph.D. in information technology/information system (IT/IS) Management and obtained the degree (2003) at University Putra Malaysia. Her current research interest includes Information Security and Ethics, Security Management, and IT/IS Adoption/Acceptance using Structural Equation Modelling Approach. She has published, among others, in Information & Management, Oxford Journal, International Journal of Cyber Society and Education, Journal of US-China Public Administration, and ACM SIGCAS Computers and Society.

Mirella M. Moro is an assistant professor at the Computer Science department at UFMG (Belo Horizonte, Brazil). She holds a Ph.D. in Computer Science (University of California Riverside - UCR, 2007), and MSc and BSc in Computer Science as well (UFRGS, Brazil, 2001, 1999). She is the Education Director of SBC (Brazilian Computer Society) and is the editor-in-chief of the new electronic magazine SBC Horizontes, which focuses on career in Computer Science. She is also a member of ACM, ACM SIGMOD, ACM SIGCSE, IEEE, IEEE WIE, and MentorNet. Mirella has been working with research in Computer Science in the area of Databases since 1997. Her research interests include hybrid XML/relational modeling, XML query optimization, stream processing, content-based dissemination systems, temporal databases, versioning management, and schema evolution.

Raymond Papp is an associate professor in the Information Technology and Management department in the John H. Sykes College of Business at the University of Tampa. Dr. Papp completed his doctorate in Information Management at Stevens Institute of Technology. His dissertation, "Determinants of Strategically Aligned Organizations: A Multi-industry, Multi-perspective Analysis" is an empirical investigation into the determinants of strategic alignment. His research interests include Strategic Alignment, IT for Competitive Advantage, and Pedagogical Issues in IT. His book Strategic Information Technology: Opportunities for Competitive Advantage highlights the use of information systems to achieve competitive advantage and contains numerous cases and research on strategic information systems. He has published in numerous journals including Annals of Cases on Information Technology, Communications of the AIS and Industrial Management and Data Systems as well as presented at national and international conferences.

Julie Prescott is studying for a PhD in the School of Population, Community and Health, at the University of Liverpool, UK. The PhD study is looking at the career aspirations and motivations of women working in the computer games industry. Prior to undertaking her PhD, Julie was a researcher on the Breaking Barriers in the workplace project at the University of Liverpool. Julie's background is in psychology and women's studies and she has a particular interest in women's careers, especially in terms of barriers and drivers, occupational segregation and the experiences of women working in male dominated occupations.

Natalie Sappleton (MA) is a doctoral researcher at Manchester Metropolitan University. Natalie graduated from The University of Glasgow in 2002 with First Class Honours in Economics and Politics, and began a career in Policy and Research. In 2003, she began working at the Equal Opportunities Commission on an Investigation into occupational segregation and gender stereotyping in science and engineering. Since then, she has worked at Manchester Metropolitan University and the University of Salford on research projects relating to gender and the labour market. Her doctoral research is investigating whether the networks of male and female entrepreneurs operating in gender atypical industries differs from those of business owners in gender typical sectors, and whether this has an impact on their performance of their businesses. Natalie is currently a short-term scholar at the Kluge Center, Washington DC.

Ruta Sevo served as Senior Program Director for the program Research on Gender in Science and Engineering at the U.S. National Science Foundation from 1998 through 2005. Dr. Sevo managed funding at the level of $10 million per year and oversaw about 150 grants in amounts from $100,000 to $1.5 million. She initiated the publication of New Formulas for America's Workforce: Girls in Science and Engineering. She served as staff to the Commission on the Advancement of Women and Minorities in Science, Engineering and Technology (CAWMSET) in 1999. Before that, she was an executive level information technology manager and specialist at the National Science Foundation, and a systems librarian and systems planner at the Library of Congress. Dr. Sevo has published in science and education policy and is currently consulting and serving on project advisory boards. Her education includes an M.A. in library and information systems and an M.A. and Ph.D. in South Asian area (civilization) studies, all from the University of Chicago.

Taisy Silva Weber received her PhD degree from the Karlsruhe University, Germany, in 1986. She is an associate professor at Universidade Federal do Rio Grande do Sul. During 7 years, she was part of the board of directors of the Brazilian Computer Society, SBC. After that, she served as a member of the council during 4 years. Her research has focused on dependable distributed systems and especially on dependability evaluation using fault injection, subjects on which she has authored and coauthored more than 100 papers. She is currently vice-coordinator of the Computer Science Undergraduate Course at UFRGS. Recently, a secondary thread of her research is the little participation of girls in graduation courses in all fields of Computer Science and means to widen this participation.

Haifa Takruri-Rizk (MBE, PhD, MIET, MBCS, FHEA) is a senior lecturer in the School of Computing, Science and Engineering at the University of Salford, UK. Her current technical research is within the Computer Networking and Telecommunications Research group, focusing on the development of wireless sensor networks for varied applications. Haifa has recently been awarded an MBE for her services to women in science, engineering and technology (SET). For many years she has been an advocate for the promotion of the contribution of women in ST, and in particular encouraging young women to study SET subjects. She is currently leading research investigating the factors contributing to the under-representation and the leaky-pipeline phenomena of women in SET and their lack of progression into SET careers. In the last a few years, she led a number of projects in this area including the Women Audio Video Engineers (WAVE) project, the Women in North West Engineering (WEWIN) project and the Developing Female Engineers (DFE) project.

Natasha F. Veltri is an Assistant Professor of Information and Technology Management in the John H. Sykes College of Business at the University of Tampa. She completed her MBA with concentration in Finance and Ph.D. in Management Information Systems at the University of Central Florida. Her research interests include Information Technology sourcing and governance, adoption of information systems and societal impacts of IT. Her current research efforts are focused on social networking, gender issues in IT and IT-enabled interorganizational collaboration. She has presented at national and international conferences and her research appeared in California Management Review, International Journal of Human-Computer Interaction and DATABASE for Advances in Information Systems.

Harold W. Webb is an Associate Professor and Chair of the Information and Technology Management Department of the John H. Sykes College of Business at the University of Tampa. He earned his Ph.D. from Texas Tech University in the area of Management Information Systems. Dr. Webb specializes in research on software testing, electronic commerce, and the impact of information technology on learning. His software testing research has focused on behavioural issues associated with the testing process and the resulting implications for managers. With electronic commerce, his interests are directed at the measurement of quality in business to consumer Web sites. With information technology and learning, he has two interest areas: the impact of technology on the individual learner and academic–industry collaboration in the adoption of advanced information technologies. His publications include journal articles, book chapters, numerous conference proceedings, and a technical report for the Department of Defence.

Gwen White is an Associate Professor of Computer Literacy and Information Technology at Gateway Community and Technical College in Northern Kentucky. In addition she teaches Information Technology courses at Xavier University in the Williams College of Business and at DeVry University. She is an entrepreneur and former owner of SMA Computers, Inc. Her interests include educational technology, Six Sigma and developing inroads for women in technology. Mrs. White has degrees from Miami University ('86) Political Science, University of Cincinnati ('90) Master of City Planning/Economic Development, Morehead State ('03) MBA in Computer Information Science and is currently enrolled at Walden University in the Applied Management and Information Science Ph.D. program.

Vicki Winstead holds a B.A. in social psychology and is currently a second year doctoral student in Medical Sociology at the University of Alabama in Birmingham, Alabama. Her research interests include women in STEM fields, neighbourhood context and health, and technology use by senior adults.

Alexandra Winter is a Principal Policy Officer with the Queensland Government's Office for Women. She has worked on delivering the Queensland Government's Smart Women – Smart State Strategy and the Women in Hard Hats policy initiative, as well the Women on Boards Strategy.

Fran Wright has a BA (Honours) degree in History from the University of Stirling, an LLB from the University of Auckland and an LLM from Victoria University of Wellington. She has taught law at Worcester College, Oxford, University of Waikato, Victoria University of Wellington and, most recently, Bradford University Law School. Fran teaches modules in Criminal Law, Equity and Trusts, Legal History and Law and Society. Her research interests include criminal law and law and disability.

Elizabeth Yost is a fourth year doctoral student in Medical Sociology at the University of Alabama, Birmingham. Her research interests focus on the effects of gender and education on mental health, and the impact of technology on health.

Index

Copyright © 2010, IGI Global, distributing in print or electronic forms without written permission of IGI Global is prohibited.

T

U

V

W

Y